Stars

of the

First

People

Other Books by Dorcas S. Miller:

Berry Finder: A Guide to Fleshy Fruits, Eastern North America

Good Food for Camp and Trail: A Guide to Planning and Preparing Delicious All-Natural Meals Outdoors

Track Finder: A Guide to Mammal Tracks, Eastern North America

Winter Weed Finder: A Guide to Pods, Seeds, and Dried Flowers

The Maine Coast: A Nature Lover's Guide

STARS of the First People

Native American Star Myths and Constellations

DORCAS S. MILLER

PRUETT PUBLISHING COMPANY
BOULDER, COLORADO

Printed in the United States
10 9 8 7 6 5 4 3 2 1

The following institutions have generously granted permission to reproduce text excerpts or illustrations from copyrighted works. From the *Journal of American Folklore,* "Legend of the North Star," "The Summer Brought North," "The Origin of the Constellation Ursus Major" (38:147, © 1925), and "The Origin of Cassiopeia's Chair" (46:182, © 1933); not for further reproduction. Reprinted with permission of the American Folklore Society. From *The Celestial Bear Comes Down to Earth,* by Frank G. Speck and Jesse Moses. Illustration of Delaware drum and beaters. Scientific Publications, vol. 7, © 1945. Reproduced with permission of the Reading Public Museum and Art Gallery. From *Prehistoric Rock Art of California,* by Robert F. Heizer and C. W. Clewlow. Illustration from petroglyph showing mountain sheep and hunters, © 1973. Reproduced with permission of Ballena Press. From *Hopi Journal of Alexander M. Stephens,* by Elsie Clews Parsons, ed. Design from a Hopi kiva showing Orion, Pleiades, Morning Star, and Moon, © 1936 by Columbia University Press. Reproduced with permission of the publisher. From *Sandpaintings of the Navajo Shooting Chants,* by Franc Johnson Newcomb and Gladys Reichard. Sandpainting showing Father Sky, © [1937] 1975. Reproduced with permission of Dover Publications.

Library of Congress Cataloging-in-Publication data

Miller, Dorcas S., 1949–
Stars of the first people : Native American star myths and constellations / Dorcas S. Miller.
p. cm.
Includes bibliographical references and index.
ISBN 0-87108-858-4 (pbk.)
1. Indian astronomy—North America. 2. Indian mythology—North America. 3. Stars—Mythology. I. Title.
E98.A88M56 1997
520'.97—DC21 97-20052
CIP

Cover and book design by Jody Chapel, Cover to Cover Design
Book composition by Lyn Chaffee
Cover painting "Night Warriors" (Yei Bi 'Chai dance, Navajo), by Wayne Beyale
Maps by Jerry Painter
All illustrations by Dorcas S. Miller except where noted otherwise.

To Ben

Contents

Preface

Twenty-five years ago I bought H. A. Rey's book *The Stars: A New Way To See Them* and began teaching myself the constellations. I was an Outward Bound instructor in those days and took the book with me on canoeing trips in the Boundary Waters Canoe Area in Minnesota and the Quetico Provincial Park in Canada and on backpacking trips in Big Bend National Park in Texas. I carried the book with me on subsequent trips in Oregon, Idaho, and Utah and used it at home in Maine. At each place I sat under the great sky, tracing out the constellations—the easy ones at first, and then the fainter and more complex groups of stars—and I read the Greek myths about them.

One night I realized that the people who lived in North America must have had, like the Greeks, their own star names and stories, and I became curious about those Native American celestial counterparts. How much information was available about Native American constellations? Do Native people here "see" the same star groups as did the ancient Greeks? Are any of the concepts the same—is the Big Dipper a bear in North America as well as in Greece?

Those initial questions led me on a long journey, one that has culminated in the writing of this book, a survey of Native American constellations in North America. I have arranged the material so that it will be, I hope, accessible and useful. Chapter 1 presents a brief introduction to star lore in Native American beliefs and culture. Chapter 2 describes and provides illustrations of classical constellations, those whose roots are in Greece and its related civilizations. These constellations are the reference points I have used throughout the text when describing and locating Native American constellations. This section also presents tips for stargazing.

Chapters 3 through 12 present information about the cultures and star lore of Native American tribes. These chapters are arranged by culture area, that is, regions where tribes share similarities in customs and beliefs. Each chapter is arranged by tribe or culture group. A chart at the end of each chapter presents the constellations, by tribe, in the area.

The Conclusion presents a few closing comments. Appendix A contains a list of classical constellations and their Native American counterparts. Tribes are listed by culture group in the order in which they appear in this book. Appendix B presents a list of the twenty brightest stars in the sky.

I have confined this book to the constellations and star myths of the Native Americans of North America and limited my research to material available in books and journals. Given the scope of this work, which covers dozens of tribes across the continent, it was impossible to visit each tribe and establish the kind of personal contact necessary for the task of collecting oral material.

When I write in the text, "Only a few star stories from this tribe have been recorded," I mean that my research has yielded few star stories. There is doubtless more material in the oral tradition and in published and unpublished texts not included in the reference list at the end of the book. "No other recorded stories," then, does not mean that there are are no other stories—anywhere—on this topic, but rather is shorthand for "This is all that I could find after a concerted search." The point is an important one, because although there is a great deal of star material here, there is certainly more elsewhere.

When introducing each section and tribe, I have described how their people historically secured food and shelter, activities that are often reflected in the constellations. I have also included, when possible, cultural information that helps explain unusual or important aspects of the stories. Although I have used the past tense when describing sustenance and shelter in order to portray the culture as it was before or during Euro-American contact, I do not mean to imply that these cultures are dead or lost. Many tribes have sustained and nurtured traditional skills, ceremonies, and beliefs about the world. Therefore, when presenting stories and beliefs and describing the practice of some rituals, I have used the present tense, "The Pawnee tell this story . . ." or "The BIG DIPPER is . . . ," because the stories and identifications continue to the present. I have also used the present tense when introducing story characters and summarizing stories, because these characters still exist and their actions are believed to have consequences even today. I have, however, used the past tense in narrating stories when the original narrators did so.

Myths do not serve up celestial information in neat, uniform parcels. Sometimes the original sources mention a constellation but do not identify it; some constellations are named but no story accompanies them. Sometimes there may be several quite different explanations for the creation of a constellation, or the events in one story may contradict those in another. Events may be "true" in a way that does not conform to Western culture's scientific laws. In several Zuni Ahayuta stories, for instance, the two brothers create the Morning Star and Evening Star, while in others they *are* these two stars. To the Ojibway, the Milky Way is the path of the soul, the path of migrating birds, and the spray from a turtle's tail. Whereas Euro-American logic may say, "The Ahayuta can either create the stars or be them, and the Milky Way can be one thing but not all three," in myth these apparent contradictions are acceptable and are, in fact, common.

Indeed, one Blackfoot narrator explained the differences by pulling a ragweed plant out of the ground. "The parts of this weed all branch off from the same stem. They go different ways, but all come from the same root. So it is with the different versions of a myth." No two stories, even ones from the same tribe, are identical.

The references I consulted did not always give the exact name of a Native American star or constellation. Although information about the Maricopa shows that they call Orion *Amŏ's* (Mountain Sheep), there are many examples in which the sources only recount that certain people (or animals) go into the sky and become stars. If the sources do not state an exact name, then I have drawn the name from the appropriate characters of the story. The Assiniboin story about Wise-One and his brothers, for example, says that they became the Pleiades; I have assigned that constellation the name "Wise-One and His Brothers."

Although the celestial beliefs of a few tribes are well documented, in many cases the information is slim at best. Not only have there been taboos about sharing this material with strangers, but many of the ethnologists who studied these tribes were not familiar with the stars and did not inquire about them. Even those who took care to record information could be frustratingly vague about identification. One ethnologist who studied the culture of the Southwest carefully describes a Pueblo constellation but never mentions the classical equivalent.

Some Native American constellations presented in this book, therefore, are unaccompanied by myths for the simple reason that the materials with which I worked contained no stories about them. It is likely that there was once a myth about each Native American star group. It is also possible that even though the myth has not been recorded in the sources I consulted, somewhere someone still tells the myth today. I have included these "empty" constellations, those without translation, identification, or stories, because they still offer insights into the life of a tribe. Although it is not known with what classical stars the Makah constellations Halibut, Shark, Skate, and Whale are correlated, their very names underscore the Makah's orientation to the ocean. The Dakota constellation Sweat Tipi, though not identified, suggests that the ritual sweat bath was important enough to merit a place in the sky.

The myths are presented here in written form, but it is important to remember that they come from—and are still part of—an oral tradition. The written word cannot convey the facial expressions, the gestures, the pauses, and the inflection of a skilled storyteller.

Moreover, the quality of the documentation, translation, and editing of myths has varied from researcher to researcher. With even the best of efforts, the process of recording and publishing a myth not only removes it from its context in the oral tradition but also puts the reader at a great distance from the original narration. There is an inevitable filtering process. Storytellers may have deleted sacred material, material they thought the translator would not understand, or material that they feared would be considered "superstitious." Translators were perhaps not fluent in the native language or in English, so the translation may have been imperfect. People who recorded the narration may have added their own emphasis or phrases. Editors—some of whom were not present when the narrator told the story—may have smoothed rough spots and rounded out an image.

My own editing has provided further distance from the original. I have, regretfully, summarized and retold many of the following stories in order to emphasize the star-related material and to present the information with some consistency. When there are several versions of a story I have given a representative narration or told one story and summarized others. Some narratives quoted from original sources have been slightly edited for easier reading.

Though I have presented some cultural information, a survey book such as this one comes with limitations, and it has been impossible to convey each tribe's extensive and complex beliefs. When I write that the customs and language of one tribe are similar to those of another tribe, I am providing general background information and do not mean to suggest that the two tribes are identical or that the differences are negligible. Each tribe and culture group covered in this book has its own distinctive beliefs and customs, and I urge readers to consult the reference list at the end of the book and to investigate areas of interest. It is also important to note that I have discussed some beliefs and practices, such as the vision quest, when they are relevant to the star lore of a tribe, but I do not discuss the vision quest in every tribe that embraces it. Simply because I do not mention a practice does not mean it is absent from a culture group.

When possible, I have included information about the storytellers, for these are their tales and the tales of their people. Most of the narratives were recorded between the 1880s and the 1930s, and many of the storytellers had lived a traditional prereservation life in their youth.

For star identification, I have relied primarily on information in the original texts given either by the storyteller or the recorder. I have also benefited from the work of anthropologists and astronomers who, in the last twenty years, have turned their attention to star myths and have identified or suggested possibilities for many unidentified groups. Occasionally I have presented my own thoughts concerning the identity of a constellation. These discussions regarding identity can usually be found in the Notes section.

Acknowledgements

I have relied on the Smithsonian *Handbook of North American Indians*, original sources, and several other texts for cultural information. The Smithsonian series has also been my general guide for the spelling of tribal names and the location of tribal territories.

I would like to thank the librarians at the Maine State Library in Augusta for their invaluable assistance in locating in-house material and in securing books and articles from other libraries across the country. The cooperation of these other libraries through the Inter-Library Loan Program has made this project possible. I also used the facilities at the Hawthorne-Longfellow Library at Bodoin College in Brunswick, Maine, and the library at the R. S. Peabody Museum of Archaeology in Andover, Massachusetts.

I am deeply grateful for the comments I received from Anne Hazard, Mary McClintock, Sheila McDonald, Dr. Trudy Griffin-Pierce, Dianne Russell, and my editor, Marykay Scott. These individuals offered excellent guidance in revising the text. Any mistakes are, of course, my own.

Thanks to Rebecca Stanley, who said, "Write this book so I can read it to my kids!" Even though Seth and Nina learned to read years ago, her comment provided an impetus to keep going and reminded me of the oral nature of the stories. I also appreciate the efforts of other friends who cheered me on.

I extend heartfelt thanks to my husband, Ben Townsend, for encouragement and support throughout the long process of writing this book.

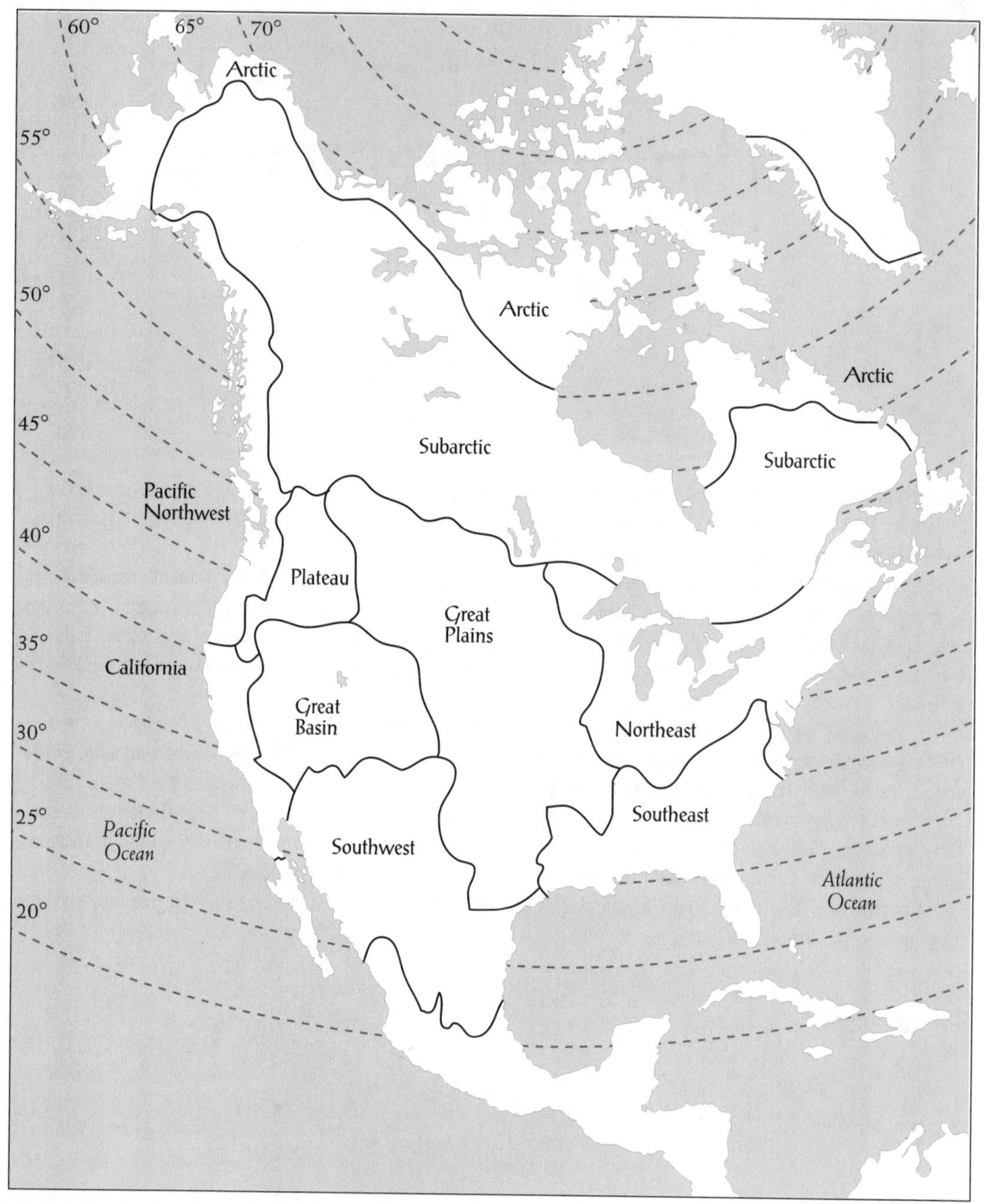

Native American culture areas in North America.

Chapter one

Stars of the First People

A Brief Introduction to Native American Star Lore

The leaves have turned brilliant colors and fallen to the ground. Winter has settled in. Perhaps a storm has layered the earth with snow. The baskets and leather storage sacks are filled with dried meat, roots, and berries. The rhythm of life has slowed and the sun sets early, leaving long nights to be filled with companionship and storytelling. People close the entrance to the lodge and gather around the fire. An elder, skilled in the art of narration, begins, "This is not a story. This really happened, back when animals were people . . ." The smooth narration, the timely pause, and the gesture for emphasis all recreate past times when the earth was made and when people were learning how to live in this world.

And so it has happened for hundreds of years. In the eastern woodlands, in the swaying prairie grass, in the dense boreal forest, along the treeless tundra, in the snow-frosted mountains, in the rock-rimmed desert, in the Northwest rain forests, and along the shimmering Pacific, indigenous people of North America have recounted their beliefs through the long winter nights. These narrations instruct listeners in the customs, rituals, and values of their community, and help them understand their surroundings as well as relate their past to their present lives. These narrations, together with the rituals themselves, are the conduit for culture.

Although some writers use the term "myth" only when referring to sacred material (including narrations about the creation), it is sometimes difficult to say exactly where the sacred myths end and nonsacred stories begin. Therefore, I have followed the lead of Native Americans and Native American scholars who consider myth "any story that reflects the quality and character of a specific Native American culture, or of Native American cultures more generally." Because all of the narrations included in this book reflect beliefs of particular cultures, and because it would be repetitious to use "myth" when referring to each and every narration, I have used the terms "myth," "narrations," and "stories" interchangeably.

In doing so I do not intend to demean the narrations or to imply that they are merely legend, amusing stories, or fantasy. Just as Christians and Jews believe that Genesis, the first book of the Bible, relates either the actual or figurative story of creation and believe that God created the world and all life therein, each Native American tribe believes that its own creation myth is real or close to real.

The narrations presented in this book are those that involve the stars. They range from complex cosmologies to elemental tales. Taken as a whole, and cutting across cultures and geographic regions, these myths contain several common elements.

Everything Has a Spirit

The stories abound with descriptions of animals talking, interacting, and otherwise engaging in human behavior. Some of the stories also attribute these actions to what Euro-Americans would call inanimate objects. The myths do not, however, simply project human qualities. Instead, they portray a natural world that is radically different from that of the Western scientific universe. The myths describe a realm in which humans, animals, stars, planets, stones—in fact, all things that exist—have spirits. These spirits take part in and influence life in many ways.

In a time long ago, the boundaries between worlds were fluid. Sky People came to earth and animals went to the sky in search of adventure, game, or fire. There was an easy flow between the two arenas and things that might now be considered impossible took place. Even today, when animals no longer talk out loud as they did then, humans may communicate with their spirits, and it is important to do so. Hunters who kill a bear must thank the bear's spirit and treat it with reverence so that it will return. The bear, representing all game, is part of the natural world and the natural cycle of life. Earthly bears are descended from the celestial bear, and just as the spirit of the celestial bear retreats to its den (the NORTHERN CROWN) each winter and then reappears in the spring (as the bowl of the BIG DIPPER), earthly bears do not die but return each spring after a long winter's sleep.

Power

To Native peoples, the concept of power is central to living a meaningful life. Unlike the Western world's idea of power, which involves dominance, the Native American concept involves spirituality. Power involves the ability to call upon guardian spirits, to have visions, and to practice ritual through song, story, and prayer. The theme of power appears in several celestial myths, including the Micmac bear hunt, in which only hunters with power can see the celestial bear in its den.

The Creation

Many of the creation myths of Native Americans account for the formation of the stars. The Navajo describe how the creators gather in a hogan and Black God makes the constellations. The Pueblo people say that when they emerged from a world far below into the present world, the stars came with them. In some cultures the stars are directly responsible for establishing this world. The Pawnee Bright Star, for example, gives Great Star a pebble, which he drops into water to create the Earth. The daughter of Bright Star and Great Star joins the son of the Moon and Sun, and together they populate the Earth.

Values

Myths convey values, and many of the star stories provide examples or moral lessons. Sometimes the stars themselves embody the rules of proper conduct. When the Twin War Gods fight with each other, they accidently spill stars out of a container. These spilled stars (the Milky Way) are a visual reminder to the Mescalero Apache that it is wrong to quarrel with family and that such dissension will upset the balance in the universe.

In the same way, the Navajo First Woman uses the stars as a tablet upon which to write the rules for proper behavior. The Milky Way, for example, symbolizes the cornmeal she scatters when she prays, and it reminds people that they should pray, too. Turning to another area of the sky, some Navajo chanters interpret Male One Who Revolves (the BIG DIPPER) and Female One Who Revolves (CASSIOPEIA) as a married couple who stay home and fulfill their traditional responsibilities to the family. Navajo couples are to emulate this example.

The idea of following the stars is evident in other areas of the continent as well. Pawnee chiefs arranged themselves in a circle like the Council of Chiefs (NORTHERN CROWN), and when someone fell ill, the Pawnee carried the sick person on a stretcher like the stretcher in the sky (the BIG DIPPER). When the Cheyenne gathered to camp, they placed their tipis in a circular pattern just like the Camp Circle (also the NORTHERN CROWN).

Still other star stories come with a moral. In a Pueblo tale, two little girls ignore the counsel of their grandmother and as a result rise to the sky, where everyone will recall their imprudence. And in a story of the Tohono O'odham, a group of women enjoy dancing the adolescence dance to such excess that they lose their homes and are changed into stars. They remain in the sky to show others on earth that although it is fitting to celebrate a woman's arrival at puberty, it is improper to celebrate *too* much.

Ritual

Stars are important in ritual. Star symbols appear on religious objects such as altars, *kiva* walls, drums, and masks, and on clothing used in vision quests. Many Native Americans used—and still use—the appearance of certain stars to begin or end ceremonies. The Hopi, for instance, carefully track the movements of the PLEIADES and ORION to decide when to begin the winter solstice ceremony. The appearance of the Pawnee constellation Swimming Ducks (two stars in the tail of SCORPIUS), in conjunction with the prolonged rumbling of thunder heard in the spring, provides the signal for the Thunder (or Creation) Ritual. The Mescalero Apache use the movement of the BIG DIPPER to time the girls' puberty ceremony.

In some cases stars are the vehicle through which people receive a ritual. In a Blackfoot narration, a woman who marries a star and lives in the sky world returns to her people with sacred objects and instructions for the Sun Dance.

Culture Heroes and Tricksters

In these stories there is usually someone who establishes the outlines of the physical world and introduces food, shelter, and rituals. Sometimes these gods (or "culture heroes," as ethnologists call them) slay ancient monsters or give animals their attributes. They make mountains and shape the flow of rivers. They introduce bison, corn, salmon, fire, fishnets, weaving, and other necessities of life to the earth. They even introduce death. Monster Slayer and Born for Water (Navajo), the Ahayuta (Zuni), Old Woman's Grandchild (Crow), and Breath Maker (Seminole) are but a few of the figures who fulfill these heroic duties.

There are also figures who are both great and foolish, a combination culture hero, trickster, and buffoon. The trickster embraces the great range of human emotion, from well-intentioned to thoughtless, from valiant to craven. The trickster may show people how to farm and hunt game and may give them the sun and fire. He may also kill a friend because he is jealous of him, or lust after his own daughters because they are beautiful. His behavior may be laudable or reprehensible, just as the actions of humans may be laudable or reprehensible. Coyote (the Great Plains, Great Basin, Plateau, and California); Raven, Mink, and Bluejay (the Pacific Northwest); Great Rabbit (the Southeast); Hare or Rabbit and Spider (the Great Plains); and Gluskap (the Northeast) all serve as both tricksters and culture heroes.

These tricksters and culture heroes may create or arrange the stars. Gluskap names the constellations. Breath Maker creates the Milky Way. Coyote scatters the stars, spoiling the plans of Black God and the creators. Raven walks across the sky in snowshoes, and his tracks become the Milky Way. Coyote's family is the PLEIADES, his cane the BIG DIPPER, he himself the star Aldebaran.

Taboos

In the past, Native Americans told sacred stories during the long winter nights but not at other times of the year. Native Americans of the Pueblos say that the Earth sleeps for a month in midwinter, at the time of the winter solstice, and this is the proper time for telling the sacred stories. A storyteller who breaks the taboo might be bitten by a snake or become ill or incur some other calamity. The Pawnee, Shasta, and other Native peoples say that a person who ignores the taboo will be bitten by a snake; according to the Pawnee, Coyote does not like people to talk about him, so he has instructed Rattlesnake Star to bite anyone who does so. In general, the spirits and animals are abroad in summer and don't like to be discussed.

Everyday Life

The stars are part of the fabric of everyday life in Native American culture, just as religion is inseparable from everyday action. Not only do the stars mark rituals, but they also signal the arrival of seasons and changes in the natural world. The Navajo constellation Man with Legs Ajar indicates the arrival of Parting of the Seasons (November), and the Milky Way is the sign for Crusted Snow (January). For the Cahuilla, "The old men used to study the stars very carefully and in this way could tell when each season began. They would meet in the ceremonial house and argue about the time certain stars would appear, and would often gamble about it. This was a very important matter, for upon the appearance of certain stars depended the season of the crops. They never went to the mountains until they saw a certain star, for they knew they would not find food there previously."

The appearance and disappearance of constellations set the time for planting, harvesting, and hunting. The Skidi Pawnee and Cherokee watched the PLEIADES to determine when to plant; the Zuni called the PLEIADES the Seed Stars. When the PLEIADES rose above the western horizon at dusk in May, the Yokut prepared for the salmon run. The Navajo did not hunt for rabbits until the constellation Rabbit Tracks (stars in the tail of SCORPIUS) had moved into a horizontal position, which meant that young rabbits were able to survive on their own and the older rabbits could be hunted. For many tribes, the movement of the PLEIADES, ORION's belt, and the BIG DIPPER also marked the divisions of the night.

These various elements—a worldview in which animals and "inanimate" objects have spirits, the active roles played by culture heroes and tricksters, the integration of the celestial and the spiritual into everyday values and actions—all express important aspects of Native American culture.

The celestial narrations presented in this book are more than entertaining stories about how a star or group of stars appeared in the sky. Each narration embodies—and contributes to—the particular Native American culture from which it comes. If these sacred stories stretch the awareness and consciousness of those from other cultures, so much the better, for new insights can provide richness to the human experience.

Chapter two

Greek Constellations

A Classical View of the Night Sky

In 1930, astronomers from around the world officially divided the sky into eighty-eight sectors, each with a constellation. Most of these constellations are figures from ancient Greek mythology, though European astronomers named several dozen others for non-mythological subjects in the 1500s and 1600s. In this book, these ancient and modern constellations are together called classical constellations to differentiate them from Native American constellations.

Recorded Native American constellations involve fewer than three dozen classical constellations. These classical star groups are given here with brief descriptions and illustrations. Only the classical constellations that appear in Native American stories are described, though some nearby constellations are presented in the illustrations. The sky maps at the end of the chapter show how the constellations move through the sky throughout the year.

It is easier to "see" some classical constellations than others. For instance, bright stars mark ORION's shoulders, belt, and feet, so it requires little imagination to envision these stars as a person. It is somewhat harder to discern a bull in TAURUS. Often people will say, "That bunch of stars doesn't look like a bull (or a herdsman, or whatever) to me!"

In this chapter, and throughout the book, I have drawn on H. A. Rey's interpretations of the classical constellations, which show them as actual figures. His TAURUS looks like a bull, and his HERDSMAN has a head, a body, and feet. Rey's book, *The Stars: A New Way To See Them* is an excellent nontechnical guide to learning the classical constellations.

I have used English names rather than the official Latin for constellations. My reasoning is that people who are becoming acquainted with the stars find English names easier to learn. Because most people are familiar with the Latin names for the constellations of the zodiac, however, I have used Latin in those instances. The classical constellations and some *asterisms* (figures within a constellation) are designated with uppercase letters to distinguish them from Native American constellations, which are capitalized. The names of individual stars, some asterisms, planets, and other astronomical features are capitalized, regardless of the language in which they appear.

Stars rise in the east, move through the sky, and set in the west. A constellation rises in Charleston, South Carolina, at the same time that it does in Pittsburgh, Pennsylvania, which is due north of Charleston. Latitude can, however, determine how fully observers can see a constellation. In the panhandle of Florida, at about thirty degrees north latitude, part of the BIG DIPPER swings below the horizon, while in Winnipeg, at about fifty degrees north latitude, the BIG DIPPER remains well above the horizon. On the other hand, in the panhandle, SCORPIUS shows in all of its glory, but in Winnipeg the scorpion's tail does not entirely clear the horizon. Essentially, the farther north you are located, the better you will be able to see constellations close to the North Star. If you were standing at the North Pole, for instance, the North Star would be directly overhead. The BIG DIPPER, LITTLE DIPPER, CASSIOPEIA, and nearby star groups circle the North Star but never set.

The constellations move through the sky in an annual cycle. As the cycle progresses, each constellation rises about four minutes earlier on each successive evening. Thus, at one time of the year a constellation may rise just after dark, at another just before dawn, and at still another during daylight hours when it cannot be seen.

This progression means that most constellations are best viewed at certain times of the year. ORION is conspicuous in the evening sky in winter and early spring, and SCORPIUS is prominent in late summer. (If you are looking at the predawn sky, you will find the opposite to be true: ORION is visible before dawn from August through November, and SCORPIUS is visible before dawn in March, April, and May.) The times given in this chapter for best sighting are in the evening, from whenever it gets dark to about 9 P.M. standard time. If it is later than 9 P.M. when you are viewing, check the subsequent star map.

Although the human eye cannot gauge the difference between, say, the fifth brightest star and the sixth brightest star, it does help to become familiar with the brightest stars because they can serve as beacons, or orienting markers. Of the twenty brightest stars in the sky, fifteen appear in Native American constellations.

Tips for Watching the Stars

For those who are unfamiliar with the night sky, it is easiest to begin by locating the most distinctive star groups, like the BIG DIPPER, CASSIOPEIA, and ORION. Then, use these groups as landmarks for locating nearby figures and bright stars.

A clear night with no moon and a location well away from artificial light provide ideal conditions for stargazing. The more light there is around you, either from the moon or from cities, the more difficult it will be to see the stars. Smog and other pollution also diminish visibility. If you plan to use this book while star-watching, cover your flashlight with a bit of red cloth. The red light will allow you to trace out figures on the page but has less impact on your eyes' ability to adjust to the dark.

Even if conditions are less than ideal, you can still see some of the major constellations. Keep in mind that you don't have to dedicate an entire evening to star-watching. Glancing at the sky whenever you step outside can help you learn what constellations are visible at different times of the year and at different times of the night.

Although it is fun to look at the PLEIADES and the Great Nebula of Orion through binoculars or a telescope, you are better off using your eyes for viewing constellations because you can more easily scan the sky for key stars and see the entire constellation.

Planets

The planets are not shown in these drawings because they, like the moon, move through the band of twelve constellations known as the zodiac. If you see what looks like an extra star in or around SAGITTARIUS, SCORPIUS, LEO, CANCER, GEMINI, TAURUS, or other constellations of the zodiac, then it is likely a planet.

Venus is the brightest planet and it is also brighter than all true stars. Venus appears in the west as the Evening Star and in the east as the Morning Star. Jupiter and Saturn are the largest planets and so are very bright; Mars appears reddish. Mercury is difficult to find because it is always close to the horizon and thus is seen at sunrise or sunset.

BIG DIPPER in GREAT BEAR (URSA MAJOR)

The BIG DIPPER is one of the best-known star figures in the sky. It is big, it is easy to see, and it circles the North Star. Technically, the BIG DIPPER is an asterism, a figure within the GREAT BEAR. Because Native American constellations involve the BIG DIPPER but not the other stars included in the GREAT BEAR, the star maps and illustrations in later chapters show only the BIG DIPPER.

The two stars on the far edge of the dipper—the side of the bowl opposite the handle—are called the Pointer Stars because they point to the North Star. Mizar, the middle star in the handle, is of note because it is paired with the fainter Alcor.

The Greek bear myth involves both GREAT BEAR and LITTLE BEAR. The Greek god Zeus fell in love with the beautiful mortal Callisto. When Zeus's wife, Hera, learned of this turn of events, she became intensely jealous and changed Callisto into a bear. Callisto, though, retained her human emotions, and when she saw her son, Arcas, she reached out to greet him. The son thought he was being attacked and started to kill the bear. At that moment Zeus intervened and placed Callisto in the sky as GREAT BEAR; he later placed Arcas near her as LITTLE BEAR. But jealous Hera was not quite finished. She made sure that the god of the sea, Poseidon, would never allow bears into his realm, which meant that the two constellations are never allowed to set. They must remain forever high in the sky.

The BIG DIPPER is visible all year in northern latitudes, but in the southern United States it swings below the northern horizon from about October to January (see Fig. 2.1).

LITTLE DIPPER or LITTLE BEAR (URSA MINOR)

The LITTLE DIPPER, also called LITTLE BEAR, includes the North Star (also called Polaris or the Pole Star), the star that does not move. The easiest way to locate the North Star is to find the BIG DIPPER and follow the Pointer Stars up from the bowl (see any of the star maps).

The farther north you are, the higher in the sky you will find the North Star. On the equator, it is on the horizon. At thirty degrees north latitude, it is thirty degrees from the horizon, at forty degrees north latitude it is forty degrees from the horizon, and so

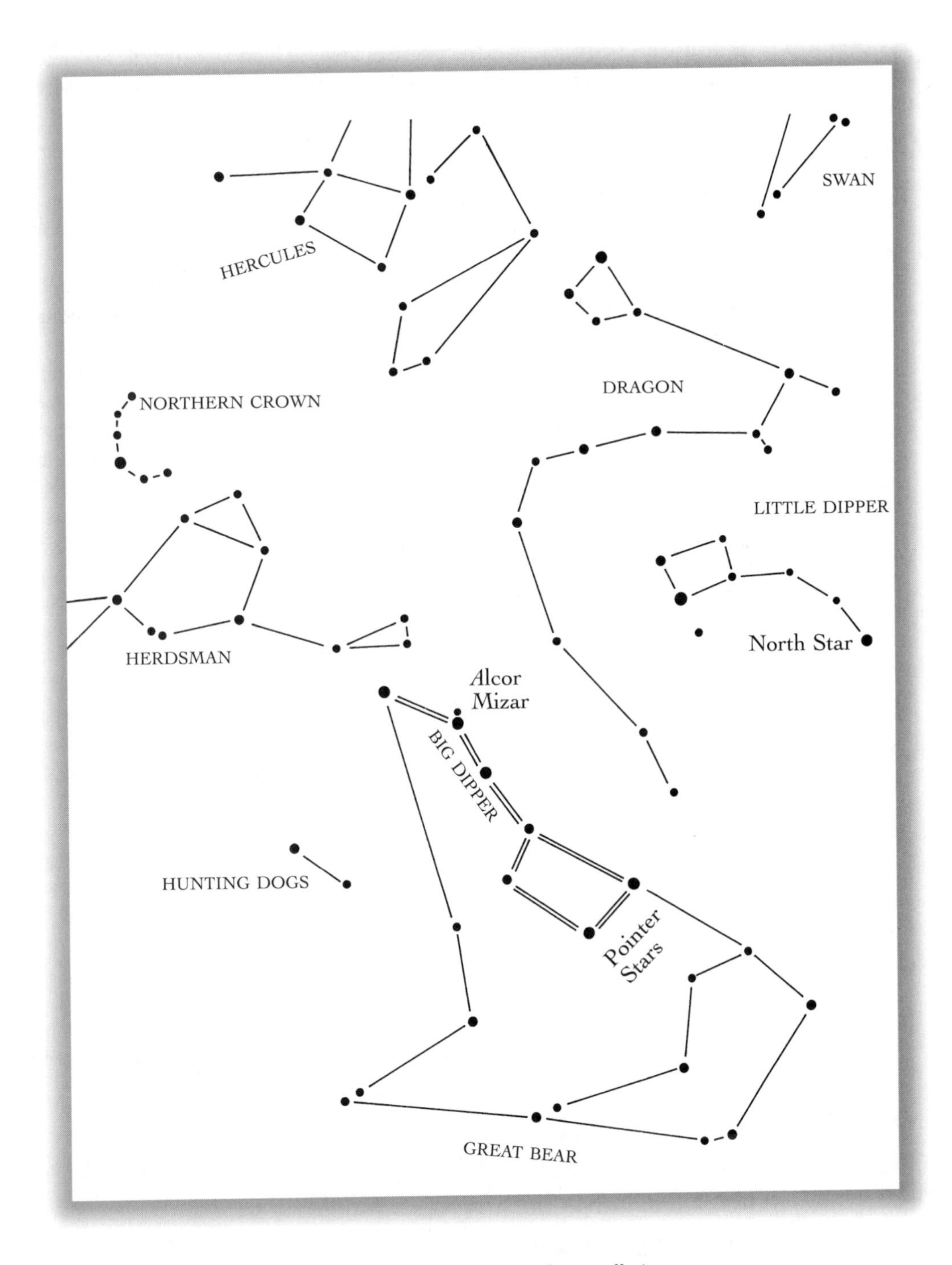

Fig. 2.1. BIG DIPPER, LITTLE DIPPER, and nearby stars and constellations.

on until at the North Pole it is ninety degrees from the horizon, or directly overhead. In the geographical area covered by this book, the LITTLE DIPPER can be seen throughout the night and throughout the year (see Fig. 2.1).

CASSIOPEIA

CASSIOPEIA is a group of stars, the brightest of which form a large W in the northern sky. This constellation lies opposite the BIG DIPPER, with the North Star in between (see Fig. 2.2 or any of the star maps).

The mythic Cassiopeia was an Ethiopian queen. Her husband was named Cepheus and their daughter Andromeda. When the queen boasted that she was more beautiful than the sea nymphs, the nymphs asked Poseidon to send a monster to destroy Ethiopia. The sea monster ravaged the coast until Cepheus sacrificed Andromeda by chaining her to a rock. But just as the monster was about to reach Andromeda, the hero Perseus set her free and killed the monster. In one form of the myth, Perseus then went to Cassiopeia and Cepheus asking for Andromeda's hand; Perseus and Andromeda married and lived happily thereafter. In another version of the myth, Perseus had agreed to set Andromeda free on the condition that he and Andromeda marry, but Cepheus and Cassiopeia reneged on their promise when Cassiopeia called in another of Andromeda's suitors during the marriage ceremony. Poseidon put both Cepheus and Cassiopeia among the stars. Because of her duplicity, Cassiopeia was tied into a market basket and—for part of each night—is upside down, a humbling position for a queen. Athene, one of Zeus's daughters, later elevated Andromeda to the sky as well. (See Fig. 2.9 for an illustration of Andromeda.)

Because CASSIOPEIA is close to the North Star, the constellation can be seen throughout the year except in southern areas of the United States, where it disappears along the horizon in April, May, and June. Fig. 2.2 shows the location of the BIG DIPPER, the North Star, and CASSIOPEIA relative to one another.

CEPHEUS

Look for King CEPHEUS next to Queen CASSIOPEIA. CEPHEUS is visible when CASSIOPEIA is visible (see Fig. 2.2).

PERSEUS

PERSEUS is a winter constellation best seen in the winter sky from November to March. PERSEUS lies between CASSIOPEIA and CHARIOTEER (see Figs. 2.2 and 2.3). The Perseid meteor shower, whose meteors appear to come from this part of the sky, takes place each August. The best display is after midnight.

LIZARD (LACERTA)

LIZARD is an inconspicuous constellation between CEPHEUS (see Fig. 2.2) and the GREAT SQUARE OF PEGASUS (see Fig. 2.9).

CHARIOTEER (AURIGA)

CHARIOTEER stands out because one of its stars, yellowish Capella, is the sixth brightest in the sky. Capella lies in an arc of bright stars circling ORION; these stars also include

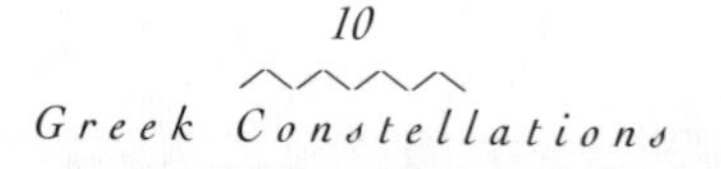

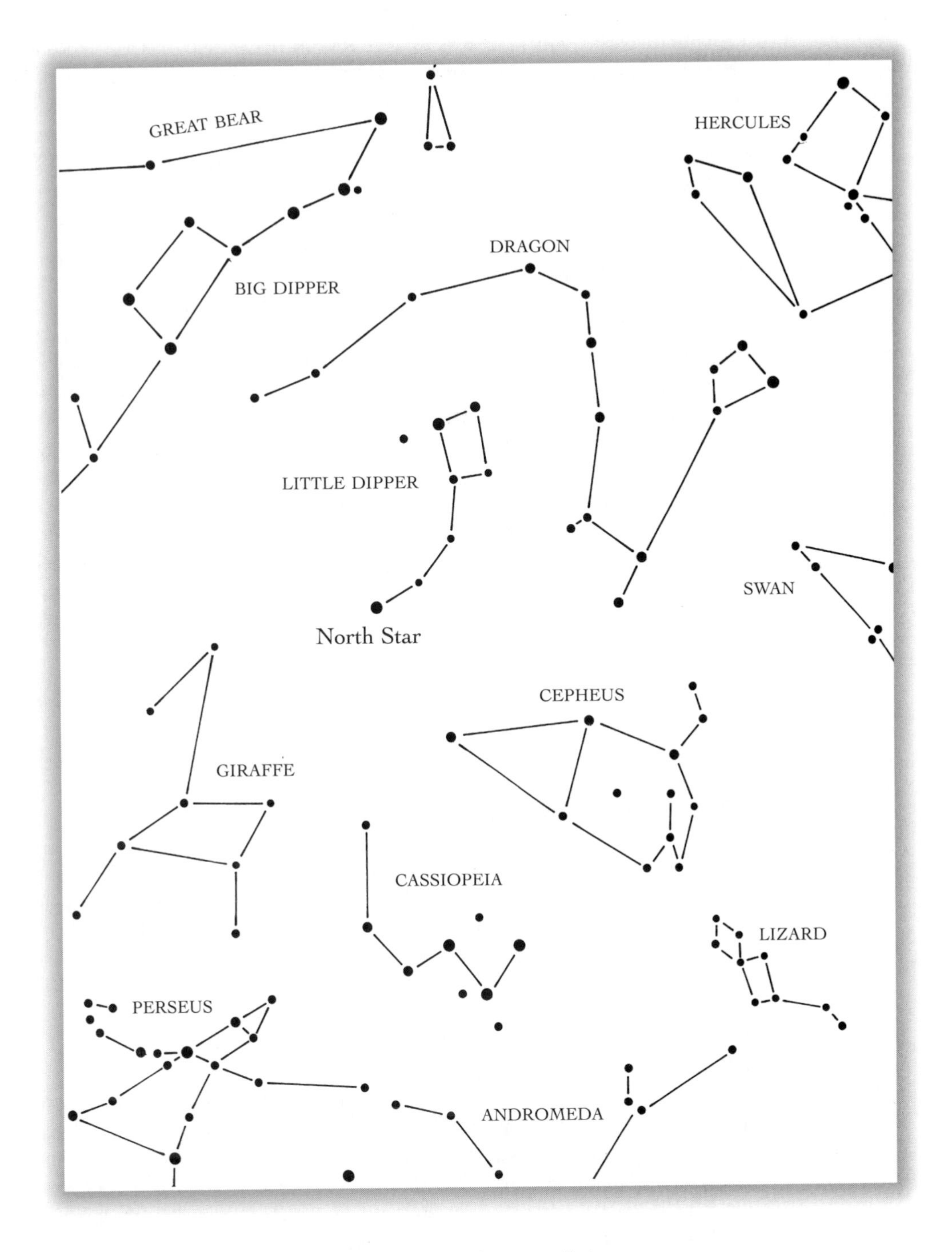

Fig. 2.2. CASSIOPEIA, PERSEUS, CEPHEUS, and nearby constellations.

Castor and Pollux, Procyon, and Sirius. (See Star Maps 1, 2, and 6, at the end of this chapter.) Officially, El Nath is in TAURUS, but Rey's drawings show El Nath as one of the horns of the bull, and his interpretation is used here.

In Greek myth, CHARIOTEER invented the chariot, which changed the face of warfare because it afforded a vast tactical advantage over foot soldiers.

Look for CHARIOTEER from October through April. See Star Maps 1, 2, 5, 6, and Fig. 2.3.

TWINS (GEMINI)

Castor, which rises first, and Pollux, which follows, do indeed look like twin stars, and the pair has been so identified by many ancient civilizations. Yellowish Pollux is somewhat brighter, though, and is the seventeenth brightest star in the sky. GEMINI is part of the zodiac, so there may be triplets when a planet passes through this constellation.

Stories about Castor and Pollux (also called Polydeuces) vary. In some, Castor was the son of the king of Sparta and Pollux was the son of Zeus, and Leda was the mother of both. They became fast friends and were twins in action, if not by birth. But when Castor was killed in a fight, he had to go to Hades because of his mortal birth. Pollux was inconsolable because he, as the son of a god, would go to Olympus and never see his brother again. Pollux convinced Zeus to allow Castor and him to remain together by spending one day in Hades and the next in Olympus throughout eternity. In one account, the two stars in the sky belonged to these two brothers. In another, Zeus set the images of Castor and Pollux among the stars.

In other Greek myths, Castor, Pollux, and Helen hatched from an egg Leda laid after she had lain with Zeus, who had taken the form of a swan. Later, Castor and Pollux fought bravely in the Trojan War, which was triggered by their sister's abduction.

Look for GEMINI from December through May. See Star Maps 1, 2, 6, and Fig. 2.3.

LITTLE DOG (CANIS MINOR)

Procyon, Greek for "before the dog," is one of the two stars in the constellation LITTLE DOG. LITTLE DOG does indeed rise before BIG DOG. Procyon is a yellowish white star that is the eighth brightest in the sky. You can find Procyon by following a curved line from Capella through Castor and Pollux in GEMINI, or draw a line east through ORION's shoulders. See page 18 (HERDSMAN) for a Greek myth that accounts for the presence of Procyon in the sky.

Look for Procyon from December through May. See Star Maps 1, 2, 6, and Fig. 2.3.

ORION

ORION is a large, beautiful winter constellation that rises on its side and marches across the southern sky. Look for three bright stars in a row, ORION's belt, below which hangs ORION's sword. The middle "star" of the sword is the Great Nebula of ORION, an immense gas cloud that appears as a hazy star (shown in Fig. 2.3 as a circle that is not filled in). Bluish white Rigel, the seventh brightest star in the sky, lies below the belt, and reddish Betelgeuse, the twelfth brightest star in the sky, lies above it.

In Rey's depiction of ORION, the hunter also holds a shield and a club (shown in

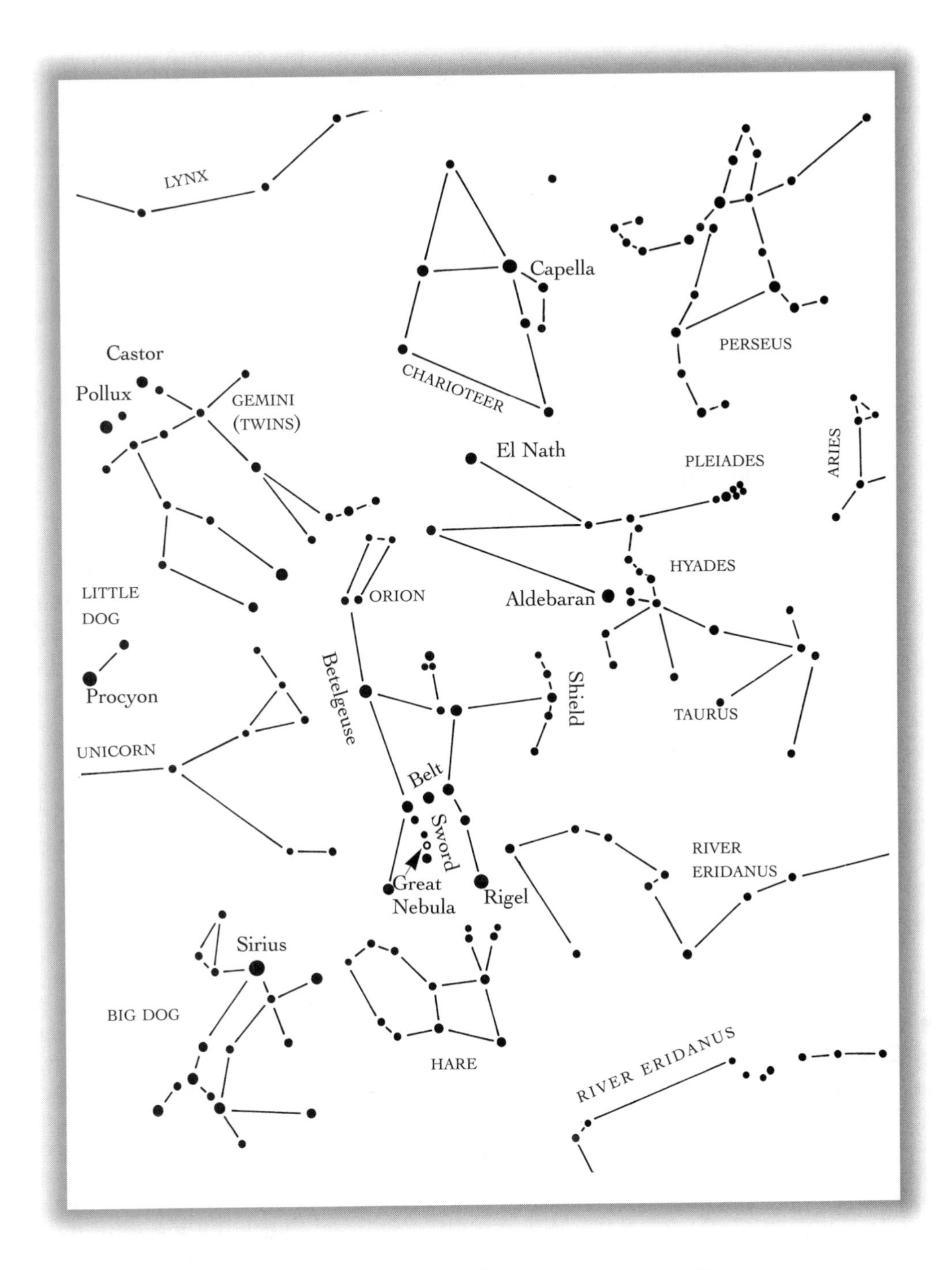

Fig. 2.3. CHARIOTEER, TWINS, ORION, TAURUS, and nearby stars and constellations.

Fig. 2.3), but because these stars are faint and do not generally appear in Native American constellations, they are not shown on the star maps and later illustrations.

Orion was a handsome and mighty hunter who fell in love with Merope—one of the PLEIADES—and promised her father that he would kill all the dangerous animals that inhabited their island. Though Orion was successful, the father went back on his word and blinded Orion. To regain his sight, Orion went on a long journey, after which he visited Crete. Here he met the beautiful goddess Artemis.

In one version of the myth, Apollo, Artemis's brother, feared that Artemis would succumb to Orion's charms and so tricked Artemis into killing Orion. She called upon the healer Asclepius to help revive Orion, but Asclepius was struck by a thunderbolt and died. Artemis lifted Orion into the sky, where he remains to this day. In another version of the myth, a scorpion stung Orion to death.

ORION appears in midwinter and dominates the heavens through April. See Star Maps 1, 2, and 6.

BIG DOG (CANIS MAJOR)

BIG DOG follows its master, ORION, into the sky. Bluish Sirius, the brightest star in the sky, easily stands out. Locate Sirius by drawing a line left from ORION's belt, or by drawing an arc from Capella through Castor, Pollux, and Procyon on to BIG DOG. In Greek myth Orthrus, a two-headed dog who fathered the Sphinx and the Nemean lion, is Sirius.

January through March is the best time to see BIG DOG. Look for Sirius on Star Maps 1, 2, 6, and Fig. 2.3.

HARE (LEPUS)

BIG DOG not only follows ORION but also hunts for HARE, which is located below ORION. HARE does not have any bright stars, so it is not prominent in the sky.

Look for HARE next to BIG DOG from January through March (Fig. 2.3).

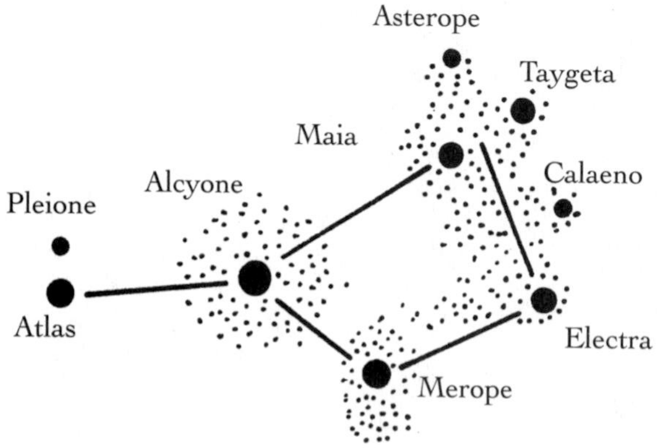

Fig. 2.4. With magnification, the brightest stars of the PLEIADES appear in a tiny dipper, and faint luminosity shrouds some of the stars.

BULL (TAURUS)

TAURUS is a large zodiacal constellation that contains three eye-catching elements: Aldebaran, the HYADES, and the PLEIADES. Orange Aldebaran, part of the head of the bull, is the thirteenth brightest star in the sky. Immediately next to Aldebaran is the V of stars that forms the HYADES. There are about two hundred stars in this cluster, though only a dozen or so are visible to the unaided eye. In Greek myth, the six Hyades were daughters of Atlas (who holds the world upon his shoulders) and half-sisters of the Pleiades. The HYADES were called the Rainy Stars because their appearance in the fall in ancient Greece coincided with the rainy season.

The PLEIADES, also called the Seven Sisters, were seven daughters of Atlas and Pleione: Electra, Maia, Taygeta, Alcyone, Merope, Celaeno, and Asterope (or Sterope). The sisters and their parents form a distinctive hazy cluster that was at one time a constellation of its own but is now folded into TAURUS. Six or seven stars are visible to the unaided eye. According to Greek myth, Electra disappeared because of her grief over the fall of Troy. In another story, it was Merope who left her sisters in the sky; she was embarrassed because she married Sisyphus, who lived in the underworld instead of the Olympic heavens. The "missing" star is probably Pleione, which varies in brightness over time.

A small telescope reveals about a hundred stars in the cluster, and a large telescope shows dozens more (see Fig. 2.4).

Aldebaran, the HYADES, and the PLEIADES are part of TAURUS, or the bull, an animal that is well known in myth. Zeus, in the form of a bull, carried the maiden Europa away over the sea. He also loved Io, whom his jealous wife turned into a heifer. Zeus named the Bosporus (which means "ox ford") in Io's honor. The Cretan queen Pasiphaë mated with a bull and gave birth to the Minotaur.

TAURUS is easy to spot because it is next to, and rises just before, ORION (see Fig. 2.3). The best time to look for this constellation is from October through March. On Star Maps 1, 2, 5, and 6, only the component parts that appear in Native American myths—Aldebaran, the HYADES, and the PLEIADES—are shown.

LYRE (LYRA)

LYRE is a small constellation from which shines forth Vega, a bluish white star that is the fifth brightest in the sky. LYRE is located near the SWAN; if you see a bright star near a cross (part of the SWAN), it is likely Vega. Or, if you see three stars in a row (part of EAGLE) in the summer sky, use them as pointers to Vega. The "summer triangle" of Vega, Deneb, and Altair is an important landmark in the night sky. (See Star Maps 3, 4, and 5 as well as Fig. 2.5.)

In Greek myth, Arion, a poet and musician who was celebrated for his skill with the lyre, was given many valuable gifts for winning a musical contest. When Arion learned that the crew of the ship he was on intended to kill him and steal his wealth, he asked to be allowed to sing one last song. He then jumped (or the crew threw him) into the sea. But his song had brought a school of dolphins to the ship, and one of the dolphins bore him to the surface and then safely to land. The lyre represents Arion and his lyre.

Look for LYRE from May through November (Fig. 2.5).

SWAN (CYGNUS)

Next to LYRE is the SWAN, a large bird with an outstretched neck that lies in the Milky Way. The NORTHERN CROSS (drawn with a double line in Fig. 2.5) lies within the SWAN. The "cross" stars form the wings of the bird, and several smaller stars stretch out in front to form the neck and head. Deneb is the tail. As the nineteenth brightest, Deneb just barely makes it onto the top twenty star list.

In classical mythologies this constellation is variously called a swan, hen, vulture, ibis, and eagle. In one ending of the Phaëthon myth (see SCORPION, page 20), when Phaëthon fell from the sky his sisters mourned him; in another version, his friend Cygnus was so distraught that he swam back and forth in the river—like a swan—looking for the body. For his efforts, Cygnus was placed among the stars in the form of a swan.

The best months to see SWAN are from June through November. Also see Star Maps 3 through 6, in which the swan is represented by its brightest stars, those in the NORTHERN CROSS.

FOX (VULPECULA)

FOX is a modern constellation that consists of two dim stars—not much to look at—near the head of the SWAN (see Fig. 2.5).

ARROW (SAGITTA)

The ARROW is a small but distinct constellation that lies near the head of the EAGLE. The celestial arrow has been linked with many arrows in Greek mythology, including those belonging to Eros (Cupid) and Heracles (Hercules). See Fig. 2.5.

DOLPHIN (DELPHINUS)

DOLPHIN, like the PLEIADES, is a small figure that stands out in the sky not because it has bright stars but because the stars are so clearly clustered. The five brighter stars look like a kite with a tail; if you include the lesser stars, the figure becomes a crescent. Greek sailors revered the dolphin, telling stories about how dolphins saved shipwrecked castaways.

DOLPHIN is visible from July though November. For its position in the sky, see Star Maps 3 through 6, where it is represented by the brighter stars. See Fig. 2.5.

EAGLE (AQUILA)

If you see three stars in a row in the evening sky from December to March, it's the belt of ORION, but if you see three stars from July through October, it's the head of the EAGLE.

Zeus often dispatched his beautiful and faithful eagle as an omen to the human world that important events would follow. The eagle also abducted Ganymede (see story about AQUARIUS on page 24) to be the Olympian cupbearer.

Altair, which is the translation of an Arabic word meaning "the flying eagle," is the brightest star in this constellation and the eleventh brightest in the sky. In Rey's depiction of the eagle, this yellowish white star is the main star in the head. You can use Altair and the two flanking stars in the head to locate other constellations. DOLPHIN and ARROW lie in EAGLE's path; if you follow the line of the three stars north toward the North Star, you will cross the bottom of the NORTHERN CROSS and arrive at Vega (see Fig. 2.5).

Altair, Vega, and Deneb in the SWAN form a large triangle called the Summer Triangle because the trio is high in the sky at this time of the year. See Fig. 2.5 and Star Maps 3, 4, and 5.

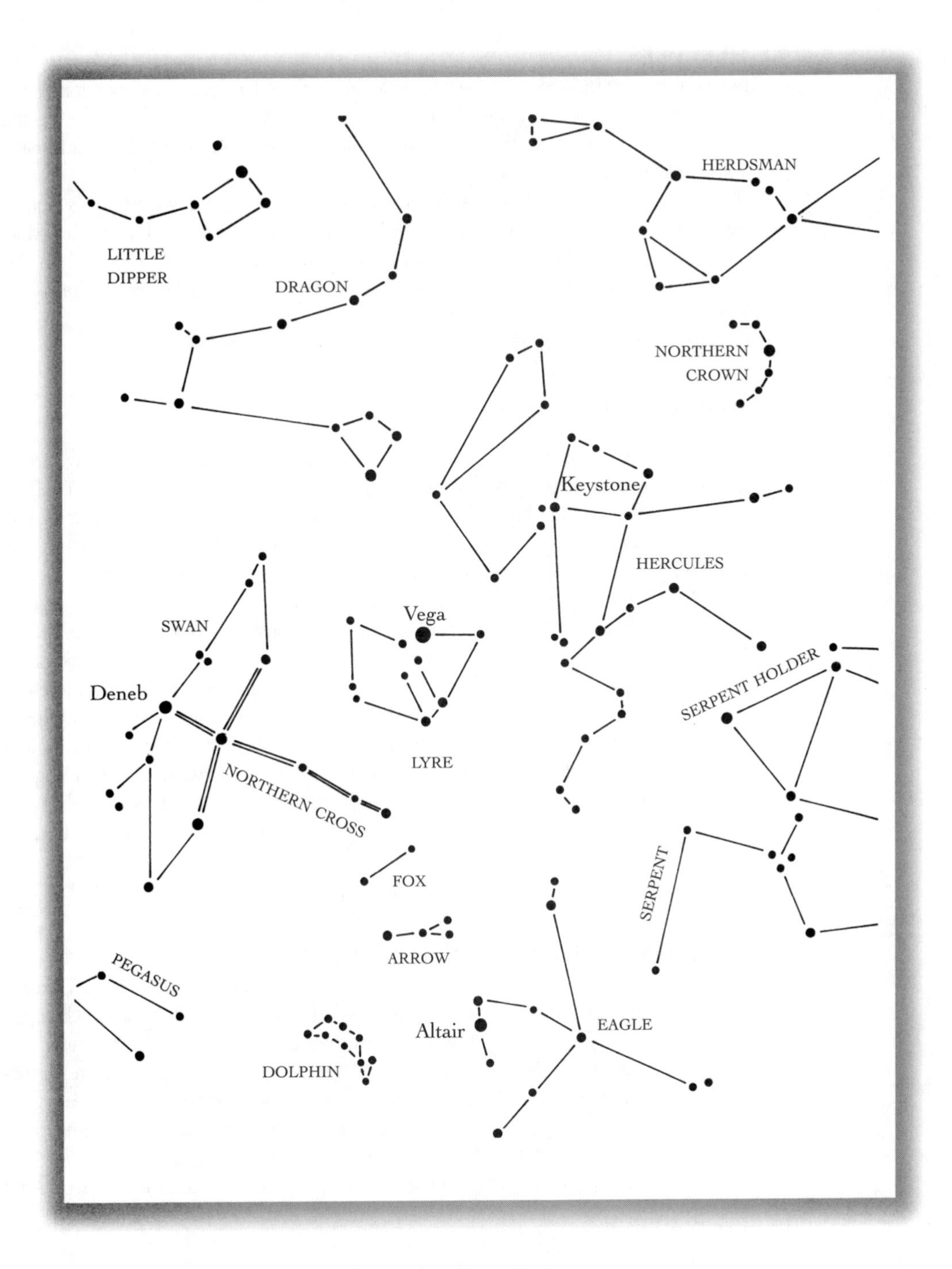

Fig. 2.5. LYRE, SWAN, FOX and nearby stars and constellations.

HERCULES

HERCULES, located between Vega and the NORTHERN CROWN, has lots of history but no large stars. First locate the "keystone," which in Fig. 2.5 is the head, and then locate the smaller stars that make up the rest of the body. Hercules is shown ready for adversity, with bent knee and club upraised.

Hercules is the Latin form of the Greek Heracles. In Greek myth, Heracles, son of Zeus, killed his own children while under the influence of a spell. To atone for his crime he performed the Twelve Labors, which included slaying the Nemean lion. In the process of performing the second labor he killed a crab; Hera set the crab in the zodiac to reward it for its bravery.

Look for HERCULES from May to October.

HERDSMAN (BOÖTES)

The HERDSMAN lies just beyond the tip of the handle of the BIG DIPPER. In Fig. 2.6 he is calmly smoking his pipe (though this image is hardly classic), the triangular configuration of stars projecting from his head. In this portrayal, the herdsman's hip is orange Arcturus, the fourth brightest star in the sky. Locate Arcturus by starting at the BIG DIPPER's handle and tracing an arc away from the bowl (see the star maps).

The HERDSMAN is so called because in ancient times the nearby stars were considered a flock of sheep and he was their keeper. The constellation was also called Bear-Watcher or Bear-Guard because it stands next to the GREAT BEAR, and it was called Wagoner, because in some cultures the GREAT BEAR was considered a wagon. Greek legend says that the HERDSMAN represents Icarius, the first human to make wine. After drinking undiluted wine and getting quite drunk, a group of shepherds set upon Icarius and killed him. Icarius's dog, Maera, led his daughter, Erigone, to his grave. When Erigone found out that her father was dead, she hanged herself, calling upon other maidens to do likewise until her father was avenged. In some versions of the myth, Icarius became BOÖTES, Erigone VIRGO, and the dog Procyon in LITTLE DOG.

Look for HERDSMAN from April through August. See Star Maps 2, 3, 4, and Fig. 2.6.

NORTHERN CROWN (CORONA BOREALIS)

The NORTHERN CROWN lies next to the HERDSMAN. The stars in the crown are not exceptionally bright, but the distinctive semicircle is easy to spot. In ancient Greek myth, Ariadne, the daughter of King Minos of Crete, helped Theseus kill the Minotaur and escape from its maze. They fled together, but when Theseus learned in a dream that Ariadne was to marry a god, he deserted her. Then the god Dionysus came along, gave Ariadne a beautiful crown, and married her. They lived happily, and when Ariadne died, Dionysus placed her crown in the sky, where it became the NORTHERN CROWN.

Look for the NORTHERN CROWN at the same time of year you look for HERDSMAN, April through August. See Star Maps 2 through 5 and Fig. 2.6.

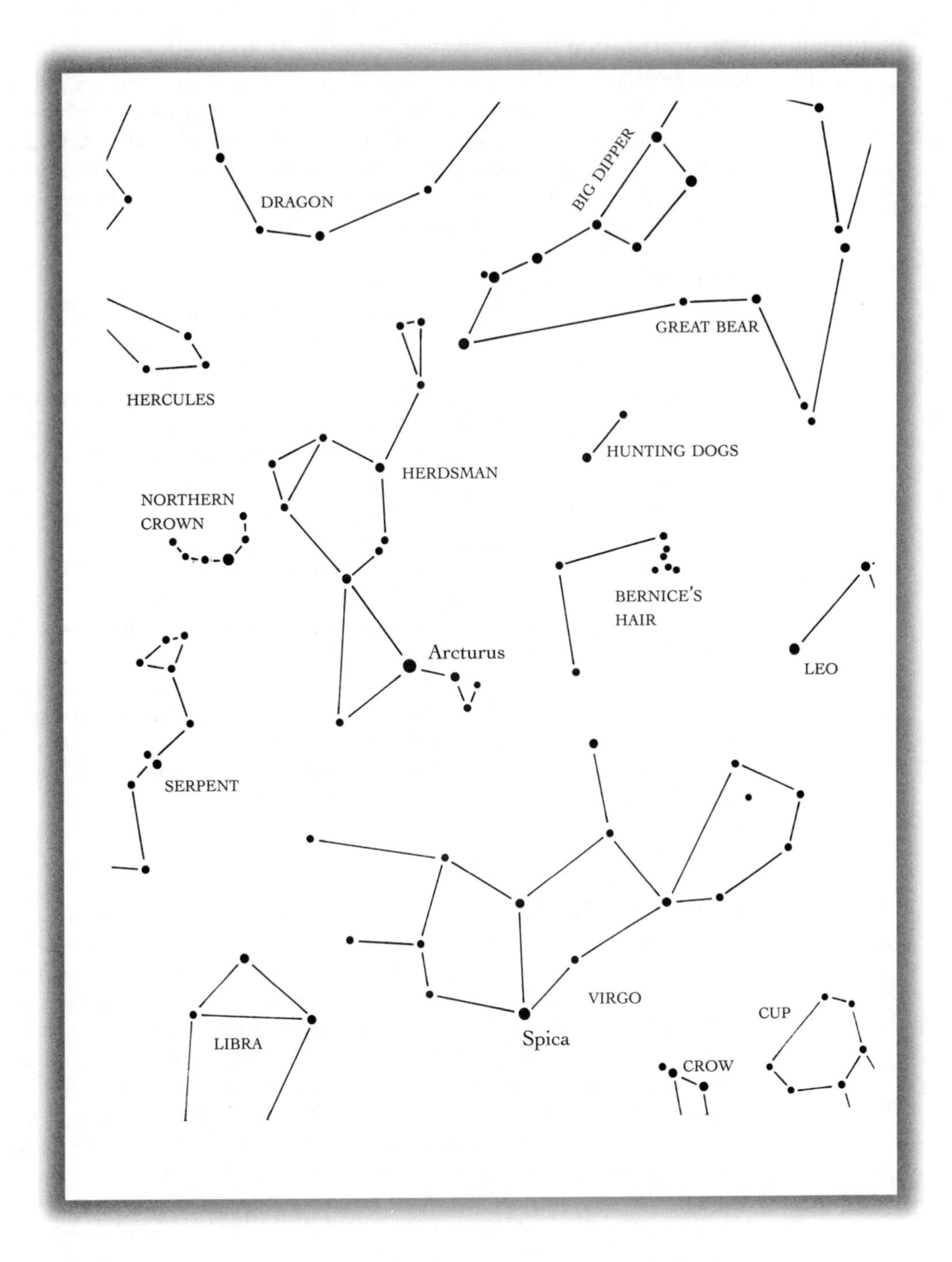

Fig. 2.6. HERDSMAN, NORTHERN CROWN, VIRGO, and nearby stars and constellations.

VIRGIN (VIRGO)

South of the HERDSMAN lies bluish Spica, the fifteenth brightest star in the sky and the only bright star in the zodiacal constellation VIRGO. Follow the arc from the BIG DIPPER'S handle out through Arcturus and on to Spica. VIRGO owes its prominence to its place in the zodiac: It lies between LEO and LIBRA. Erigone, who killed herself from grief over her slain father (see HERDSMAN, page 18), became the constellation VIRGO (see Fig. 2.6).

Look for VIRGO in April, May, and June.

SCORPION (SCORPIUS)

SCORPIUS is a majestic constellation that dominates the southern sky in July and August. The farther south you are, the more you can see of its striking, sweeping curves, and in the southern United States you can see it in all of its glory.

SCORPIUS features reddish Antares, the sixteenth brightest star in the sky. The only other objects of this color (excluding lights on planes) are reddish Betelgeuse in ORION, orange Aldebaran in TAURUS, and reddish MARS. Because Betelgeuse and Aldebaran are winter stars and have left the sky when Antares appears, if you see a red star in summer, it's Antares—unless it is the planet MARS, which travels through the zodiac, of which SCORPIUS is a part. SCORPIUS also contains Shaula and Lesath, a distinctive pair of stars called the Stinger Stars that are located at the tip of the scorpion's tail.

In one version of the Orion myth, a scorpion stung and killed Orion.

In other myths, the scorpion and other members of the zodiac are portrayed as living creatures in the sky. When young Phaëthon begged his father, Apollo, to allow him to drive Apollo's chariot—the sun—through the sky, Apollo tried to dissuade his son by describing the perils he would face: "You will have to pass beasts, fierce beasts of prey, and they are all that you will see. The Bull, the Lion, the Scorpion, the great Crab, each will try to harm you." But Phaëthon insisted, and the ride ended in disaster. The horses ran the chariot up against SCORPIUS and nearly ran over CANCER. Phaëthon was killed and fell from the sky.

See Fig. 2.7 and Star Maps 3 and 4.

ARCHER (SAGITTARIUS)

SAGITTARIUS follows SCORPIUS into the southern sky. At forty degrees north latitude—the latitude of Denver, Dayton, and Philadelphia—the archer's legs are often lost in haze, but in the southern United States the whole constellation is visible on a clear night. The most distinctive part of SAGITTARIUS is the Milk Dipper, or Teapot (double lines in Fig. 2.7), a small rectangle that looks like the bowl of the BIG DIPPER but is only about one-fourth as large. The Milk Dipper gets its name because it lies along the Milky Way.

SAGITTARIUS is generally described as a *centaur,* a being with the head, chest, and arms of a man and the body of a horse, though in Rey's interpretation the archer appears as a man with two legs, a bow, and a feather in his head. There are many centaurs in Greek mythology, most of whom were loud and loutish. The centaur Chiron, however, was kind, learned, and wise, and when he was killed by accident, Zeus placed him in the stars.

Look for SAGITTARIUS, which lies in the zodiac next to SCORPIUS, in July and August. See Fig. 2.7 and Star Maps 3 and 4.

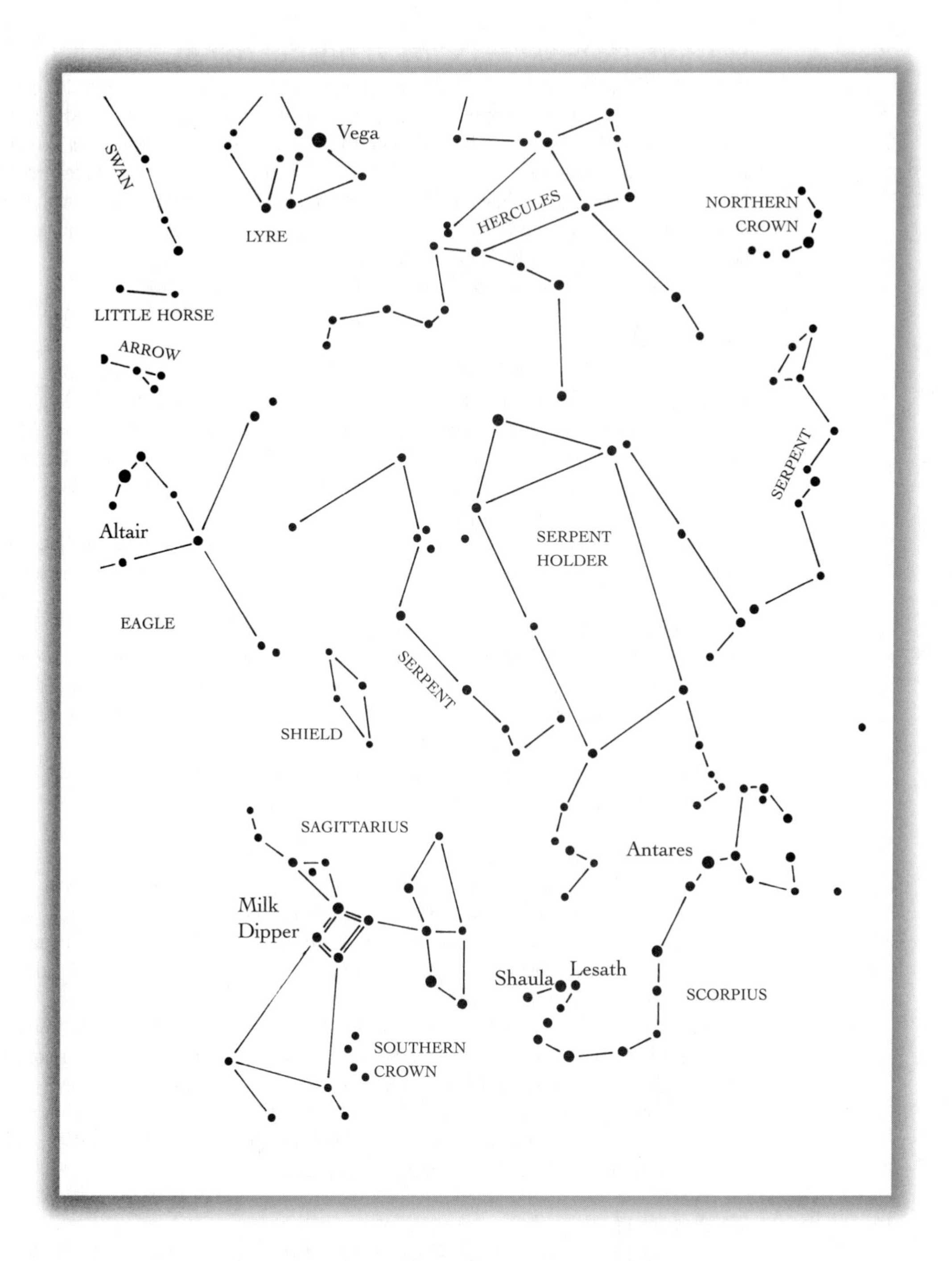

Fig. 2.7. SCORPIUS, SAGITTARIUS, SERPENT, and nearby stars and constellations.

SERPENT (SERPENS) and SERPENT HOLDER (OPHIUCHUS)

SERPENT and SERPENT HOLDER are two large intertwined constellations that require concentration to trace in their entirety. Apollo and Coronis, a mortal woman, had a son, Asclepius. Asclepius became a healer; Chiron the centaur was his first teacher. When Asclepius tried to save Orion (see page 14), Zeus killed the healer with a thunderbolt but later brought Asclepius back to life. Zeus placed the image of the healer—with the serpent that Asclepius considered sacred—among the stars. The SERPENT is the only classical constellation that is divided into two parts.

The SERPENT and SERPENT HOLDER are visible in July and August (see Fig. 2.7).

LION (LEO)

LEO, a member of the zodiac, strides across the sky below the BIG DIPPER. The constellation's brightest star, bluish white Regulus, is the twenty-first brightest in the sky. Regulus is part of the Sickle, a figure within LEO that is marked in Fig. 2.8 with double lines. The most famous lion in Greek myth is the Nemean lion, which Hercules killed as the first of his Twelve Labors. It is not known, however, whether LEO represents the Nemean lion or some other lion in Greek myth.

To locate Regulus, start at the BIG DIPPER's Pointer Stars and look south. See Fig. 2.8 and Star Maps 1, 2, and 3. Look for LEO below the BIG DIPPER from February through June.

CRAB (CANCER)

CANCER is an undistinguished constellation that lies in the zodiac between LEO and GEMINI. Hercules killed a crab that hindered him during the second of his Twelve Labors (see page 18). This constellation contains the Beehive (also called M44), a cluster of about fifty stars that appears as a faint haze.

You need good viewing conditions to see CANCER; look for it from January through May (see Fig. 2.8).

CROW (CORVUS)

The crow is a member of the Corvidae family, which also includes ravens, blue jays, and magpies. The crow in the sky has no bright stars, but it is quite recognizable because it lies in a rather empty section of the sky. You need to use a little imagination to see a crow in these stars, which otherwise appear as a parallelogram. The upper side is aligned with Spica (see Star Map 2).

In Greek myth, Corvus was a pet raven who did errands for Apollo. In Roman myth, the bird was a crow. Apollo sent the bird—either raven or crow—to get a goblet of water, but the bird loitered until some fruit became ripe enough to eat and then returned with a snake instead. Apollo sent the bird, cup, and snake to the sky.

The CROW moves through the southern sky from April through June and is more prominent in the southern United States. See Fig. 2.8.

CUP (CRATER)

CUP is an undistinguished constellation. It is the goblet or cup that Apollo asked Corvus to fetch. See Fig. 2.8.

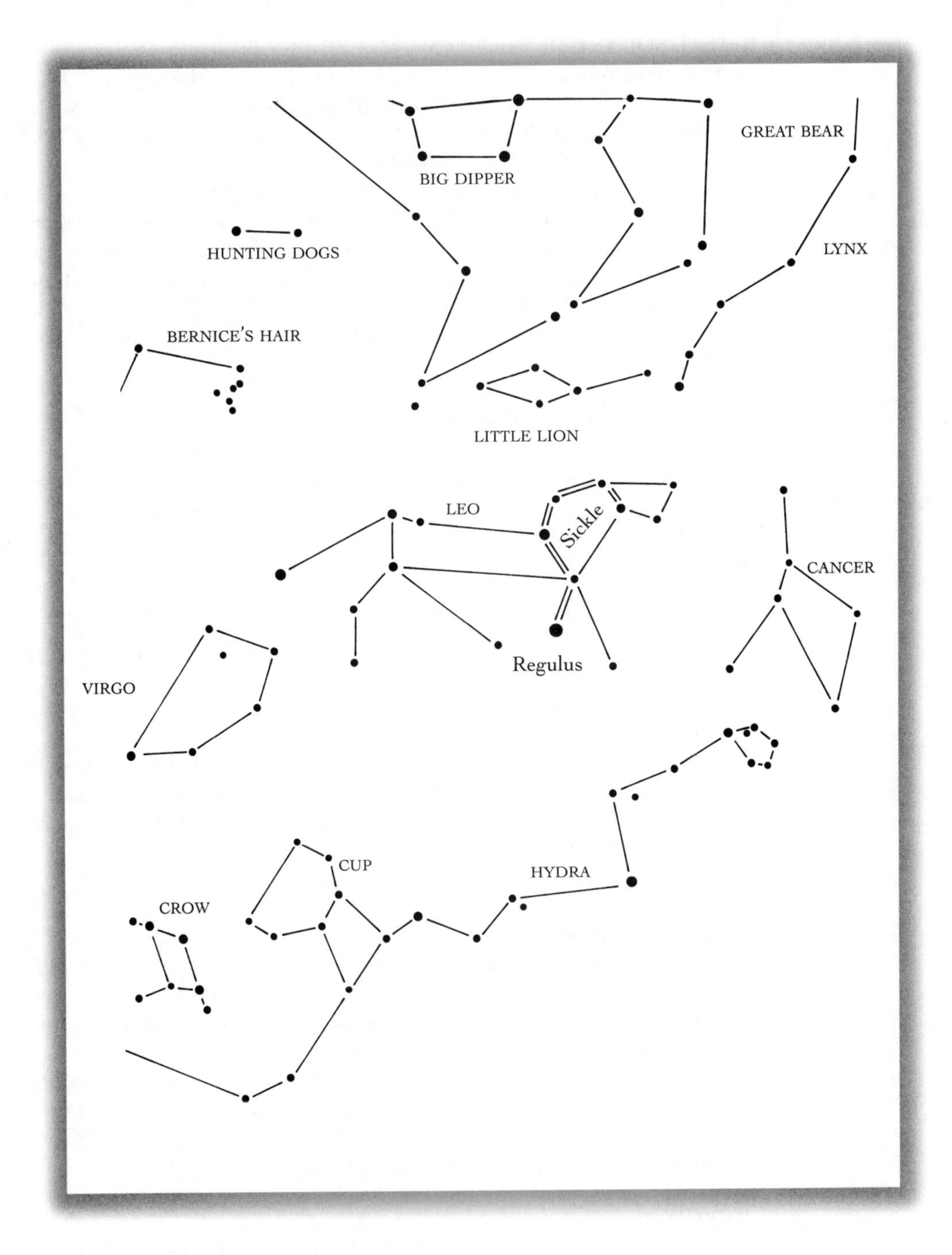

Fig. 2.8. LEO, CANCER, CUP, CROW, and nearby stars and constellations.

GREAT SQUARE OF PEGASUS

PEGASUS and ANDROMEDA contain stars that together make up the GREAT SQUARE OF PEGASUS. In Rey's version of PEGASUS and ANDROMEDA, the wings of PEGASUS and ANDROMEDA's head form this square (double lines in Fig. 2.9). The GREAT SQUARE stands out because it occupies an area that is otherwise devoid of distinctive features.

The winged horse Pegasus sprang from the womb of Medusa just as Perseus cut off her head. A handsome youth, Bellerophon, charmed Pegasus with a golden bridle and the two set off on many adventures. When Bellerophon desired to visit Olympus, however, Zeus sent a gadfly that stung Pegasus and made him throw his rider. Pegasus then went to live in the Olympian stables.

Look for the GREAT SQUARE from August to October. It is shown in Fig. 2.9 and in Star Maps 4, 5, and 6.

SOUTHERN FISH (PISCIS AUSTRINUS)

The dimmer stars of SOUTHERN FISH are largely lost in haze, but dazzling Fomalhaut stands out as the brightest star in the region and is the eighteenth brightest in the sky. SOUTHERN FISH should not be confused with PISCES (FISHES), a constellation of the zodiac.

Look for Fomalhaut from September through November. See Fig. 2.9 and Star Map 5.

WATER CARRIER (AQUARIUS)

The gods chose Ganymede, a handsome mortal, to serve as the cupbearer to Zeus. The choice angered Hera, because her daughter Hebe had held that job. Hera so annoyed Zeus over the matter that Zeus placed Ganymede among the stars as the constellation AQUARIUS, the water-carrier.

AQUARIUS, like CANCER, gets its distinction from its place in the zodiac rather than from brilliant stars. AQUARIUS lies between PISCES and CAPRICORNUS, two other rather dim zodiacal constellations. Locate AQUARIUS by drawing a line diagonally through the GREAT SQUARE from ANDROMEDA's head through PEGASUS's tail (see Fig. 2.9).

Look for AQUARIUS from August through October.

SHIP'S KEEL (CARINA)

In Greek myth, the ship *Argo* carried Jason and the Argonauts on their adventures. In the sky, the *Argo* is composed of several constellations, including SHIP'S KEEL. Yellowish white Canopus in SHIP'S KEEL is the second brightest star in the sky, but it is not visible in most of North America.

SHIP'S KEEL is not illustrated, but see Star Map 1, which shows Canopus just below the horizon at forty degrees north latitude. Canopus can be seen south of about 35 degrees north latitude in the evening in late February and early March. For reference, Los Angeles, Dallas, Houston, Atlanta, and Miami fall within this area.

Milky Way

All the stars that we see in the sky are part of a galaxy, an assemblage of stars, dust, and gas held together by gravitational forces. (Our home galaxy is designated in this

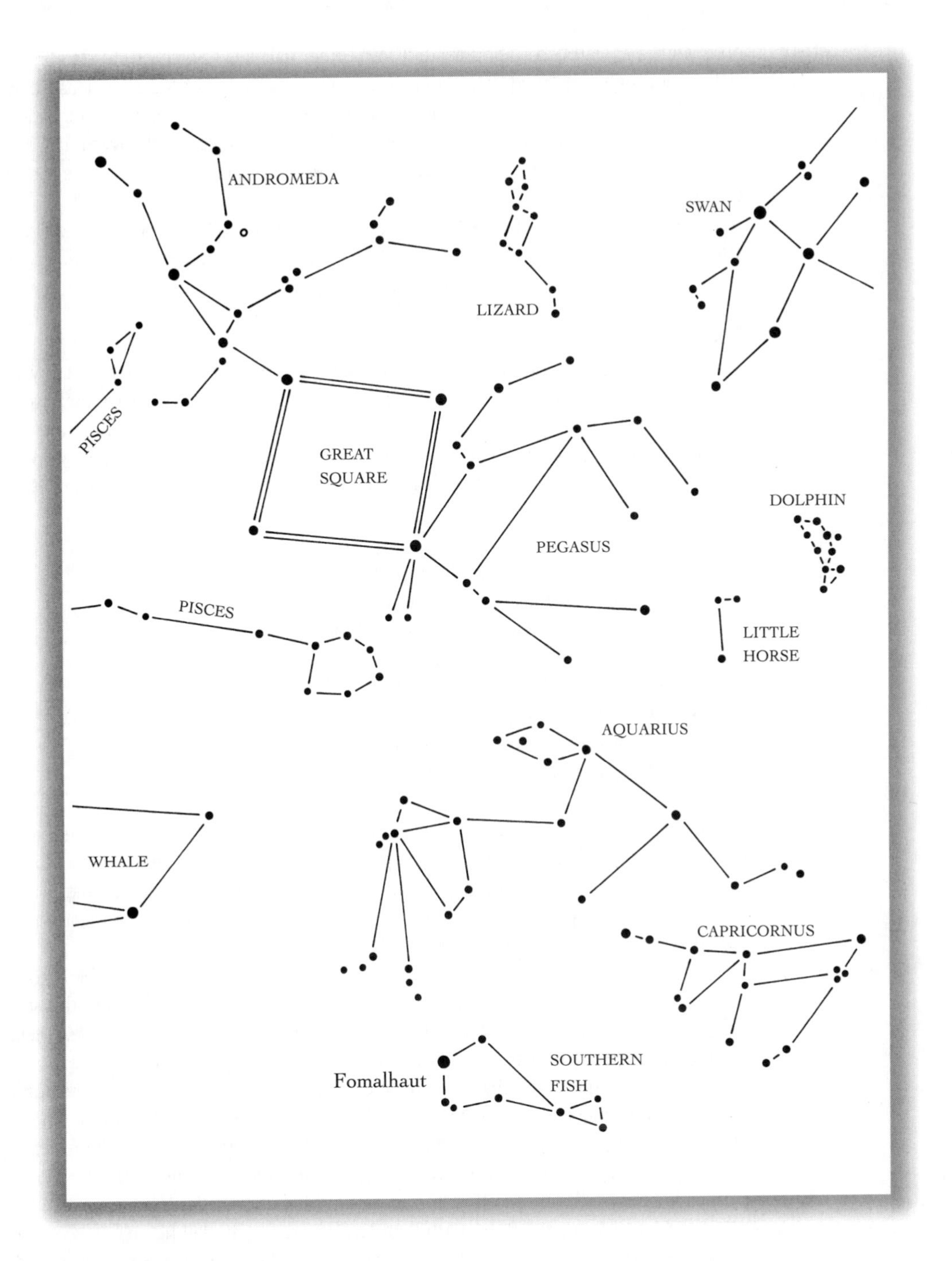

Fig. 2.9. The GREAT SQUARE OF PEGASUS, AQUARIUS, SOUTHERN FISH, and nearby stars and constellations.

book with a capital G to differentiate it from other galaxies.) Some of the stars in our Galaxy appear to us singly while others appear to be concentrated in a dense band called the Milky Way. This band moves through the night sky just as the constellations do. With good viewing conditions you can see it most of the year, although in late May and early June, when it sweeps along the northern horizon, it may become lost in ground haze.

The Milky Way appears as a backdrop to CHARIOTEER, PERSEUS, CASSIOPEIA, CEPHEUS, SWAN, EAGLE, SAGITTARIUS, SCORPIUS, and constellations of the Southern Hemisphere. See Star Maps 1 through 6 for the positioning of the Milky Way.

Star Maps

The six star maps that follow provide an overview of the night sky throughout the year. These maps show only the brightest stars and those constellations that appear most often in Native American stories. Use these maps to become familiar with how the brightest stars are positioned relative to one another and with how they move through the sky. These maps, like the constellation figures, are based on the drawings of H. A. Rey.

For **winter evenings,** use Star Maps 1 and 6
For **spring evenings,** use Star Map 2
For **summer evenings,** use Star Maps 3 and 4
For **fall evenings,** use Star Map 5
For **winter mornings,** use Star Maps 2 and 3
For **spring mornings,** use Star Map 4
For **summer mornings,** use Star Maps 5 and 6
For **fall mornings,** use Star Map 1

If the time at which you are stargazing is not shown next to the map, then refer either to the preceding or subsequent map, depending upon the time that most closely matches that of your observation. The times on the maps are for standard time. If you are observing in daylight savings time, subtract one hour from your observation time (i.e., if it is 9 P.M. daylight savings time, it is 8 P.M. standard time).

The maps show the sky as seen from forty degrees north latitude, roughly the latitude of Philadelphia, Columbus, Indianapolis, Denver, and Reno. Although the stars shown within the map circles are technically above the horizon, even on a clear night those stars along the horizon will not be visible because of ground haze.

If you are farther south than forty degrees, you will lose some stars along the northern rim of the circle and will be able to see more stars along the southern rim. On Star Map 1, for example, Canopus appears outside the horizon circle in the southern sky. At thirty degrees north latitude, however, Canopus rises above the horizon. The Florida panhandle, New Orleans, and Houston lie along the thirtieth latitude.

Likewise, if you are farther north than forty degrees, you will be able to see more of the northern stars and fewer of the southern stars. On Star Map 1, Deneb is outside the horizon circle. But at fifty degrees north latitude, Deneb is above the horizon. The northern part of Newfoundland, Winnipeg, Medicine Hat, and the upper section of Vancouver Island lie on the fiftieth latitude; the line runs just north of Minnesota, North Dakota, Montana, Idaho, and Washington.

On the star maps, east lies on the left and west lies on the right side of the map because astronomical maps are configured opposite of land-navigation maps. If you turn toward the south, you will notice that the sun rises to your left, in the east, and sets to your right, in the west. In the same way, when you "face" the southern sector of the star map, you will have east on your left and west on your right. The star maps are drawn so that you can hold them above your head and compare them to what is in the sky.

Star Map 1

The stars as seen at forty degrees north latitude.

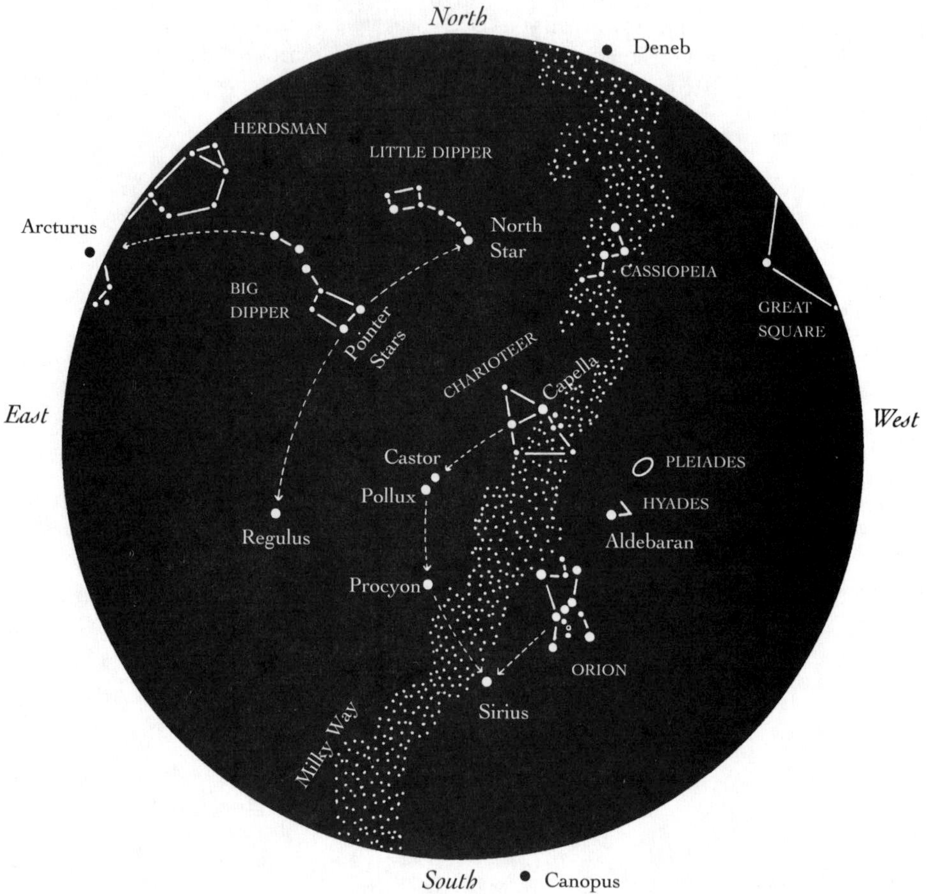

Evening viewing times:
January, from 10 P.M. to 1 A.M.;
February, from 8 P.M. to 11 P.M.;
March, from dusk to 9 P.M.

Morning viewing times:
October, from 4 A.M. to dawn.

The major sky features at this time:

The BIG DIPPER points to the North Star; CASSIOPEIA lies on the opposite side. Regulus is high in the sky; follow the Pointer Stars in the BIG DIPPER to the south. The bright stars Capella, Castor, Pollux, Procyon, and Sirius form an arc around ORION; ORION's belt points toward Sirius. Aldebaran, the HYADES, and the PLEIADES are in view. In the southern U.S., Canopus barely makes it above the horizon.

Star Map 2

The stars as seen at forty degrees north latitude.

North

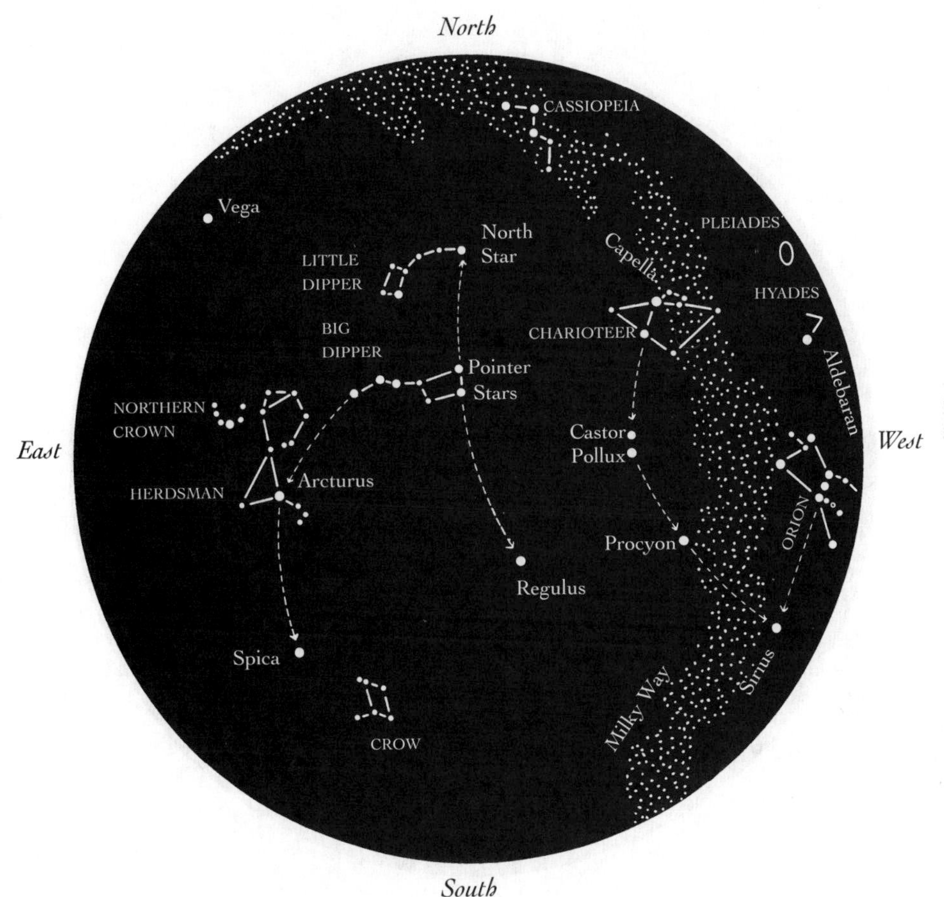

East | West

South

Evening viewing times:
March from 10 P.M. to 1 A.M.;
April from 8 P.M. to 11 P.M.;
May, at dusk.

Morning viewing times:
December, from 4 A.M. to 7 A.M.

The major sky features at these times:

HERDSMAN and the NORTHERN CROWN have risen; follow the BIG DIPPER's handle to locate orangish Arcturus in the HERDSMAN, then continue the arc to Spica in VIRGO; locate the CROW nearby. Vega has risen in the northern sky. Regulus is high in the sky; follow the Pointer Stars in the BIG DIPPER to the south. The BIG DIPPER points to the North Star; CASSIOPEIA lies on the opposite side. The bright stars Capella, Castor, Pollux, Procyon, and Sirius form an arc around ORION; ORION's belt points toward Sirius. Aldebaran, the HYADES, and the PLEIADES are setting.

Star Map 3

The stars as seen at forty degrees north latitude.

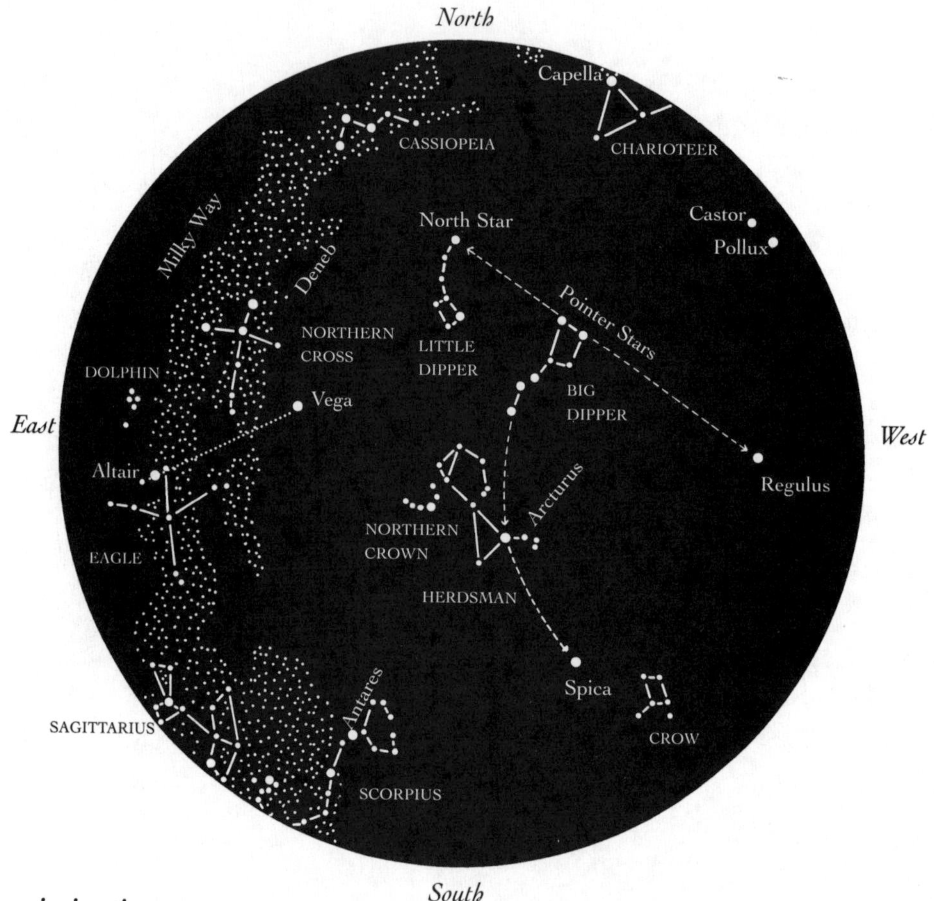

Evening viewing times:
May, from 10 P.M. to 1 A.M.;
June, from 8 P.M. to 11 P.M.;
July, at dusk.

Morning viewing times:
February, from 4 A.M. to 7 A.M.

The major sky features at these times:

Vega, SWAN (shown here as the NORTHERN CROSS), EAGLE, and DOLPHIN have risen; Vega, Deneb, and Altair form the "summer triangle." In the southeastern sky SCORPIUS (with bright red Antares) has appeared, and SAGITTARIUS follows. The HERDSMAN and the NORTHERN CROWN are high in the sky. Follow the BIG DIPPER's handle through orangish Arcturus in the HERDSMAN to Spica in VIRGO; locate the CROW nearby. Regulus is still in the sky; follow the Pointer Stars in the BIG DIPPER toward the west. Castor, Pollux, and CHARIOTEER are setting.

Star Map 4

The stars as seen at forty degrees north latitude

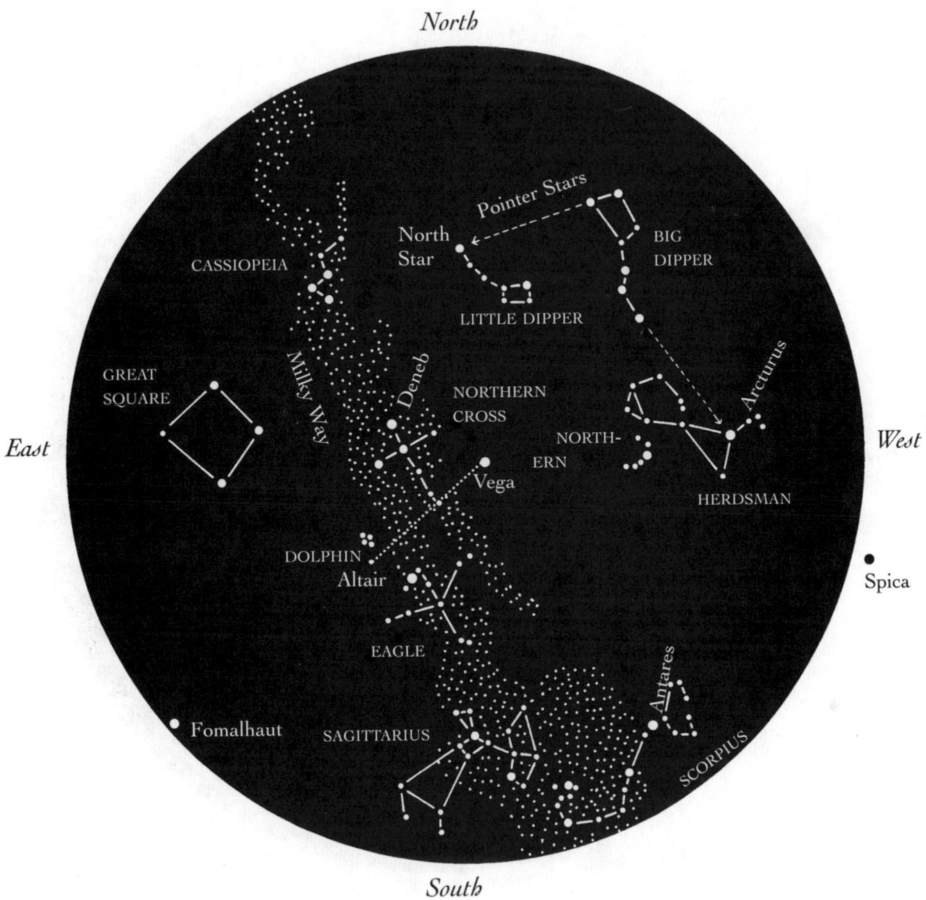

Evening viewing times:
July, from 10 P.M. to 1 A.M.;
August, from 8 P.M. to 11 P.M.;
September, from 6 P.M. to 9 P.M.

Morning viewing times:
April, at dawn.

The major sky features at these times:

The GREAT SQUARE has risen in the east; Fomalhaut is barely above the southeastern horizon. Vega, SWAN (shown here as the NORTHERN CROSS), EAGLE, and DOLPHIN are high in the sky; Vega, Deneb, and Altair form the "summer triangle." SCORPIUS, with red Antares, and SAGITTARIUS dominate the southern sky. In the northwestern sky HERDSMAN, with orange Arcturus, the NORTHERN CROWN, and the BIG DIPPER are prominent.

Star Map 5

The stars as seen at forty degrees north latitude.

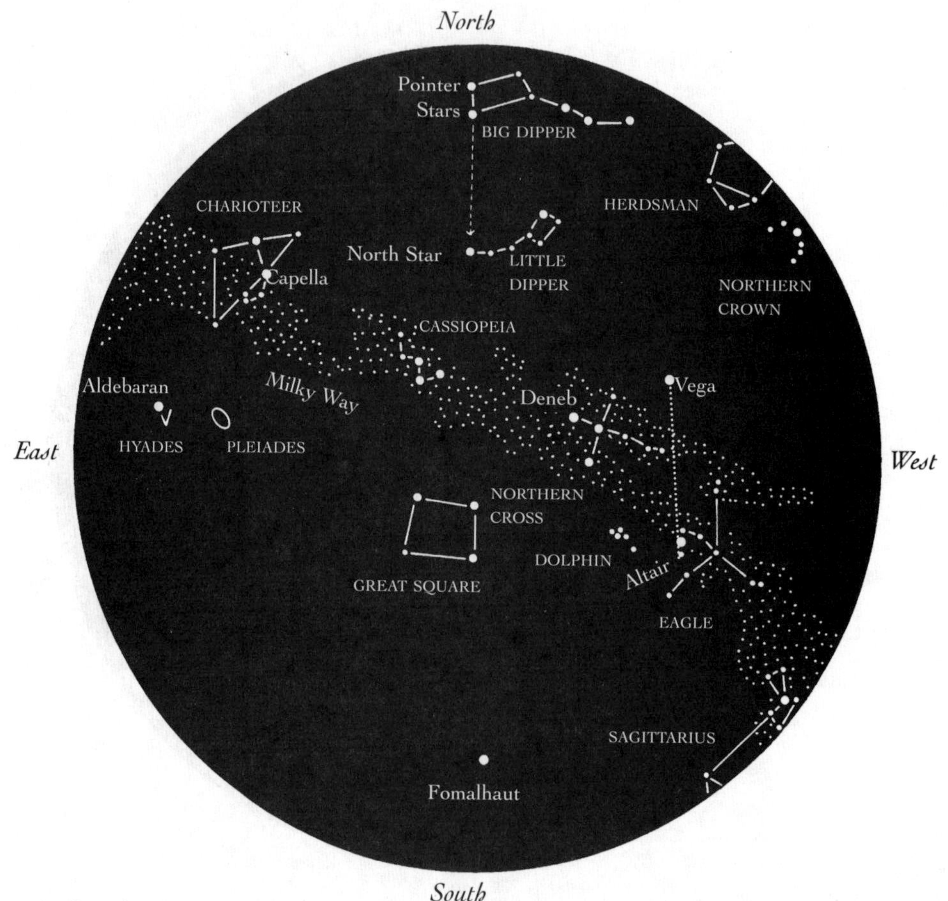

Evening viewing times:

September, from 10 P.M. to 1 A.M.;

October, from 8 P.M. to 11 P.M.;

November, from 6 P.M. to 9 P.M.;

December, from dusk to 7 P.M.

Morning viewing times:

July, from 2 A.M. to dawn.

The major sky features at these times:

Aldebaran, the HYADES, and the PLEIADES (all in TAURUS) have risen, as has yellowish Capella in the CHARIOTEER. The GREAT SQUARE is high in the sky. Fomalhaut stands out in the southern sky. Vega, SWAN (shown here as the NORTHERN CROSS), EAGLE, and DOLPHIN are moving toward the western horizon. Vega, Deneb, and Altair form the "summer triangle." HERDSMAN and the NORTHERN CROWN are setting. The BIG DIPPER sinks toward the north.

Star Map 6

The stars as seen at forty degrees north latitude.

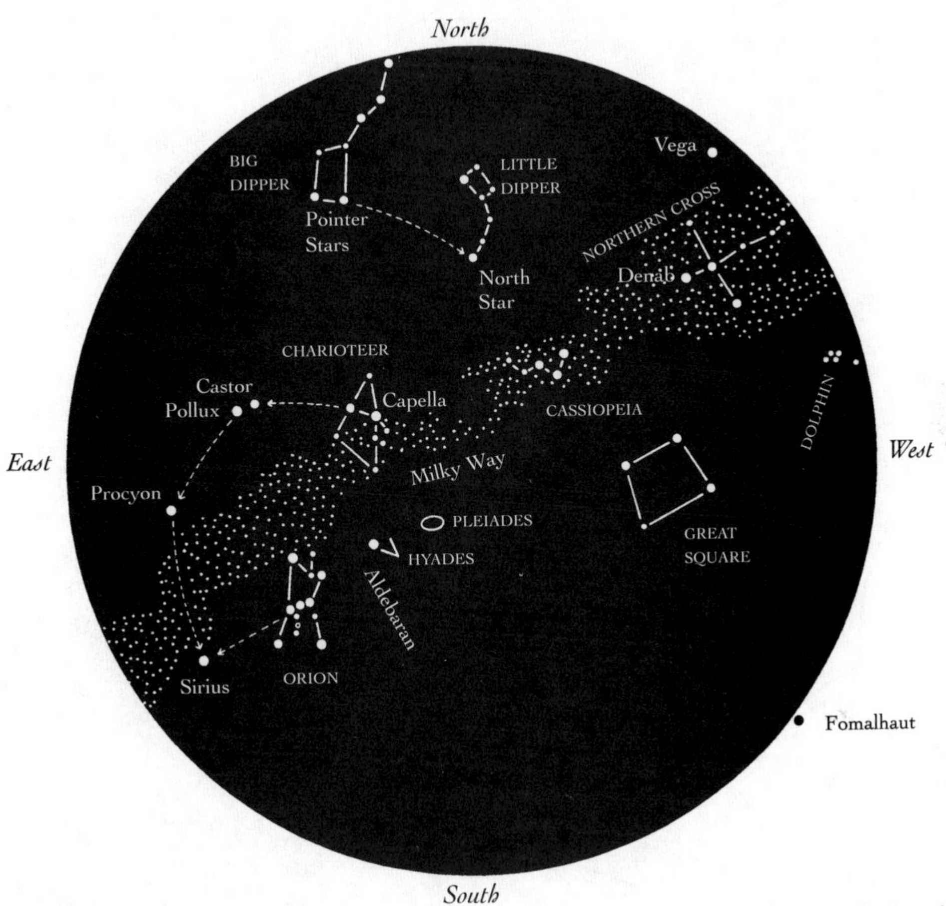

Evening viewing times:
November, from 10 P.M. to 1 A.M.;
December, from 8 P.M. to 11 P.M.;
January, from 6 P.M. to 9 P.M.;

Morning viewing times:
August, at dawn.

The major sky features at these times:

The bright stars Capella, Castor, Pollux, Procyon, and Sirius form an arc around ORION; ORION's belt points toward Sirius. Aldebaran, the HYADES, and the PLEIADES are in view. The GREAT SQUARE moves toward the west, and the SWAN (shown here as the NORTHERN CROSS) is on the horizon. Fomalhaut has just set.

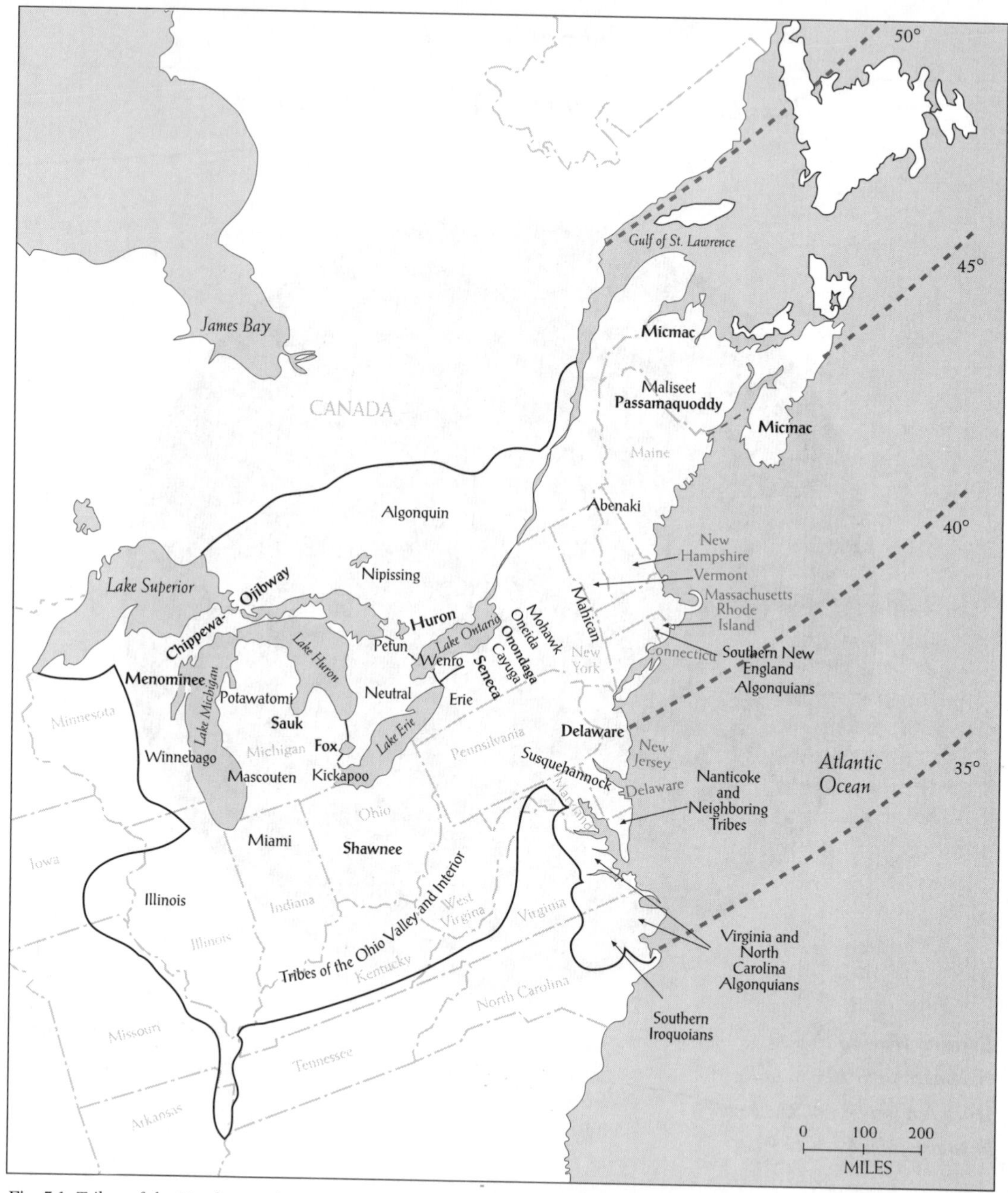

Fig. 3.1. Tribes of the Northeast culture area and their approximate territories at the earliest time for which information is available. The Mohawk, Oneida, Onandaga, Cayuga, and Seneca formed the Iroquois Confederacy. The tribes in **boldface type** are included in this chapter. Not all tribes that inhabited this region are shown. (After Trigger, *Handbook of the North American Indians*, Vol. 15, ix.

Chapter three

The Celestial Bear Hunt

Stars of the Northeast

The Northeast culture area stretches from the Atlantic Ocean to the Great Plains, from the woodlands of the Southeast to the boreal forest of the subarctic. The area encompasses southern sections of Ontario and Quebec as well as the northeastern sector of the United States, including New England, the mid-Atlantic states, and the area west to Illinois and Wisconsin.

The tribes of this region are characterized by their dependence on hunting deer and other game as well as on harvesting corn or wild rice. To the north of the Saint Lawrence River the growing season is too short for corn, so Subarctic and Arctic tribes relied primarily on hunting and gathering. To the south there was an even greater reliance on crops because the population was denser and it was harder to secure adequate amounts of game. To the west bison provided a major source of food.

The tribes of the Northeast spoke dialects of the Iroquoian language family (Iroquois, Huron, and nearby tribes) and the Algonquian language family (all other tribes, including Micmac, Passamaquoddy, Penobscot, Delaware, Fox, Menominee, and Ojibway).

In hunter-gatherer societies in the Arctic and the Great Basin, individuals lived with, hunted with, and owed their allegiance to their extended family, or *band*. In contrast, individuals in the Northeast lived, secured food, found protection, and practiced their spiritual beliefs within a larger, more organized group, the *tribe*.

Tribes along the Atlantic, such as the Delaware in the Northeast and the Cherokee in the Southeast, bore the brunt of early European exploration and settlement and were the first to see their cultures erode. Many tribes were dispersed or squeezed onto the territories of inland tribes.

Power, which is fundamental to living a spiritual life, is an important element in several star stories of the Northeast. In order to gain or maintain his or her own personal power, a man or woman may have visions, call on guardian spirits, and practice certain rituals. Only Micmac hunters with great power, for example, can see the celestial bear when it is invisible in the sky, and only Delaware shamans with special power can "know" the Bunched Up Stars (PLEIADES). In the story of the Fisher Stars, both the husband and the wife have power, but in the end the wife's power is greater; elders use this story to instruct young girls about how they, too, can gain power and live a spiritual life.

In the Northeast, the most widespread celestial image is that of the never-ending bear hunt. The Passamaquoddy, Seneca, Delaware, and Fox all identify the BIG DIPPER as a bear and hunters, but it is the Micmac narrative that describes the hunt most carefully and lays out the underpinnings of the tale. It is important that the celestial bear does not die, that its spirit lives on after each hunt and comes back in the form of another bear each spring. Because bears on Earth are thought to descend from the bear in the sky, earthly bears do not die: They sleep in the winter and reappear each spring. "The sky is just the same as the earth, only up above, and older," says the Micmac narrator. The story of the never-ending bear hunt celebrates the miraculous annual rebirth of both the celestial and the earthly bears and affirms the belief that other game will be continuously reborn as well.

Micmac

The derivation of the word *Micmac* is uncertain, but it is thought to mean "our kin" or "people of the red earth." The Micmac hunted and still live in the forests and along the shores of what is now Nova Scotia, Prince Edward Island, eastern New Brunswick, and the Gaspé Peninsula of Quebec. The short growing season and variety of wildlife in this region encouraged hunting rather than farming, and each season provided food: seal, moose, caribou, bear, and beaver in winter; Canada geese and waterfowl eggs in the spring; shellfish in the summer; large game again in the fall; and fish all year long. October is called "the month of fat, tame animals." (Indeed, in the following star story of a bear hunt, the bear is killed in the autumn.)

The Micmac hunted in small groups during winter, but in spring they gathered in larger groups along the coast to take advantage of the abundant food there. They built cone-shaped structures and covered them with birch bark, animal skins, and evergreen branches. They also made distinctive birchbark canoes for plying the waters and decorated clothing and containers with quillwork.

Gluskap (also spelled Glooscap) is the Micmac culture hero, the mythical trickster and transformer who showed the people how to hunt and fish. He also named the constellations and stars, shaped the landscape, and gave animals their attributes.

BIG DIPPER: The Celestial Bear

Native peoples in the Northeast, the Southeast, the western Subarctic, and the Plateau all relate stories of the BIG DIPPER as a celestial bear and hunters. The Micmac of Nova Scotia told the following story to the ethnologist Stansbury Hagar in the late 1800s.

The narrative describes how the passage of the seasons on Earth is illustrated in the night sky. The bear (the bowl of the BIG DIPPER) emerges from its den (NORTHERN CROWN) in the springtime. Throughout the summer, seven hunters (stars in the handle of the BIG DIPPER and in HERDSMAN) pursue the bear across the northern horizon; one of the hunters (Mizar) carries a cooking pot (Alcor). The bear stands upright at this time. In

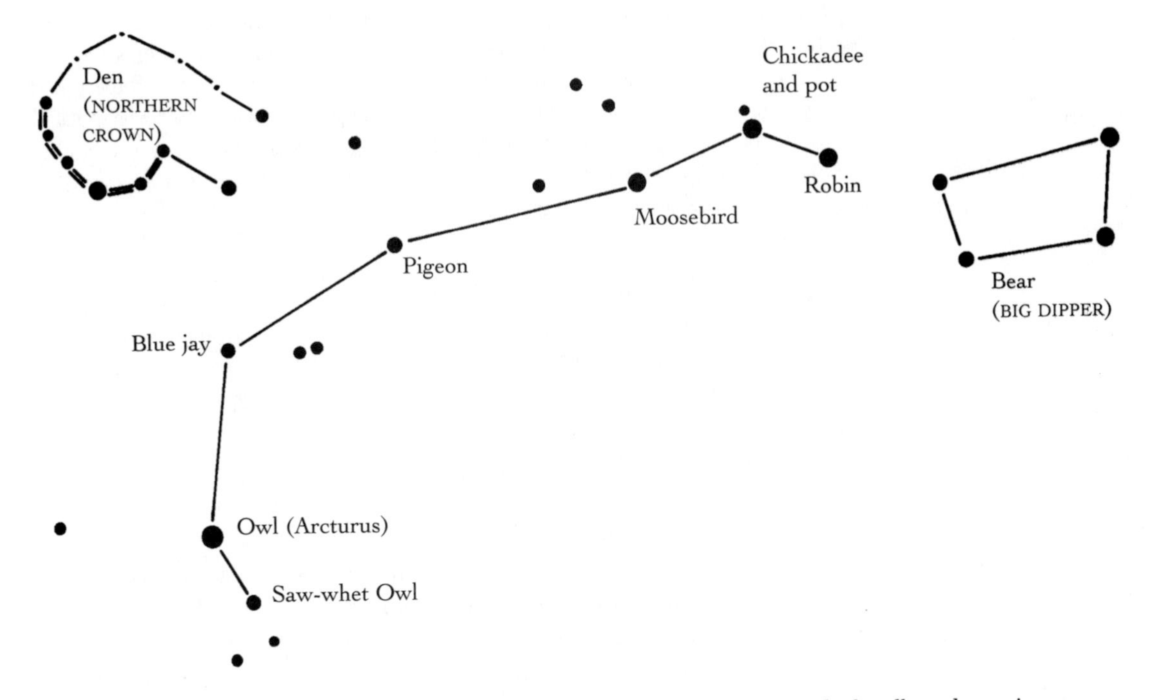

Fig. 3.2. The Never-Ending Bear Hunt. The bowl of the BIG DIPPER is the bear, the handle and stars in HERDSMAN are the hunters, and the NORTHERN CROWN is the bear's den. In addition to the six stars of the NORTHERN CROWN (indicated with double lines), the den includes an entrance, consisting of two stars in HERDSMAN with several fainter stars.

the fall, several of the hunters drop out of the chase (these stars drop below the horizon for a few hours each night). The bear now lies on its back, preparing for hibernation. The remaining hunters kill the bear, whose spirit goes to its den.

It is important to note the faithfulness with which the Micmac narrator presents detail about the characteristics of the animals and the positions of the stars. The story accurately portrays the relative positions of the stars as well as their appearance and disappearance in the sky at forty-five degrees north latitude in Nova Scotia. The bird hunters represent actual birds: The chickadee is a black-capped chickadee, the moose bird is a gray jay, the pigeon is a passenger pigeon, and the smaller owl is a saw-whet owl. The larger owl is probably a barred owl. The characterization of the moose bird is accurate, for gray jays are known as "camp robbers" that will land on a picnic table and demand or simply take their share. The call of the saw-whet owl, which Hagar describes as "rasping," is a whistle: *too, too, too, too, too*. (The aside about why one should not imitate the saw-whet owl is particular to the Micmac narration and is not found in other versions of the myth.)

The Micmac story explains other aspects of nature—why the robin has a red breast and why the trees turn red in fall, for example. (The Housatonic, also in the Northeast, say that the bear's fat makes the snow, whereas the Cherokee in the Southeast believe that the fat becomes honeydew, a sweet deposit secreted on plants by aphids.)

It is ironic that even as Hagar recorded this story, Euro-American settlers had hunted almost to extinction the great flocks of passenger pigeons that once populated North America. In 1914 the last known passenger pigeon was captive in a zoo, and it, too, died. The only passenger pigeon that can be seen now is the one in the never-ending bear hunt. The story goes like this:

Late in spring, the bear—waking from her long winter sleep—leaves her rocky hillside den and descends to the ground in search of food. Instantly the sharp-eyed chickadee perceives her and, being too small to undertake the pursuit alone, calls the other hunters to his aid. Together the seven start after the bear, the chickadee with his pot being placed between two of the large birds so that he may not lose his way. All the hunters are hungry for meat after the short rations of winter and so they pursue eagerly, but throughout the summer the bear flees across the northern horizon and the pursuit continues.

In the autumn, one by one, the hunters in the rear begin to lose their trail. First of all the two owls, heavier and clumsier of wing than the other birds, disappear from the chase. But you must not laugh when you hear how Kōpkéch, the smaller owl, failed to secure a share of the bear meat, and you must not imitate his rasping cry, for if you disregard either warning, be sure that wherever you are, as soon as you are asleep he will descend from the sky with a birchbark torch and set fire to whatever clothing covers you. Next the blue jay and the pigeon also lose the trail and drop out of the chase. This leaves only the robin, the chickadee, and the moose bird, but they continue the pursuit, and at last, about midautumn, they overtake their prey.

Brought to bay, the bear rears up on her hind feet and prepares to defend herself, but the robin pierces her with an arrow and she falls over upon her back. The robin being himself very thin at this season is intensely eager to eat some of the bear's fat as soon as possible. In his haste he leaps upon his victim and becomes covered with blood. Flying to a maple tree near at hand in the land of the sky, he tries to shake off this blood. He succeeds in shaking all of it off save a spot on his breast. "That spot," says the garrulous chickadee, "you will carry as long as your name is Robin."

But the blood that he does shake off spatters far and wide over the forests of Earth below, and hence we see each autumn the blood-red tints on the foliage. It is reddest on the maples, because trees on Earth follow the appearance of trees in the sky, and the sky maple received most of the blood. The sky is just the same as the Earth, only up above, and older.

Some time after these things happen to the robin, the chickadee arrives on the scene. These two birds cut up the bear, build a fire, and place some of the meat over it to cook. Just as they are about to begin to eat, the moose bird puts in his appearance.

He has almost lost the trail, but when he regains it he has not hurried because he knows that it will take his companions some time to cook the meat after the bear is slain, and he does not mind missing that part of the affair so long as he arrives in time for a full share of the food. Indeed, he is so impressed with the advantages of this policy that ever since then he has ceased to hunt for himself, preferring to follow after hunters and share their spoils. And so, whenever a bear or a moose or other animal is killed today in the woods of Megumaage, Micmac Land, you will see him appear to demand his share. That is why the other birds name him Mikchăgōgwéch (He-who-comes-in-at-the-Last-Moment), and the Micmacs say there are some men who ought to be called that, too.

However that may be, the robin and chickadee, being generous, willingly share their food with the moose bird. Before they eat, the robin and moose bird dance around the fire (*neskouadijik*) while the chickadee stirs the pot. Such was the custom in the good old times, when Micmacs were brothers all to all and felt it a duty to share their food together, and to thank each other and the Universal Spirit for their present happiness.

But this does not end the story of the bear, though one might think so. Through the winter her skeleton lies upon its back in the sky, but her life-spirit enters another bear who also lies upon her back in the den, invisible and sleeping the winter sleep. When the spring comes around, this bear will again issue forth from the den to be pursued and slain by the hunters, and she will send her life-spirit to the den to issue forth yet again, when the sun once more awakens the sleeping Earth.

And so the drama keeps on eternally. And so it is, the Micmacs say, that when a bear lies on her back within her den she is invisible even to those who might enter that den. Only a hunter gifted with great magic power can perceive her then.

(Reprinted from Hagar, *The Celestial Bear*, 93–95.)

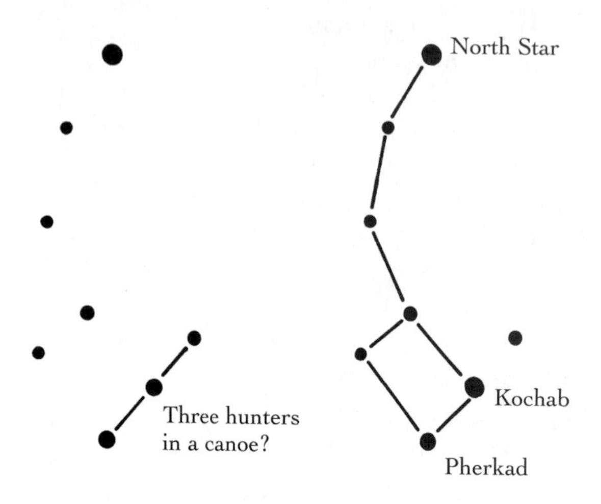

Fig. 3.3. The Micmac tell of three hunters in a canoe chasing a bear that hides near the North Star. One suggestion regarding the identity of the hunters is that they are Pherkad, Kochab, and a small star next to Kochab, all in the LITTLE DIPPER.

LITTLE DIPPER?: Three Hunters in a Canoe

There is a second hunt underway in the northern sky. Father Chretienne Le Clerq reported in 1691 that Native Americans of the Gaspé, who were familiar with both the BIG DIPPER and the LITTLE DIPPER, told him that "the three guardians of the North Star are a canoe in which three [hunters] have embarked to surprise a bear." A Micmac told Hagar in the late 1890s that there is a another bear (separate from the one in the BIG DIPPER) hidden near the pole and that "nearby" stars are hunters circling around to locate the bear's den. The three hunters in a canoe may possibly be Kochab and Pherkad (two stars in the bowl of the LITTLE DIPPER) and a third star that with them forms a line of three stars.

Passamaquoddy

In 1692 the French explorer Sieur de Antoine de la Mothe Cadillac sailed near the mouth of the Saint Croix River, which now forms part of the boundary between Maine and New Brunswick. Cadillac called the bay *Pesmocady,* and the people who lived on its shores became known as the Passamaquoddy. These Native people call themselves *Pestemohkatiyek* (Those of the Place Where the Pollack are Plentiful). Like the Micmac, they used the resources at hand—including fish, shellfish, eggs of seabirds, and game—but they also grew corn. The Passamaquoddy now live in eastern Maine and western New Brunswick.

Kuloskap (a variation of Gluskap) is a culture hero responsible for changing animals into their present form and shaping many of the natural features of the area. It was taboo to tell stories about this supernatural being in any season except winter.

Milky Way: *Ketagūswōt* the Spirits' Path
Morning Star: Chief M´Sūrtū

Mrs. W. Wallace Brown, the wife of a U.S. government Indian agent, originally recorded this story in rough form in 1890, and Abby Alger presented a more polished version in

1897. The story is likely Passamaquoddy because Mrs. Brown lived in Calais, Maine, near two Passamaquoddy reservations.

The wonderful game of ball that the old chief encounters in the sky may be some version of lacrosse, which the Passamaquoddy, Huron, Chippewa, Delaware, and other eastern tribes played. The ebb and flow of *wawbāban* (the northern lights) can easily be seen as a fluid game of ball, featuring players with lights on their heads and rainbow belts at their waists.

There once lived an old chief, called M´Sūrtū, or the Morning Star. He had an only son so unlike all the other boys of the tribe as to distress the old chief. He would not stay with the others or play with them but, taking his bow and arrows, would leave home, going towards the north, and stay away many days at a time.

When he came home, his relations would ask him where he had been, but he made no answer.

At last the old chief said to his wife: "The boy must be watched. I will follow him."

So Morning Star kept in the boy's trail, and traveled for a long time. Suddenly his eyes closed and he could not hear. He had a strange sensation, and then knew nothing until his eyes opened in an unknown and brightly lighted land. There were neither sun, moon, nor stars; but the land was illumined by a singular light.

He saw human beings very unlike his own people. They gathered about him and tried to talk with him, but he could not understand their language. He knew not where to go nor what to do. He was well treated by this marvelous tribe. He watched their games and was attracted by a wonderful game of ball that seemed to change the light to all the colors of the rainbow—colors that he had never seen before. The players all seemed to have lights on their heads, and they wore curious girdles, called *memquon* (rainbow belts).

After a few days an old man came to him and spoke to him in his own tongue, asking if he knew where he was. He answered, "No." The old man then said, "You are in the Land of Northern Lights. I came here many years ago. I was the only one here from the Lower Country, as we call it; but now there is a boy who visits us every few days."

At this, the chief inquired how the old man got there, what

way he had come. The old man said, "I followed the path called *Ketagūswōt* or the Spirits' Path (the Milky Way)."

"That must be the same path I took," said the chief. "Did you have a strange feeling, as if you had lost all knowledge, while you travelled?"

"Yes," said the old man. "I could not see or hear."

"Then you did come by the same path. Can you tell me how I may return home again?"

The old man said, "The Chief of the Northern Lights will send you home, friend."

"Well, can you tell me where or when I may see my son? The boy who visits you is mine."

The old man said, "You will see him playing ball, if you watch."

Morning Star was very glad to hear this, and a few moments later a man went around to the wigwams, telling all to go and have a game of ball.

The old chief went with the rest; when the game began, he saw many most beautiful colors on the playground. The old man asked him if he saw his son among the players, and he said that he did. "The one with the brightest light on his head is my son."

Then they went to the Chief of the Northern Lights, and the old man said, "The Chief of the Lower Country wishes to go home, and he also wants his son."

The chief asked him to stay a few days longer; but he longed to go home, so the Chief of the Northern Lights called together his tribe to take leave of M´Sūrtū and his son and ordered two great birds to carry them home. As they travelled over the Milky Way Morning Star had the same strange sensation as before, and when he came to his senses he found himself at his own door. His wife rejoiced to see him, for when the boy had told her that his father was safe, she had not heeded him but feared that he was lost.

(Reprinted from Alger, *In Indian Tents,* 130–133.)

BIG DIPPER: Three Hunters and a Bear

This Passamaquoddy "Song of the Stars" describes the never-ending bear hunt. The three hunters are stars in the BIG DIPPER's handle, the bear is the BIG DIPPER's bowl, and a road for spirits is the Milky Way.

We are the stars which sing,
We sing with our light;
We are the birds of fire,
We fly over the sky.
Our light is a voice;
We make a road for spirits,
For the spirits to pass over.
Among us are three hunters
Who chase a bear;
There never was a time
When they were not hunting.
We look down on the mountains.
This is the Song of the Stars.

(Reprinted from Leland, *The Algonquin Legends of New England; or, Myths and Folk Lore of the Micmac, Passamaquoddy, and Penobscot Tribes,* 379.)

The Penobscot, to the west of the Passamaquoddy, tell a somewhat different tale of the BIG DIPPER, one that is strikingly similar to Blackfoot and Arikara stories. When six brothers go off to scout for the enemy, their older sister becomes a bear that kills everyone in the village except for the two youngest children in their family. She forces these two to serve her. The young boy gets help from a great magician, and when the six brothers return they are all able to escape into the sky and become the BIG DIPPER.

Iroquois Confederacy

The Iroquois Confederacy, centered in what is now New York State, was the strongest and most powerful group of Native Americans in the Northeast. Its five constituent tribes—the Seneca, Cayuga, Onondaga, Oneida, and Mohawk—agreed sometime between 1400 and 1600 to form the Great Peace, a lasting alliance. The confederacy was also called the Five Nations. In the early 1700s, the Tuscarora, who had migrated from the Southeast, joined the alliance, making it the Six Nations.

The Iroquois lived in longhouses, which were on the average twenty-five feet wide and eighty feet long. The actual length varied according to the number of families that lived within. A longhouse generally had three to five fires along the centerline of the structure. Apartments were paired on either side of the fire; each apartment was half the width of the longhouse and about twenty-five feet long. Along the outer wall there was a wide, low platform for sitting and sleeping, with a storage shelf above. Food was kept in bays between apartments. The Iroquois used elm bark, which was more plentiful than birch bark, to cover their longhouses as well as their canoes.

When the League of the Iroquois was set up, the Seneca, to the west, became the Keepers of the Western Door and the Mohawk, to the east, became the Keepers of the Eastern Door. These tribes composed a longhouse of "five fires," the Five Nations. The people of the Iroquois Confederacy also became known as the People of the Longhouse.

Each fall, after the women had harvested corn, beans, squash, and tobacco, families left the longhouse for the fall deer hunt. They remained "in camp" until they saw the PLEIADES reach their zenith at dusk at the time of the winter solstice. Then the families returned to their villages and celebrated the New Year.

Iroquois men hunted deer, bear, wild turkey, partridge, waterfowl, and passenger pigeon (caught with nets or pushed out of their nests), and caught many species of fish. In Seneca tradition, hunters addressed the Morning Star, the moon, and the sun for assistance. Iroquois women were in charge of planting, tending, and harvesting the garden plots as well as of gathering wild edibles such as walnuts, beechnuts, chestnuts, and various berries.

The specific tribal identity for most of the following stories is not known; the individuals who recorded them simply called them "Iroquois" rather than attributing them to individual tribes in the confederation. For a few stories, the originating tribe—Onondaga or Seneca—is noted.

Nicolaes van Wassenaer, who in the early 1600s chronicled the history of Europe (including the settlement of New Netherland), observed that Iroquois women knew the constellations well: "The women are the most experienced star-gazers; there is scarcely one of them but can name all the stars; their rising and setting. The position of [the GREAT BEAR] is as well known to them as to us, and they name [other constellations] by other names." Wassenaer adds that the women observed stars in the head of TAURUS to determine when to plant. They call these stars "a horned head of a big, wild animal which inhabits the distant country, not their's, and when it rises in a certain part of the heavens, at a time known to them, then is the season for planting." Wassenaer's brief comments do not address why women were the most experienced stargazers and what place the constellations held in the spiritual or natural world.

Creation of the Stars and Constellations

There are several recorded versions of the Iroquoian creation story. In the Seneca account presented here, a woman marries the "Ancient One" in the world above, but she is with child and the Ancient One is angry. He uproots a great tree and pushes her into the hole. She falls into a watery world far below. The animals there make a place for her to land by diving for mud below the water and placing it on the back of the turtle. Sky Woman, as she is now known, makes the sun, moon, and stars, fashioning these constellations: They Are Pursuing the Bear (BIG DIPPER), It Brings the Day (Morning Star), They Are Dancing (PLEIADES), She Is Sitting (ORION), Beaver that Spreads its Skin (LITTLE DIPPER), and The Group Visible (an unidentified constellation). Her statement that men will "watch the Beaver" means that they will refer to the North Star, which is a navigational marker because it lies due north and does not move.

After the woman found a place on the turtle and grass and shrubs appeared, she stood up and said, "Now will come the sun, which shall be called *En-dek-ha* (Pertaining to the Day)." It appeared, and when it set it was dark again.

Then she said, "Now will come the stars like spots in the

sky." They came. Then she told what some should be called. Toward the north were several, and she said, "These shall be called *Ni-a-gwai had-i-she* (They Are Pursuing the Bear)."

Then she looked to the east and said, "A large star will be there, rising usually before daylight, and it shall be called *Tgen-den-wit-ha* (It Brings the Day)." She pointed to another group, saying, "That shall be called *Gat-gwa-da* (The Group Visible). That will be a sign of the coming spring."

Then she said of the Pleiades, "That group shall be called *De-hon-nont-gwen* (They Are Dancing)." Another she named *I-en-i-u-ci-ot* (She Is Sitting). Of another group she said, "These shall go with them and be called *Nan-ga-ni-a-gon Ga-sa-do* (Beaver that Spreads its Skin). When men travel by night they will watch this group." To others she gave names.

(Reprinted from Beauchamp, *Iroquois Folk Lore,* 205.)

PLEIADES: *Od-je-so-dah,* the Dancing Stars
PLEIADES: *Oot-kwa-tah,* There They Dwell in Peace

Two Iroquois tales describe the origin of the PLEIADES. In the first, a hunter is teaching his eleven sons the secrets of hunting, when they awaken one night, charmed by the weird chanting of sky witches. The brothers begin dancing and find that they cannot stop. Day after day and night after night they whirl around the sky, making the stars dizzy. At last the Moon changes them into a group of stars and directs them to dance over the council house every year during the ten days of feasting at the New Year. The stars are called *Od-je-so-dah* (the Dancing Stars). Some of the stars are not visible because they are small and because they are dancing behind the larger ones.

In the second story, an Onondaga tale, a group of children dance on purpose, despite a warning from a mysterious man. This story, like the previous one, occurs in the winter.

A long time ago a party of hunters went through the woods toward a good hunting-ground, which they had long known. They traveled several days through a very wild country, going on leisurely and camping by the way. At last they reached *Kan-ya-ti-yo* (the beautiful lake), where the gray rocks were crowned with great forest trees. Fish swarmed in the waters, and at every jutting point the deer came down from the hills around to bathe or drink of the lake. On the hills and in the valleys were huge beech and chestnut trees, where squirrels chattered, and bears came to take their morning and evening meals.

The chief of the band was Hah-yah-no (Tracks in the Water), and he halted his party on the lake shore that he might return thanks to the Great Spirit for their safe arrival at this good hunting-ground. "Here we will build our lodges for the winter, and may the Great Spirit, who has helped us on our way, send us plenty of game, and health and peace."

The pleasant autumn days passed on. The lodges had been built, and hunting had prospered, when the children took a fancy to dance for their own amusement. They were getting lonesome, having little to do, and so they met daily in a quiet spot by the lake to have what they called their jolly dance. They had done this a long time, when one day a very old man came to them. They had seen no one like him before. He was dressed in white feathers, and his white hair shone like silver. If his appearance was strange, his words were unpleasant as well. He told them they must stop their dancing or evil would happen to them. Little did the children heed, for they were intent on their sport, and again and again the old man appeared, repeating his warning.

The mere dances did not afford all the enjoyment the children wished, and a little boy, who liked a good dinner, suggested a feast the next time they met. The food must come from their parents, and all these were asked when they returned home. "You will waste and spoil good food," said one. "You can eat at home as you should," said another, and so they got nothing at all. Sorry as they were for this, they met and danced as before. A little to eat after each dance would have made them happy indeed. Empty stomachs cause no joy.

One day as they danced, they found themselves rising little by little into the air, their heads being light through hunger. How this happened they did not know, but one said, "Do not look back, for something strange is taking place." A woman, too, saw them rise and called them back, but with no effect, for they still rose slowly above the Earth. She ran to the camp, and all rushed out with food of every kind, but the children would not return, though their parents called piteously after them. But one would even look back, and he became a falling star. The others reached the sky, and are *Oot-kwa-tah* (There They Dwell in Peace). Every falling or shooting star recalls the story, but the seven stars shine on continuously, a pretty band of dancing children.

(Reprinted from Beauchamp, *Onondaga Tale of the Pleiades*, 281–282.)

Milky Way: *Dja-swĕn´-do',* the Great Sky Road

The Great Sky Road (the Milky Way) is the belt of Ga-do-wăās, a hunter. According to the Iroquois, each soul travels its own path from the soul house (the body) to the Sky Road, where it goes on to the Place of the Maker.

Ga-do-wăās dwells in the top sky, and with his four eyes watches every corner of the earth. At one time Ga-do-wăās was an Earth dweller and a hunter, but because he presumed to celestial power and destroyed all the game, he was transferred to the heavens and watches the gate through which each soul passes to immortality.

When Ga-do-wăās assumed his duty as soul watcher, he removed his hunting belt, which possessed the charm of enticing game. He decorated it with stars and cast it into space, where it spans the entire heavens and illuminates each path to which he guides a soul.

So luminous is this path that its blended light reaches down to the earth and divides its rays, stationing one at each lodge where a human is dying, that the departing soul may not lose its way as it leaves the dead.

No human has seen these rays; they are visible only to the soul. The south wind accompanies the soul till it reaches the gate where Ga-do-wăās watches, and as it passes the portal of this journey place he reaches into space and grasps a star which he fastens in the belt, thereby to guide the soul on its journey.

When the soul has crossed the entire heavens, Ga-do-wăās removes the star from his belt and returns it to its appointed place in space.

The Milky Way is a procession of stars, each guiding a soul. If there is a confusion in this procession, it is because some soul is disturbed and out of the path. But the star, which never loses its way, will search for it and return it to its course.

(Reprinted from Converse, *Myths and Legends of the New York State Iroquois,* 56–57.)

North Star: *Ti-yn-sōu-dă-go-êrr,* the Star That Never Moves

Not surprisingly, there are many Native American stories of the North Star, the Star That Never Moves. The following tale describes two Iroquois practices, burning tobacco at a time of decision and holding a council. After the hunters return from their adventure, the head chief calls a meeting of all the tribes—much like gathering the tribes of the Iroquois Confederation—to name the star. The Iroquois, like the Cherokee to the south, say that there are little people who serve as spirit helpers.

A large party of men, while moving in search of new hunting grounds, wandered on for many moons, finding but little game. At last they arrived at the banks of a great river entirely unknown to them, where they had to stop, not having the material to build boats. Lost and nearly famished with hunger, the head chief was taken very ill, and the people decided to hold a council to devise means for returning to their old homes. During the dance, and while the tobacco was burning, a little being like a child came up, saying she was sent to be their guide.

Accordingly they broke up their camp and started with her that night. Preceding them with only a *gi-wăh* (small war club), she led them on until daylight and then commanded them to rest while she prepared their food. This they did, and when awakened by her they found a great feast in readiness for them. Then she bade them farewell, with the assurance of returning to them again in the evening.

True to her word, at evening she reappeared, bringing with her a skin jug, from which she poured out some liquid into a horn cup, and told them each to taste of it. At first they feared to do so, but at last yielding they began to feel very strong. She then informed them that they had a long journey to make that night. Again they followed her, and in the early morning arrived at a great plain, where she again told them to rest for the day, with the exception of a few warriors who were to be shown where they could find plenty of game. Two of the warriors had accompanied her but a short distance when they encountered a herd of deer. She directed them to kill all they wished in her absence, and then, again promising to return at night, she took leave of them.

At nightfall she returned, saying her own chief would soon follow her to explain to them how they could reach their own homes in safety. In a short time he arrived, with a great number of his race, and immediately all held council together. He informed the hunters that they were now in the territory of the little people, who would teach them a sign, already in the sky, that would be to them a sure guide whenever they were lost. The little people pointed out the pole star [North Star] and told them that in the north, where the sun never goes, while the other stars moved about, this particular star stands still as the Indian's guide in his wanderings. They had only to follow its light and they would soon return to their tribe, where they would find plenty of game.

They thanked the good little people and traveled every night until they arrived safely in their homes. When they had recounted all their adventures, the head chief called a meeting of all the tribes and said they ought to give this star a name. So they called it *Ti-yn-sōu-dă-go-êrr* (Star That Never Moves), by which name it is called unto this day.

(Reprinted from Smith, *Myths of the Iroquois,* 39–40.)

BIG DIPPER: *Nya-gwa-ih,* the Celestial Bear

There are several legends about the BIG DIPPER, all of which describe a bear hunt. In a Seneca story, six hunters go out to hunt bear, but one of the men is lazy (just as the moose bird in the Micmac BIG DIPPER story is lazy) and claims that he is sick. His friends make a litter and carry him in addition to their own belongings. When they discover the track of a bear, they set him down and continue on. The lazy hunter realizes that he will be left behind, so he jumps up, outraces his tired companions, and kills the bear. All the while, the six hunters have been climbing into the sky, where they remain to this day. The bear is the bowl star nearest to the North Star. The other stars in the BIG DIPPER are the hunters. The star Alcor is the kettle that one hunter has been carrying.

In an Iroquois myth, a stone monster attacks a group of bear hunters, killing all but three men. These hunters (who become the BIG DIPPER's handle) and the bear (who becomes the bowl) are transported to the sky. The hunter nearest the bear carries his bow, the middle hunter carries a kettle, and the hunter farthest away gathers sticks. Only in the fall do their arrows pierce the bear's skin; its blood turns the leaves red at that time.

In another Iroquois myth, a huge bear is devouring the winter game of the Iroquois. Three brothers (stars in the handle) chase the bear (the bowl) into the sky.

Fig. 3.4. Stone Coats chasing Iroquois hunters. In an Iroquois myth, a stone monster kills all but three of a group of bear hunters. The three who remain and the bear become the BIG DIPPER. (Illustration from Henry Rowe Schoolcraft, *Indian Tribes of the United States* [Philadelphia: Lippincott, Grambo & Co., 1851].)

The dog, Ji-yeh, is Alcor, the small star next to the middle star in the handle. The net under which the bear hides is the NORTHERN CROWN. The Iroquois were known as great runners, and the tireless hunters described here run until they reach the edge of the world, then they continue running into the sky.

Numbers of hunters had banded together and plodded through all the forests in search of the bear, but to no avail. At times it would come near for a moment, but it would distance their arrows in a most mysterious way, and the blinding snow would fall fast and thick as if to cover its track.

In the darkness it frequently prowled near the villages, where the terrified people would hide from its roaring voice, and a deep snowfall always followed these visitations. Baffling all their plans for its death, the *nya-gwa-ih* (bear) continued his ravage of plunder.

The winter was fierce in its cold blasts, and the snows had drifted into mountains high in the forest. The trails were lost. The deer were vanishing, and their haunts were strewn with their bones that the bear had left behind him. Then one night each of three brothers dreamed he had found the bear and, deeply impressed by the remarkable coincidence on the following morning, they silently left the village and started on their secret hunt, accompanied by their faithful dog, Ji-yeh, whose keen nose ridged the snow down to the trail.

In their pursuit one day they saw the bear. It had pushed under a snow bank and was ravenously devouring a deer. So certain were they of its capture that they cut down a small pine and made ready the fire for cooking it, but when they resumed their hunt the bear had vanished, and there was no trail of it in the swift-falling snow that had covered its track. Chagrined that they had been so near and had failed, they decided not to stop again till they captured it.

They bundled the fire brush on the shoulders of one of the brothers, and to their belts tied their strong bags of *o-no-oh,* the roasted corn flour that would sustain them while they were running, and again set out on the chase.

At night they slept not; during the day they rested not; for the elusive shadow of the rapid running bear could be seen on the snow hills as they ran to the north sky.

As if avenging, the freezing winds pursued them, the ice weighted down their moccasins, and the pitiless snows drifted near to the skies. But, impelled by their dream, the intrepid hunters faltered not until they had reached the end of the flat earth where it edges close to the north sky. Then the shadow of the bear disappeared, and the distant paths seemed enveloped in a vaporous mist like a hiding cloud that floats over the water.

Yet the tireless hunters would not rest, but climbed higher and higher and farther away from the earth, when again they saw the bear, who was now slow in its path yet mighty as it pushed the white clouds before it, weaving an invisible net that it cast over the skies and crawled under to rest.

Astray in the strange place, the untiring hunters, who knew not fatigue nor hunger, rejoiced when they came near the bear to find him sleeping. "We will not lose it now, and will carry it back to our people," was their victorious cry.

The listening bear slowly opened its sleepy eyes, and rising in its giant height, lifted the net with its huge paws and, dragging the hunters under it, drove them far away to roam the broad skies forever! And the hunters and their faithful dog, Jiyeh, unknowing their imprisonment under the invisible net, are ceaselessly following the snow bear, who ever eludes them.

(Reprinted from Converse, *Myths and Legends of the New York State Iroquois*, 57–59.)

Morning Star: Star Woman
Morning Star: *Ga^{n}dewitha*, It Brings the Day

The Morning Star was once an important celestial being, but perhaps because of the cultural erosion caused by Euro-American contact, by the early 1900s the Iroquois reportedly no longer gave the star special honors. Morning Star figures in two recorded stories—as a beautiful woman in one and as a man who can control the sun in another.

A mighty hunter caught sight of the Sky Elk and followed it into the heavens. Dawn rescued the hunter from falling back to earth and gave him the job of guarding her lodge in the eastern sky. His other duty was to watch the earth's forests and escort the sky night hunters there. On one of his journeys to the earth with the night hunters he saw a beautiful woman, with whom he fell in love. He could approach her only in the form of a bird, however, and after several unsuccessful attempts, he changed himself into a hawk, plucked her from the earth, and took her to the sky. When Dawn found him she was angry at him for neglecting his duties. She bound his arms and turned the maiden into a star, which she placed on his forehead.

The star holds a torch; should the hunter try to escape, the fire will consume him. Thus, the hunter must forever yearn for the maiden, Star Woman, whom he cannot see and who has never known him. It is said that before the Sun appears on the horizon, he lights his council fire by the torch of Star Woman, the Morning Star.

(Adapted from Converse, *Myths and Legends of the New York State Iroquois,* 60–63.)

In the early 1900s, John Armstrong, a Seneca, told the following story, in which Morning Star has the power to hurry the sun's rising but chooses not to use this power for every mortal who asks. At the beginning of the tale, a man learns that his wife has become a cannibal and wants to eat him. (Cannibals often appear in Native American myths. Seneca stories describe not only Cannibal Woman but also Stone Coats, cannibal giants who wear coats of flint.) The man escapes and finds another wife, but the new wife's grandmother also plots his death. The beleaguered man calls on the Morning Star for assistance.

The old woman said to her grandson, "We must go to the island and hunt." The island was low and in the center of a deep lake. They landed and drew up the canoe, then the old woman said, pointing in a direction away from the landing, "Take your place over there. I will drive the game toward you."

When the man had gone some distance, he turned and saw that the old woman was in the canoe and paddling as fast as she could go. He called to her, but she didn't answer. He stayed all day on the island; there was no way of escape. After a while he noticed water marks very high on the trees and then he knew that at times the island was almost under water.

When night came the water began to rise, and the man climbed the tallest tree he could find. The water kept rising and he kept climbing. About halfway between midnight and morning, when all the smaller trees were covered, the man was at the very top of the high tree, and around was a crowd of creatures waiting to devour him.

All at once the man saw the Morning Star shining brightly. Then he remembered that in his youth the Morning Star had promised him, in a dream, to help him in time of trouble or peril, and he thought, "If the Morning Star will hurry the day and make light come quickly, then water will go down and I will be saved." And he called out, "Oh, Morning Star, hurry on the day! Oh, Morning Star, hurry on the day! When I was young you promised to help me if I were ever in great peril."

The Morning Star lived in a beautiful house and had a small boy as servant. Hearing the voice, he called to the boy, "Who is that shouting on the island?"

"Oh," said the boy, "that is the husband of the little old woman's granddaughter. He says that when he was young you promised in a dream to help him."

"Yes, I did promise," said Morning Star. "Let day come right away!" Day came immediately, and the water on the island went down at once. When the ground was dry the man slipped down from the tree and, going to the landing place, buried himself in the sand, leaving only his nostrils and eyes exposed.

Early in the forenoon the little old woman came in a canoe, and pulling it up on the beach she said to herself, "The flesh of my granddaughter's husband is eaten up, but maybe his bones are left. They are young and full of nice marrow. I'll find them and eat the marrow." And she began to search for the bones. When she was far enough away, the man crawled out of the sand, sprang into the canoe, pushed out, and paddled away.

When he was some distance from the island the old woman turned, and seeing him cried out, "Come back, my grandson, come back! I'll never play another trick on you. I will love you."

"Oh, no!" called the man. "You'll not play another trick on me," and he hurried on. When night came and the water began to rise, the old woman climbed the tall pine tree. Halfway between midnight and morning, when the water was near the top of the tree and the creatures in the water were waiting to eat her, she screamed out to the Morning Star, "When I was young, you promised to help me if ever I were in distress. Help me now."

The Morning Star heard the voice and called to his boy, "Is that man on the island yet?"

"Oh, no!" answered the boy. "He got off yesterday. That is the little old woman herself. She says that, when she was young, you promised in a dream that if ever she were in trouble you would help her."

"Oh, no!" said the Morning Star. "I never had any conversation with that old woman. I never made her any promise." The Morning Star went to sleep and let day come at its own time.

The water rose till it reached to top of the pine tree, then the creatures of the lake seized the little old woman and ate her up.

The man went home to his wife and then they lived happily.

(Reprinted from Curtin, *Seneca Indian Myths,* 79–81.)

?: The Old Man

Tribal elders tell this tale and ones like it to children to instill values. Although we do not know what the old man did, clearly it was a terrible deed to merit expulsion from the tribe, and he stands in the sky to remind others of his fate. This constellation has not been identified.

An old man, despised and rejected by his people, took his bundle and staff and went up into a high mountain, where he began singing the death chant. Those below, who were watching him, saw him slowly rise into the air, his chant ever growing fainter and fainter, until it finally ceased as he took his place in the heavens, where his stooping figure, staff, and bundle have ever since been visible, and are pointed out as *Nă'gê-tci* (The Old Man).

(Reprinted from Smith, *Myths of the Iroquois,* 39.)

Delaware

The Delaware call themselves *Lenni Lenape* (Common People, Real Men). These Native peoples inhabited the Delaware River basin and adjacent areas, including what is now New Jersey and parts of eastern Pennsylvania, eastern Delaware, and southern New York. There were thirty to forty autonomous Delaware communities, each with its own chief. Individuals from different bands often intermarried, hunted together, and helped protect one another.

The Delaware, like the Iroquois, used a longhouse during the winter and several other structures at other times of the year. They grew corn, squash, beans, pumpkin, and tobacco in the summer. They hunted year-round, principally for deer but also for bear, moose (inland), wolf, raccoon, turkey, passenger pigeon, and other game, and they fished for shad, sturgeon, striped bass, eels, and other species. Coastal groups took advantage of shellfish and fish from the ocean.

The Delaware recognized a Creator or Great Spirit as well as various *manitous* (spirits), including stars. The stars are spirits that help the moon to light up the night. These *manitous* regulate various aspects of Delaware life. The Delaware used the position of the stars, for example, to time the harvest of medicinal herbs and crops. The stars also told them when fish were spawning and when animals were breeding. The Delaware honored the stars in various ceremonies.

BIG DIPPER and NORTHERN CROWN: The Celestial Bear

Historically, the major Delaware religious ceremonies commemorated events in the annual cycle of securing food: the Bear Sacrifice, the Maple Sugar Dance, the Planting Ceremony, the Strawberry Dance, and the Green Corn Ceremony. These festivals were held in the Big House, a ceremonial longhouse. The Bear Sacrifice reflected events in the sky; each year a bear was sacrificed in the ceremony, just as the celestial bear (the BIG DIPPER and the NORTHERN CROWN) was slain in the heavens. Indeed, the earthly bear was considered to contain a fragment of the celestial bear, and the ceremony renewed the relationship between the beings on Earth and the beings in the sky.

To the Delaware, the BIG DIPPER is a bear's body and the NORTHERN CROWN is its head. The body is trying to catch up to the head. When this event occurs, the world will come to an end. A large unidentified constellation nearby is the bear's den or nest. Still another nearby constellation, also unidentified, is a beast pursuing the bear. "At a certain season of the year you will find the leaves of the forest lying on the ground, appearing to be greasy and covered with spots of water. If at this time of year the beast has caught the bear the greasiness of the leaves is believed due to the bear-grease," said Wi'tapano´xwe (Walking with Daylight), who described Delaware customs to the ethnologist Frank Speck in the early 1900s.

PLEIADES: Bunched Up Stars

Walking with Daylight says that only shamans or individuals who possess supernatural power can "know" the Bunched Up Stars, which are holy men or prophets. The following story describes how these stars came into being.

> Because so many people came to a group of holy men for assistance, the prophets changed themselves into seven stones. Still the people came, and they changed themselves into beautiful pine trees, though some appeared as cedars. [Cedars, to the Delaware, are holy trees.] But still the people came. At last the prophets changed themselves into seven stars that move around in the sky during the year. They became the Bunched Up Stars, Pleiades.
>
> (Adapted from Speck, *A Study of the Delaware Indian Big House Ceremony,* 171–173.)

It is not unusual for a tribe to tell several different stories about one constellation. In another Delaware myth, a group of boys who are fasting become a cluster of stars, presumably PLEIADES.

> A shaman took ten young men to an island in the river in order to help them gain spiritual power. The boys fasted and prayed; they were allowed to eat only what the medicine man gave them. In the evening, the shaman visited the island and

threw each youth a corn cake, which he could eat if he caught it on a pointed stick. For nine days none of the boys was able to catch a corn cake, but on the tenth day the first boy succeeded, and he immediately rose into the air, singing a sacred song. Each boy thereafter also caught a cake and rose singing.

(Adapted from Speck and Moses, *The Celestial Bear Comes Down to Earth,* 79–81.)

Fig. 3.5. Folded hide drum and beaters used in Delaware Bear Sacrifice Ceremony. (Illustration from Speck and Moses, *The Celestial Bear Comes Down to Earth,* fig. 3.) Reprinted with permission.

Other Stars and Constellations

The Delaware call Morning Star the Great Star or Red Star. The Evening Star is *Kweᵉtci´pən˙nᵉs* (untranslated). The Milky Way is the path along which the soul travels into the spirit world.

Huron (Wyandot)

"Huron" generally refers to Native peoples who lived in an area between Lake Ontario and Lake Huron in Canada. The Wyandot are Hurons who migrated to the United States. The Huron spoke an Iroquoian language, and their customs were similar to those of the nearby Iroquois even though the Huron were not part of the Iroquois Confederacy and the two groups historically were enemies.

PLEIADES: Seven Dancing or Singing Maidens

One Huron story of PLEIADES is similar to the Shawnee tale about the NORTHERN CROWN (see page 69). In the Huron version, a young man is fasting when he hears voices nearby. The voices belong to seven young women who are dancing. When they realize he is near, they spring into a basket and disappear into the sky. The young man eventually catches one woman, who tells him that she cannot marry him unless he goes with her to Sky Land. "We know this is true," the story ends, "because even today the seven sisters are in the sky. Six are clearly seen, but the seventh sits back in the shadow with her husband. The seventh star may be Pleione, which varies in brightness. During its less bright phase, it can be described as sitting in a shadow.

The Wyandot describe the stars of PLEIADES as Singing Maidens:

The Sun and the Moon had many children, including six girls born at the same time. These girls sang sweetly and danced beautifully. One day they decided to visit the Wyandot, but their father the Sun forbade them to go. They went anyway and sang for the people of the villages, who had never heard songs so lovely. When they returned to the sky, their father was very angry. He promised to send them to a place so distant that they can hardly be seen—and there they are, the Singing Maidens, barely showing in the sky.

(Adapted from Connelley, *Wyandot Folk-Lore,* 109–111.)

Chippewa (Ojibway)

The Chippewa, or Ojibway, were originally centered north of Lakes Superior and Huron but later expanded into what is now Michigan, Wisconsin, Minnesota, North Dakota, southern Ontario, and western Saskatchewan. "Chippewa" is preferred in the United States and southern Ontario, and "Ojibway" is used in other parts of Canada. The Ojibways' name for themselves is *Aniššina·pe·* (Human Being).

This central part of the continent, once covered with glaciers, is crowded with lakes that provide fish and wild rice. The annual cycle included making maple sugar in the spring, farming on small plots and gathering wild berries and nuts in summer, gathering wild rice in the fall, and hunting and fishing all year long.

The Chippewa commonly made dome-shaped wigwams and covered them with cattail mats and birch bark, though hunters sometimes built peaked structures. They also constructed the *mite´win* or conjuring lodge, used for *Midewiwin* (Medicine Dance) ceremonies, and sweat lodges. The Chippewa used birch bark for canoes, containers, and decorative items.

Spirits live in the sun, moon, wind, thunder, and lightning, as well as in other animate and inanimate objects. Nanabozho, the culture hero, secured tobacco for the people and showed them how to hunt and grow food. Tobacco plays an important part in ceremonies; it is offered as a gift and smoked at the beginning of all ceremonies.

Creation of the Stars

A Chippewa tale about how a fox scatters the stars around the sky is similar to the Navajo creation myth in which Coyote blows the stars into the heavens, spoiling the patterns that Black God had planned (see pages 187–188).

Each night Ojishonda, the ruler of Starland, carried a sack of stars around with him and placed the stars in the sky. But being early one evening, he dropped the sack on the ground and took a short rest. His pet red fox, a crafty fellow, decided to have some fun, so he grabbed the bag and took off with it. The stars spilled out everywhere, all over the sky, and Ojishonda was never able to gather them together again. Now Ojishonda must walk around and light the stars one by one, a job that takes the entire night. Sometimes, when the snow is deep or when he is tired, he is not able to light them all. That is why you cannot see some stars in the sky and why some stars disappear in winter. It is just that Ojishonda is getting older and older and cannot complete his rounds.

(Adapted from Cappel, *Chippewa Tales*, 45–46.)

Star Husband

The Temagami Ojibway in Canada narrate this tale, which shows the fluid interaction between people on the Earth and people in the star world; this interaction appears in many Native American stories. Over forty tribes across the continent tell some type of star husband story. In the Blackfoot myth, when the woman returns from the sky she brings with her sacred items and knowledge of rites and ceremonies that she shares with her people. The Ojibway tale does not describe similar ceremonial events but seems to be saying that you should be careful what you wish for, because you might get it.

Although the girls are called foolish for sleeping outdoors in the winter and—apparently—for wishing for something they should not desire, they are at least cleverer than the wolverine, whom they trick into rescuing them. The wolverine is an uncommon but ferocious member of the weasel family with glands, like those of a skunk, that produce an unpleasant odor. The wolverine inhabits the northern boreal forest and Arctic wilderness.

The hole in the sky over which the old woman presides and through which the girls escape is, in Ojibway sacred lore, PLEIADES. Even modern Ojibway shamans believe that the PLEIADES are the point of exchange between our world and the star world. Traditionally, shamans with certain powers built small structures with seven poles to represent the seven stars of PLEIADES for the shaking-tent (also called conjuring-lodge) ritual. In this ritual, which is held at night, the shaman enters the one-person *mite´win* lodge and communicates with invisible spirits. The spirits shake the lodge, giving the ritual its name. The shaman's spirit may leave the body and go through the hole in the sky to the spirit world. The shaking-tent rite is also important to the Cree and the Menominee.

At the time of which my story speaks people were camping just as we are here. In the wintertime they used birchbark wigwams. All the animals could then talk together. Two girls, who were very foolish, talked foolishly, and were in no respect like the other girls of their tribe, made their bed out-of-doors and slept right out under the stars. The very fact that they slept outside during the winter proves how foolish they were.

One of these girls asked the other, "With what star would you like to sleep, the white one or the red one?" The other girl answered, "I'd like to sleep with the red star." "Oh, that's all right," said the first one, "I would like to sleep with the white star. He's the younger; the red is the older." Then the two girls fell asleep. When they awoke, they found themselves in another world, the star world. There were four of them there, the two girls and the two stars who had become men. The white star was very, very old and was grey-headed, while the younger was red-headed. He was the red star. The girls stayed a long time in this star world, and the one who had chosen the white star was very sorry, for he was so old.

There was an old woman up in this world who sat over a hole in the sky, and whenever she moved, she showed them the hole and said, "That's where you came from." They looked down through and saw their people playing down below, and then the girls grew very sorry and very homesick.

One evening, near sunset, the old woman moved a little way from the hole. The younger girl heard the noise of the *mite´win* down below. When it was almost daylight, the old woman sat over the hole again and the noise of the *mite´win* stopped; it was her spirit that made the noise. She was the guardian of the *mite´win*.

One morning the old woman told the girls, "If you want to go down where you came from, we will let you down, but get to work and gather roots to make a string-made rope, twisted. The two of you make coils of rope as high as your heads when you are sitting. Two coils will be enough." The girls worked for days until they had accomplished this. They made plenty of rope and tied it to a big basket. They then got into the basket and the people of the star world lowered them down. They descended right into an eagle's nest, but the people above thought the girls were on the ground and stopped lowering them. They were obliged to stay in the nest because they could do nothing to help themselves.

Said one, "We'll have to stay here until someone comes to get us." Bear passed by. The girls cried out, "Bear, come and get us. You are going to get married sometime. Now is your chance!" Bear thought, "They are not very good-looking women." He pretended to climb up and then said, "I can't climb up any further." And he went away, for the girls didn't suit him.

Next came Lynx. The girls cried out again, "Lynx, come up and get us. You will go after women some day!" Lynx answered, "I can't for I have no claws," and he went away.

Then an ugly-looking man, Wolverine, passed and the girls spoke to him. "Hey, Wolverine, come and get us." Wolverine started to climb up, for he thought it a very fortunate thing to have these women and was very glad. When he reached them, they placed their hair ribbons in the nest. Then Wolverine agreed to take one girl at a time, so he took the first one down and went back for the next. Then Wolverine went away with his two wives and enjoyed himself greatly, as he was ugly and nobody else would have him. They went far into the woods, and then they sat down and began to talk.

"Oh!" cried one of the girls, "I forgot my hair ribbon." Then Wolverine said, "I will run back for it." And he started off to get the hair ribbons. Then the girls hid and told the trees, whenever Wolverine should come back and whistle for them, to answer him by whistling. Wolverine soon returned and began to whistle for his wives, and the trees all around him whistled in answer. Wolverine, realizing that he had been tricked, gave up the search and departed very angry.

(Reprinted from Speck, *Myths and Folklore of the Timiskaming Algonquin and Temagami Ojibwa,* 47–49.)

Milky Way: The Road of Souls
Milky Way: Pathway of Birds

The Ojibway of Parry Island in Georgian Bay, part of Lake Huron, recounted several different stories about the Milky Way in 1929, when these tales were collected. Francis Pegahmagabow, a chief at Parry Island, says that the Milky Way holds up the Earth "like a curving bucket-handle." If the Milky Way were to split apart, the world would end; if the Milky Way were to turn in another direction, life in the world would also change.

Another Ojibway describes the Milky Way as the road on which souls travel to arrive at the land of the dead. Along the way, each soul meets an old man, the sun; an old woman, the moon; a dog; frogs who eat those who cannot offer them tobacco; and a man who removes the soul's brain and directs the shadow to return to the grave. At last the soul enters the land of the dead and reunites with its relatives, with great rejoicing.

Temagami Ojibway and Parry Island Ojibway both call the Milky Way a birds' path. The Parry Islanders relate that the Great Spirit ordered turtle to make a path, the Milky Way, for migrating birds. Jim Nanibush, an Ottawa who grew up on the island, recounts a second turtle story in which the Great Spirit has not assigned any duties to the turtle, so it stays at the bottom of the water.

Whenever it sees people canoeing, the turtle swims to the surface and eats them. The Great Spirit tells two boys to shoot the turtle. The turtle thinks it will escape by swimming north, but when it surfaces the boys shoot it and their arrow lodges in its tail. The turtle waves its tail in the air, splashing water through the sky and creating the Milky Way. From that time on, the turtle has been helpful to the Ojibway.

(Adapted from Jenness, *The Ojibwa Indians of Parry Island, Their Social and Religious Life,* 28.)

BIG DIPPER: *Ojeeg Annung,* the Fisher Stars

Several myths, including one from the Cree and the following Chippewa tale, describe how the people of the cold northern climate obtained summer. The main character in this tale is Fisher. The fisher is an elusive member of the weasel family that can travel on the ground as well as climb trees. The Fisher Stars are the BIG DIPPER. The handle of the dipper is the fisher's tail and the arrow is the little star Alcor that lies next to the middle star in the tail.

Henry Rowe Schoolcraft, an ethnologist and U.S. government agent sent to the Chippewa, married the granddaughter of a Chippewa chief and wrote extensively about Native Americans. He published a long Ojibway tale about *Ojeeg Annung* (the Fisher Stars) in 1856. Ojeeg (the Fisher) is a great hunter—some consider him a *manitou*—who lives on the southern shore of Lake Superior. He rarely returns from hunting without meat, bringing home choice venison or other delicacies. His son begins to hunt with him, but the continual cold freezes his fingers and makes him aim poorly. Snow has covered the country for years. A squirrel convinces the boy to ask his father to find and bring back summer, a great undertaking.

The father cannot refuse his son. Fisher calls together his friends Otter, Beaver, Lynx, Badger, and Wolverine, and they roast a bear for a feast. Fisher says goodbye to his wife and son, for he knows that he will not return, and he and his companions set out on the search for summer.

After a time they come to the lodge of a *manitou,* who gives Fisher advice about where to find the birds of summer and about how they should act once they are there. Then:

They pursued their way, and traveled twenty days more before they got to the place of which the *Manitou* had told them. It was a most lofty mountain. They rested on its highest peak to fill their pipes and refresh themselves. Before smoking, they made the customary ceremony, pointing to the heavens, the four winds, the earth, and the zenith; in the meantime, speaking in a loud voice, they addressed the Great Spirit, hoping that their object would be accomplished. They then commenced smoking.

They gazed on the sky in silent admiration and astonishment, for they were on so elevated a point that it appeared to be only a short distance above their heads. After they had finished smoking, they prepared themselves. Ojeeg told the Otter to make the first attempt to try and make a hole in the sky. He consented with a grin. He made a leap, but fell down the hill, stunned by the force of his fall; and the snow being moist, and falling on his back, he slid with velocity down the side of the mountain. When he found himself at the bottom, he thought to himself, it is the last time I make such another jump, so I will make the best of my way home. Then it was the turn of the Beaver, who made the attempt, but fell down senseless; then of the Lynx and Badger, who had no better success.

"Now," says Fisher to the Wolverine, "try your skill; your ancestors were celebrated for their activity, hardihood, and perseverance, and I depend on you for success. Now make the attempt." He did so, but also without success. He leaped the second time, but now they could see that the sky was giving way to their repeated attempts. Mustering strength, he made the third leap and went in. The Fisher nimbly followed him.

They found themselves in a beautiful plain extending as far as the eye could reach, covered with flowers of a thousand different hues and fragrance. Here and there were clusters of tall, shady trees separated by innumerable streams of the purest water, which wound around their courses under the cooling shades and filled the plain with countless beautiful lakes, whose banks and bosom were covered with waterfowl basking and sporting in the sun. The trees were alive with birds of different plumage warbling their sweet notes and delighted with perpetual spring.

The Fisher and his friend beheld very long lodges and the celestial inhabitants amusing themselves at a distance. Words cannot express the beauty and charms of the place. The lodges were empty of inhabitants, but they saw them lined with *mocuks* (baskets or cages) of different sizes, filled with birds and fowls of different plumage. Ojeeg thought of his son and immediately commenced cutting open the cages and letting out the birds, who descended in whole flocks through the opening that they had made. The warm air of those regions also rushed down through the opening and spread its genial influence over the north.

When the celestial inhabitants saw the birds let loose and the warm gales descending, they raised a shout like thunder and ran for their lodges. But it was too late. Spring, summer, and autumn had gone; even perpetual summer had almost all gone; but they separated it with a blow, and only a part descended; but the ends were so mangled that wherever perpetual summer prevails among the lower inhabitants, it is always sickly.

When the Wolverine heard the noise, he made for the

opening and safely descended. Not so the Fisher. Anxious to fulfill his son's wishes, he continued to break open the cages. He was, at last, obliged to run also, but the opening was now closed by the inhabitants. He ran with all his might over the plains of heaven and, it would appear, took a northerly direction. He saw his pursuers so close that he had to climb the first large tree he came to. They commenced shooting at him with their arrows, but without effect, for all his body was invulnerable except the space of about an inch near the tip of his tail. At last one of the arrows hit the spot, for he had in this chase assumed the shape of the Fisher after whom he was named.

He looked down from the tree and saw some among his assailants with the totems of his ancestors. He claimed relationship and told them to desist, which they only did at the approach of night. He then came down to try and find an opening in the celestial plain by which he might descend to the earth. But he could find none. At last, becoming faint from the loss of blood from the wound on his tail, he laid himself down towards the north of the plain and, stretching out his limbs, said, "I have fulfilled my promise to my son, though it has cost me my life; but I die satisfied in the idea that I have done so much good, not only for him, but for my fellow-beings. Hereafter I will be a sign to the inhabitants below for ages to come, who will venerate my name for having succeeded in procuring the varying seasons. They will now have from eight to ten moons without snow."

He was found dead next morning, but they left him as they found him, with the arrow sticking in his tail, as it can be plainly seen at this time in the heavens.

(Reprinted from Schoolcraft, *The Myth of Hiawatha and Other Oral Legends, Mythologic and Allegoric, of the North American Indians,* 121–128.)

PLEIADES: The Seven Brothers

In the Star Husband story presented earlier in this chapter (see page 58), PLEIADES is the hole in the sky that connects this world and the star world. Another Chippewa tale about seven bunched stars, with one paler than the others, doubtless also describes PLEIADES. In the story, seven brothers dance and sing every evening in honor of the stars. Soon they become restless with their lives on Earth, and when they hear the stars beckoning they decide to join them in the sky. One brother, the youngest, does not really want to go, and he is the faint star that is smaller than the others. He is homesick for the Earth and his mother.

Fox

The Fox called themselves *Meškwahki haki* (Red Earths or Red Earth People), and before contact with Europeans they lived west of Lake Michigan. They grew corn, squash, beans, and other crops along river bottoms and lived nearby in long buildings covered with elm bark. Each building housed an extended family. In the winter, groups as small as a few families or as large as a band moved from place to place while hunting game. They lived in dome-shaped structures covered with cattail mats.

The Fox believe that the Great Manitou rules the sky, a region of goodness. There is also an area of evil under the Earth. Important spirits inhabit the four directions, and many lesser *manitous* have various functions. A person can make contact with a *manitou* by blackening the face, fasting, wailing, smoking, or offering tobacco.

William Jones, who wrote about the Fox, had an unusual vantage point as an ethnologist. He was born in 1871 on the Sauk and Fox reservation in Oklahoma and grew up with his paternal grandmother, a Fox woman. He worked as a cowboy, then attended preparatory school and Harvard College. He received a Ph.D. in anthropology from Columbia University in 1904. During the summers he was engaged in fieldwork with various Algonquian tribes and planned to continue these studies but took a research position in the Philippines. He was inexplicably murdered there in 1909 by one of the tribes he was studying.

The Fox had initiated him into many rites, but Jones was careful to respect the people from whom he learned secret religious material. He recorded and sealed some information—from individuals who were the last ones to know it—until those people "went to their fathers." Many of his notes were published after his death. Jones had a deep love of his people and his homeland on the prairie. In a letter written to a friend shortly before his death, he said, "You saw the stars that I used to see. Did you ever behold clear moonlight nights anywhere else? Did you hear the lone cry of the wolf and the yelp of the coyote?"

Manitou Stars

Wīsa'kä, the Fox creator and culture hero who lives in the north, made the Earth and everything in it. Cāwanōa, the great *manitou* [spirit] of the south, has power over the Thunderer Beings. A *manitou* star serves as the messenger between these two great beings.

Wīsa'kä lives in a wigwam in the north, and with him lives a great *manitou* star. Every morning Wīsa'kä directs the star to visit Cāwanōa, who lives in a wigwam in the south, and give him a message. The star travels a road that leads directly through the sky, and en route he meets the Sun. They both stop, talk, and rest. (That is why people talk and rest when the sun is high.) But the Sun cannot tarry long, for his heat might burn up the Earth and he doesn't like for people to look at him for too long. They both resume their journeys. The star arrives at the wigwam of Cāwanō[a], gives him the news from Wīsa'kä, and continues through the underworld to get back to the north by morning. This star makes the journey every day and every night.

(Adapted from Jones, *Ethnography of the Fox Indians,* 21.)

Stars and Constellations

Some other stars are also great *manitous,* such as Wâpananānāgwa (the Morning Star) and Maskuīgwäw[a] (Red-eyed), the highest star of the evening. Maskuīgwäw[a] opens his eyes to see how things are on the Earth, then descends in the west to go home.

The Fox call the Milky Way *Wâpisīpow[i]* (the White River). These stars are *manitous* who live along the banks of the river. "Some of these *manitous* are those whom Wisa'kä drove away from the earth," writes Jones. Others are *manitous* who are partly mortal who once lived on Earth. But ordinary people never go to the White River.

BIG DIPPER: They that Chase after the Bear

Like other tribes in the Northeast, the Fox say that the bowl of the BIG DIPPER is a bear and the handle is three hunters. The oak and the sumac are symbols of the trees on Earth. Every year the bear comes to life again, and the hunters pursue him once more.

It is said that once on a time long ago in the winter, at the beginning of the season of snow after the first fall of snow, three men went on a hunt for game early on a morning. Upon a hillside, into a place where the bush was thick, they trailed a bear. One of the men went into the brush and startled the bear into running. "The bear is speeding away to the place where the cold comes from!" the hunter said to his companions. "He is speeding toward the place of the noonday sky, and now toward the place of the going-down of the sun!"

Back and forth among themselves they kept the bear fleeing. Finally it went up into the sky. "Oh, River-that-Joins-Another, let us go back! We too are being carried up into the sky," cried one hunter, but his friend did not heed him. His friend was the middle hunter and had a little puppy, Hold-Tight, for a pet.

In the autumn they overtook the bear and slew it. Then they cut boughs of sumac and oak and lay the bear on top of the leaves. They cut up the bear and began flinging and scattering the meat in every direction. Toward the place of the coming of the morning they flung the head and the backbone. In the wintertime, when the morning is about to appear, some stars usually rise. It is said that they came from the head of the bear and the backbone. The backbone stars lie huddled close together.

And they say that the four stars in the lead were the bear, and the three stars at the rear were they who were chasing after the bear. In between two of them is a tiny star, which hangs nearby another; they say that it was the puppy, the pet Hold-Tight of River-that-Joins-Another.

Every autumn the oaks and sumacs redden in the leaf because it is then that the hunters lay the bear on top of the leaves and cut it up; the leaves become red with blood. This is the reason why every autumn the leaves of the oaks and sumacs turn red.

(Adapted [with some original material] from Jones, Fox Texts, 71–75.)

Menominee

Menominee comes from the Ojibway *Mano'mini* (Wild Rice People). The Menominee inhabited an area west of Lake Michigan, in what is currently northeastern Wisconsin and part of the Upper Peninsula of Michigan. Today some members of the tribe live on a reservation in northeastern Wisconsin. Although the Menominee tended small gardens, the long winter and short growing season made it necessary to secure most of their food by hunting, fishing, and gathering. Deer, buffalo, sturgeon, and wild rice were important components of their diet. Wild rice, which is actually a grass, required no planting or cultivation, provided a significant source of food, and was useful as a trade item. Each fall as the rice was ripening, the Menominee poled their canoes through the grasses, and harvesters knocked the grains into the canoes with wooden paddles. The seeds were then dried, pounded, and winnowed.

At puberty, Menominee children underwent a fast of about ten days in length, during which they received a dream from their personal dream guardian, a spirit. The vision usually came in the form of a supernatural animal. A shaman later interpreted the vision, clarifying both the spiritual power and the responsibilities that the individual had received during the dream.

Ethnologists Alanson Skinner and John V. Satterlee gathered stories from the Menominee during the summers of 1910–1914. Satterlee, himself a Menominee, translated for Skinner and collected material independently. Skinner writes that although stories came up in ordinary conversation, he had to "pay" for the telling of myths as a Menominee youth would pay, probably with a gift of tobacco or some other desirable item. He also notes that although the Milky Way was known as the road of the dead among the Menominee, he did not hear any tales involving this road.

BIG DIPPER: *Ut'cikanäo,* the Fisher Stars

The following story of the Fisher Stars (BIG DIPPER) is quite different from the Chippewa Fisher Stars story (see page 61). Instead of describing how the people secured summer, the tale juxtaposes the power of the young man with that of his wife, who becomes the Fisher Stars. Girls listening to the story learn that if they are pure when they fast, they too can receive power from this constellation.

The tale begins, "This is a sacred story concerning the power given through fasting to a young man who was clean, and pure, and free from sin, and whose name was Wânasâtakiu. No one can ever bear this name again because he was so powerful." Wânasâtakiu leaves his home and goes to another village where the chief's daughters are receiving suitors. Although the girls have turned away every man, when Wânasâtakiu sits next to one of them, she says she has been waiting for him for a long time.

Wânasâtakiu took the girl to be his wife, according to their custom, for she accepted him when he lay down with her. Then, as Wânasâtakiu was a famous hunter and great in power, he lived with his new wife for a year, killing everything that he desired, for nothing was hard to him. After this time had passed he said to his wife, "We will now go over to my house and live with my parents for a while." On their way to his parents' lodge, Wânasâtakiu said, "We will go through the village." When they were seen coming, the unsuccessful suitors called out, "Oh, here comes Wânasâtakiu on his way home with his bride!"

The new wife lived with her mother-in-law for a time. Whenever a feast occurred in other villages, her husband would attend. One time he met another nice girl whom he married and lived with as his wife, while his first spouse knew nothing of it. When the feast was over he returned to his home and said to his first wife, "Make me a pair of moccasins. I am going to join a war party."

This was only a lie, his scheme to make an opportunity to live with his paramour. There was indeed a war party about to set out, but the youth went over to dwell with the other woman. He then sent over his paramour's parents to tell his father and mother that he had been killed, and they and his wife believed it.

Every night for four nights Wânasâtakiu went out to the foot of the hill and played on his flute, and each night his discarded wife heard it and recognized the song. She knew well enough that he was not killed and was playing the flute to spite her. Every day the poor woman cut wood as usual and carried it home, until one day she went in the direction from which she heard the flute. There she saw a dead tree and chopped it down, thinking it would give a lot of wood and last a long time. When the tree fell it broke into pieces, and out ran a mouse. The woman snatched up her ax to kill the animal, but it stopped and spoke to her.

"Don't kill me! I was going to tell you something, but now I won't tell you because you want to slay me!"

"Oh then tell me!" cried the deserted wife. "If you do I will pay you well. I will give you some of my hair oil to eat. It is sweet. Come, tell me, and you may have it tonight."

Then the mouse answered, "Do you hear a flute song evenings at the top of that far hill?"

"Yes," replied the woman. "I do."

"Well then," returned the mouse, "that is your husband. It is he that plays there. The others told you a lie. He is not dead, he is alive, and he is staying with another woman." When the wife heard this she was so angry she went right home without cutting any more wood. The mouse had said to her, "On your shelf you will see a bundle. Open it, and you will find it to be a bunch of dried bones that represent your man's death." When she arrived at her lodge she opened the bundle and the bones all fell to pieces.

The mouse had also said, "You go over this evening near that hill and watch for your husband there. You will see him playing. You can see his tracks." The woman went there and found all these things were true. So she said to herself, "Well now, my man Wânasâtakiu, where in the world can you ever escape me? When you come here again you will get it, and you will know it comes from me!"

Then she sat down right there, in sight of the whole great village and began to sing her magic song, directing it against her husband, for she also had great power. Her song meant that she was offended and would not have the man live with her again, and her song told that she was a god. She sang it loud for all the villagers to hear. They listened and were frightened, for they heard her say that her name was Ut´cikasikwäo, really a god-woman.

The deserted wife sang this song four days at this place, and all the people heard and were frightened, knowing her to be a great-powered god-woman. On the fourth and last day, in the morning, she threw herself on the ground and rolled over and over like a horse, and when she had finished she had become a small animal, a fisher. Then she went into the village and at her approach the people, knowing her, ran away into this world to hide. The fisher went directly to the place where her husband was cohabiting with his paramour and killed both of them, with all their relatives. She chased them all over this earth before she caught them and bit them to death. Then she cried, "I am now so mad nothing can ever pacify me. I will never go home again, but as I am really possessed of power I will make a sign for those who are to people the world in the future and they will say of me that I did right."

So she jumped up and ascended into the northern heavens. She is now there as a female, and is called *Ut´cikanäo* (the Fisher Star), meaning the Dipper.

This is a true, sacred story pertaining to the nature of all females, as the female is the mother of mankind. It is said of the Dipper that any girl who was pure and fasting may receive some of her power and it shall be known as long as this world shall stand.

(Reprinted from Skinner and Satterlee,
Folklore of the Menomini Indians, 471–474.)

Shawnee

The *Shawnee* (People of the South) are southerners in the Algonquian language family. Although subgroups moved often, the Shawnee generally lived west of the Cumberland Mountains in and around what is now Ohio.

The Shawnee revere Our Grandmother (or the Creator), a supreme being who created the Earth and taught the Shawnee how to live, raise corn, hunt, and conduct religious ceremonies. The Spring Bread Dance and the Fall Bread Dance were the most important ceremonies, though like the Southeast tribes the Shawnee celebrated the Green Corn Ceremony in August.

NORTHERN CROWN: The Celestial Sisters

This story about a star family is thought to describe the NORTHERN CROWN, whose circular form does resemble a basket. The tale is similar to the one that the Huron relate about PLEIADES (see page 56). The woman here, like sky beings in some other Native American tales, is unhappy on Earth and wants to return to her home in the sky.

The tale starts as Waupee (White Hawk), a celebrated hunter, chances upon a worn circle in the sod in a place he has never visited. He finds no footprints leading to and from the place, so he hides and watches. He sees a basket descend from above and hears beautiful music. Twelve lovely sisters emerge from the basket and dance around the magic ring, striking a shining ball as if it were a drum.

Waupee admires the sisters, especially the youngest, and lunges toward them but they jump into the basket and disappear into the sky. He returns several times, first in the form of an opossum, and then as a mouse. He captures the youngest sister, but the others are drawn up in their basket. Then:

Waupee exerted all his skill to please his bride and win her affections. He wiped the tears from her eyes. He related his adventures in the chase. He dwelt upon the charms of life on the earth. He was incessant in his attentions, and picked out the way for her to walk as he led her gently towards his lodge. He felt his heart glow with joy as she entered it, and from that moment he was one of the happiest of men. Winter and summer passed rapidly away, and their happiness was increased by the addition of a beautiful boy to their lodge.

She was a daughter of one of the stars, and as the scenes of earth began to pall her sight, she sighed to revisit her father. But she was obliged to hide these feelings from her husband. She remembered the charm that would carry her up, and took occasion, while Waupee was engaged in the chase, to construct a wicker basket, which she kept concealed. In the meantime she collected such rarities from the earth as she thought would please her father, as well as the most dainty kinds of food.

When all was in readiness, she went out one day, while Waupee was absent, to the charmed ring, taking her little son with her. As soon as they got into the basket, she commenced her song and the basket rose. As the song was wafted by the wind, it caught her husband's ear. It was a voice that he well knew, and he instantly ran to the prairie. But he could not reach the ring before he saw his wife and child ascend. He lifted up his voice in loud appeals, but they were unavailing. The basket still went up. He watched it till it became a small speck, and it finally vanished in the sky. He then bent his head down to the ground and was miserable.

Waupee bewailed his loss through a long winter and a long summer, but he found no relief. He mourned his wife's loss sorely, but his son's still more. In the meantime his wife had reached her home in the stars, and almost forgot, in the blissful employments there, that she had left a husband on the earth. She was reminded of this by her son, who, as he grew up, became anxious to visit the scene of his birth.

His grandfather said to his daughter one day, "Go, my child, and take your son down to his father, and ask him to come up and live with us. But tell him to bring along a specimen of each kind of bird and animal he kills in the chase." She accordingly took the boy and descended. Waupee, who was ever near the enchanted spot, heard her voice as she came down the sky. His heart beat with impatience as he saw her form and that of his son, and they were soon clasped in his arms.

He heard the message of the Star, and began to hunt with the greatest activity, that he might collect the present. He spent whole nights, as well as days, in searching for every curious and beautiful bird or animal. He only preserved a tail, foot, or wing of each to identify the species; and when all was ready, they went to the circle and were carried up.

Great joy was manifested on their arrival at the starry plains. The Star Chief invited all his people to a feast and, when they had assembled, he proclaimed aloud that each one might take of the earthly gifts such as he liked best. A very strange confusion immediately arose. Some chose a foot, some a wing, some a tail, and some a claw. Those who selected tails or claws were changed into animals, and ran off; the others assumed the form of birds, and flew away. Waupee chose a white hawk's feather. His wife and son followed his example, when each one became a white hawk. Pleased with his transformation and new vitality, the chief spread out gracefully his white wings and, followed by his wife and son, descended to the earth, where the species are still to be found.

(Reprinted from Schoolcraft, *The Myth of Hiawatha and Other Oral Legends, Mythologic and Allegoric, of the North American Indians,* 116–120.)

CONSTELLATIONS OF THE NORTHEAST

Culture Group	Star or Constellation	Classical Equivalent
Micmac	Celestial Bear	BIG DIPPER's bowl
Micmac	Hunters	BIG DIPPER's handle
Micmac	Hunters	HERDSMAN
Micmac	Bear's Den	NORTHERN CROWN
Micmac	Three Hunters in a Canoe	LITTLE DIPPER
Passamaquoddy	Three Hunters and a Bear	BIG DIPPER
Passamaquoddy	Spirits' Path	Milky Way
Passamaquoddy	Chief M'Sūrtū	Morning Star
Penobscot	Six Brothers and a Sister	BIG DIPPER
Iroquois	Three Hunters and a Bear	BIG DIPPER
Iroquois	Great Sky Road; Star Belt	Milky Way
Iroquois	Star Woman	Morning Star
Iroquois	Star That Never Moves	North Star
Iroquois	Bear's Net	NORTHERN CROWN
Iroquois	Dancing Stars	PLEIADES
Iroquois	Horned Head of Wild Animal	TAURUS
Iroquois	The Old Man	Unidentified
Seneca	They Are Pursuing the Bear	BIG DIPPER
Seneca	Beaver that Spreads its Skin	LITTLE DIPPER
Seneca	It Brings the Day	Morning Star
Seneca	She Is Sitting	ORION
Seneca	They Are Dancing	PLEIADES
Seneca	The Group Visible	Unidentified
Onondaga	There They Dwell in Peace	PLEIADES
Delaware	Bear's Body	BIG DIPPER
Delaware	Spirit's Path	Milky Way
Delaware	Great Star; Red Star	Morning Star
Delaware	Bear's Head	NORTHERN CROWN
Delaware	Bunched Up Stars (seven prophets)	PLEIADES
Delaware	Boys Dancing	PLEIADES
Delaware	Beast Pursuing Bear	Unidentified
Delaware	Bear's Den	Unidentified
Huron/Wyandot	Seven Dancing Maidens; Seven Singing Maidens	PLEIADES

CONSTELLATIONS OF THE NORTHEAST (*cont.*)

Culture Group	Star or Constellation	Classical Equivalent
Chippewa	Fisher Stars	BIG DIPPER
Chippewa	Seven Brothers	PLEIADES
Ojibway	Soul's Road	Milky Way
Ojibway	Path of Migrating Birds	Milky Way
Ojibway	Spray from Turtle's Tail	Milky Way
Ojibway	Hole in the Sky	PLEIADES
Fox	They that Chase after the Bear	BIG DIPPER
Fox	White River	Milky Way
Fox	Wâpananananāgwa (a *manitou*)	Morning Star
Fox	Red-eyed (a *manitou*)	Unidentified
Menominee	Fisher Stars	BIG DIPPER
Menominee	Road of the Dead	Milky Way
Shawnee	Celestial Sisters	NORTHERN CROWN

Chapter four

The Fisher Stars

Stars of the Subarctic

The vast northern boreal forest sweeps across the southern provinces of Canada and into the Northwest Territories, the Yukon, and interior Alaska. The northern extent of this wooded area roughly separates the forested Subarctic from the unforested Arctic. Because the severe Subarctic climate is unsuitable for agriculture, the nomadic and seminomadic Native peoples of the region relied on caribou, moose, snowshoe hare and other small game, waterfowl, and fish. In many Subarctic tribes, hunters had a profound respect for the animals they killed. After butchering they placed the bones, which they took care not to break, in a protected area (such as in a tree) so that dogs could not get at the remains. Some tribes developed more elaborate rituals, such as adorning the dead animal with decorations and demonstrating reverence through a pipe-smoking rite. These actions reveal the deep connection between these Native people and the animals upon which they depended for food and, in a larger sense, for their spiritual well-being.

The tribes considered here historically used a variety of dwellings. In the eastern Subarctic, the Montagnais, Naskapi, and East Cree built conical buildings covered with birch bark that housed several families, rectangular structures with a ridgepole, and dome-shaped structures. The Chilcotin and Tahltan made rectangular buildings with a gabled roof and vertical walls, and the Tutchone used domed tents in winter and lean-tos or brush shelters at other times. The Koyukon made various structures, including semisubterranean houses for winter use.

There are two large language groups in the Subarctic. Tribes of eastern and central Canada, including the Naskapi, East Cree, Montagnais, West Main Cree, Woods Cree, and Plains Cree spoke dialects of the Algonquian language family. Tribes of central and western Canada and Alaska, including the Chipewan, Chilcotin, Carrier, Tahltan, Tutchone, and Koyukon spoke dialects of the Athapaskan language family.

Few constellations of Subarctic tribes have been recorded. Native people may have refrained from sharing star lore with people outside of the tribe, or they may not have considered the stars an important part of their lives. Richard Nelson, who lived with the Koyukon in Alaska, explains in his 1983 book why the tribe does not have extensive

Fig. 4.1. Tribes of the Subarctic culture area and their approximate territories in the mid-1800s; some tribes are shown at a somewhat earlier time. The tribes indicated in **boldface type** are included in this chapter. Not all tribes that inhabited this area are shown. (After Helm, *Handbook of North American Indians* Vol. 6, ix.)

celestial lore: "The long period of darkness and the clear skies present a matchless opportunity to watch and chart the heavens, but the Koyukon have little practical reason to do so. They navigate by landforms, and they devote their ideology to the more immediate living world, so they watch the stars just for pleasure or to mark the passage of time."

Montagnais (Innu), East Cree, and Naskapi

Three tribes with similar cultures and customs inhabit subarctic Quebec: the Naskapi in the north, the East Cree in the central portion, and the Montagnais in the south. In the past, people of this region used snowshoes and toboggans in the winter and birchbark canoes (sometimes spruce bark in the far north) for travel to and from winter hunting grounds and for local summer transportation. One or more families hunted in a single winter hunting ground. During the summer people banded together at large lakes or the mouths of rivers to escape the fierce insects of the interior forest and to take advantage of plentiful fish. Some groups, like the Cree, today follow their traditional way of life, including the movement between winter hunting territories and summer villages.

Men hunted large and small game, waterfowl, and other birds as these species were seasonally and regionally available. Moose and caribou were the main large game animals, but hunters also took beaver, bear, porcupine, squirrel, hare, and other mammals.

A person's relationship with the supernatural world was largely individual and depended on that person's ability to communicate with the world of spirits. Tribal members engaged in spiritual practices such as disposing of animal bones respectfully and using a scapula from a hare or caribou to ascertain the location of game in order to achieve "good health, good hunting, and successful birth." Individuals with particular powers became shamans; in the shaking-tent rite, a shaman converses with spirits in a small structure built for that purpose. (See the Ojibway Star Husband story on page 58, for more information about the tent-shaking ritual.)

BIG DIPPER: *Wətcə´kətək,* the Fisher Stars

Native peoples north of the Saint Lawrence River call the BIG DIPPER a fisher, a member of the weasel family. The Ojibway and Menominee in the western part of the Northeast also call the BIG DIPPER a fisher, in contrast to tribes in the eastern part of the region that identify these stars with a bear and hunters. The Ojibway version of the Fisher Stars includes descriptions of spiritual practices, such as conversing with a *manitou* (spirit), not found in the following story.

Ka´kwa, a religious authority among the Mistassini Cree, related the following legend of the Fisher Stars in 1915. The tale illustrates a widespread Native American belief that animals once talked, planned, and—in this narration—secured summer for the North Country. Ka´kwa weaves in an element of humor as one attacker disguises himself as a moose and other attackers silence the guards by pasting their mouths closed. The phrases, "So he was told" and "That's what they said," seem less repetitious when the tale is read aloud; in the oral tradition, phrases such as these serve to emphasize points and introduce the action that follows.

Animal fat, which is rich in calories, is an important source of energy for people who live in a frigid climate, and it is not surprising that grease figures prominently in the diet of these mythical animals.

The story begins with a young boy crying. He will stop only when he can hunt the Birds of Summer. The People of the North decide to go get summer for him. At first they want to leave Otter behind because he laughs too much and will spoil the attack. They discuss at length who will get what portion of fat—an important consideration in a cold land that is without summer—and then they all set out. At last they arrive at a very big, long wigwam. One of the People of the North, a bird, flies up to the tent and sees that summer (in the form of the Birds of Summer) is kept there. Muskrat then goes in to investigate.

The people of the wigwam saw a muskrat swimming far out in the water. They said, "A stranger appears to be coming. Invite him to come in." They called to him. "You swimming, who are you?"

He answered, "I am muskrat, always going around alone."

"Haven't you any neighbors?"

"No," he said, "I am always alone."

"Then swim ashore."

The muskrat swam ashore. "Come ashore," they told him.

"I never come ashore on the land, only on the rocks," said the muskrat.

"If you come ashore, we will give you some grease to eat," they said.

"Oh, throw some over here to me," said he. Then they threw him some grease and he ate it.

"I can't find any taste to it," he said. "I made a mistake, I mixed water in my mouth. Throw me some grease again," he said, "this time a bigger piece."

They threw him some grease again. "If you come ashore you will get a bigger share," they said. When he had eaten it, they asked him, "Is that good? If it is good, come ashore," they said.

"It is good," said he, "only it is too rich." Then he started to swim away. "Don't tell anybody that you have seen us," he was told.

"No," he said. Then he spoke to himself, "I am going to tell it pretty soon." Then he dove; they could not see the muskrat any more.

The muskrat searched for his companions. Together they arrived with the bird who had peeped in upon Summer. That bird explained where the Birds of Summer were located. Muskrat said to his friends, "Every morning they paddle after moose over the narrows." He referred to the habits of the people who were guarding Summer.

Then the muskrat was told, "Tonight go and gnaw their paddles, and likewise bite through the bottoms of their canoes." So he was told. "As soon as it is morning swim over there in the narrows, pushing the root of a tree in the place where the moose generally swim across."

The sucker and the sturgeon were accustomed to guard Summer, so they were told. Accordingly the muskrat swam pushing the tree so that it looked just like the antlers of a moose swimming by. As it grew light in the morning they saw the branches just like the antlers of a moose swimming by.

One of them said, "Moose is swimming over there." They all ran out and jumped into their canoes. But the sucker and the sturgeon, the keepers of Summer, stayed back. The others pursued the moose a little ways out there in the water. Thereupon the People of the North rushed in from their hiding places. Some of them dipped up some sturgeon glue and pasted up the mouths of sucker and sturgeon so that they were not able to cry out for help. Then the People of the North took hold of Summer. They ran outside and ran off with it. The sucker and the sturgeon could not call for help because their mouths were glued up. One of them picked up an arrow, punched it in his mouth, and called out, "Our Summer they have taken from us." That's what they said, that sucker and sturgeon. "They have taken from us our Summer," so they said. "Sho! Paddle back quickly."

They tried to paddle back hard but their paddles broke where Muskrat had gnawed them. Others sank to the bottom where he had made holes in their canoes. After a while some of them ran into the camp.

"What's that you said?" they asked of the sucker and the sturgeon.

"They have taken our Summer away from us."

"Then we will go after them," they said.

They started in pursuit and soon caught up to the People of the North, who cried, "Now they are overtaking us! Who will engage them and delay them?"

"You will!" the People of the North said to Otter. "Run under or inside a stump!"

He ran into one. The pursuers came up. "*Ma!*" [Hello!] they said. "Where is one of them? Oh! Here he is in this hole! Get hold of him!" They seized him.

"What will be his manner of death?" asked one.

"Throw him into the fire," said another.

Otter responded, "You will all make yourselves sick, and die when the flames spread out if you do that."

"Strike him to death!" cried still another.

"That will cause your deaths when I bleed from the wound!" replied Otter.

"Then let's drown him in the water!"

The otter cried in great terror, "*Nawe, nawe, nawe*" (Horrors!).

Then they seized him and threw him in the water. After a while over there way out in the water he emerged. He called, "This is where I live. To die here is impossible for me."

Again they started off after those who were fleeing with the

Summer. Again they almost overtook them. When they came close, the People of the North said, "Who will engage them again to delay them?" They told Fisher, "You, Fisher!" So he was told.

"What shall I do?" he said.

"Run up a tree," they told him. He ran up a tree, and the pursuers came to that place. One arrived first.

"Where is he?"

"He must be up in the tree. Look for him!"

They saw Fisher up in the tree, and they shot at him. They only ripped off a little of his tail, for he ran around behind the tree so fast that they could not do anything to him.

They said, "Where is the most expert bow-man?"

"He has not yet arrived."

At last he came along. "We cannot do anything to Fisher. It is your turn." He shot at him. Twice he shot at him, but he did not hit him once. He said, "This time I'll shoot you with an arrow right."

He shot him. Fisher flew off toward the sky with the arrow sticking in him. They did not know how that was. Then they looked in the sky and saw Fisher.

"We could kill him," they said, "But he will be a sign for man when he comes here in the future, the Fisher Stars." So they said.

(Reprinted from Speck, Montagnais and Naskapi Tales from the Labrador Peninsula, 28–31.)

North Star: *Məca´o wətce´gətək*, Great Star

The son of Ka´kwa, a religious authority of the Mistassini Cree, told this story about the North Star in 1920.

People of another, earlier world were living in a village. They knew that a new world was going to be formed. One day some of them started to quarrel. Among them was North Star. The others fell upon him and were going to kill him, but he fled and soared into the sky, all after him. When they saw that they could not get him they declared, "Well! Let him be, he is North Star, and will be of good use to serve the people of the world that is to come, as a guide by night in their travels." So the North Star became the guide of the people.

(Reprinted from Speck, Montagnais and Naskapi Tales from the Labrador Peninsula, 27–28.)

Other Stars and Constellations

To the Cree, Montagnais, or the Naskapi (or perhaps to all three) *Tcï˙pai meckənu´* (Ghost Road or Dead Persons' Path) is the Milky Way, and *Nictotcima´uts* (Three Chiefs) is ORION's belt.

Chilcotin

The historic territory and current reserves of the Chilcotin lie in along the Chilcotin River and nearby areas in what is now British Columbia. The Chilcotin fished with dip and gill nets, gaffs, spears, and harpoons; they hunted with bows, arrows, and stone-pointed spears; and they used digging sticks, sap scrapers, and bark strippers for gathering.

It was taboo to eat snakes, frogs, owls, and wolves, but the Chilcotin hunted most other animals. Men hunted caribou, deer, elk, mountain goat, sheep, and sometimes black bear, as well as smaller mammals and birds. In the 1800s, moose moved into the Chilcotin hunting territory and elk moved out. Salmon, trout, whitefish, and suckers were an important part of the diet. Roots, berries, and other plant foods supplemented meat in the diet.

Individuals with skills as hunters or shamans played leadership roles within the local bands or villages. Shamans treated illness, which was thought to be caused by the soul leaving the body or by a spirit or object entering the body. Only after coming into contact with Euro-Americans did the Chilcotin establish the formalized role of "chief."

THE BELT OF ORION?: Three Young Men
Morning Star: Old Woman with a Torch

According to ethnologist Livingston Farrand, who recorded this Chilcotin story, the three brothers are probably the stars in ORION's belt. If so, then perhaps the moose and two dogs are the stars in ORION's sword. The Carrier to the north also see game animals in the sky; for them, the PLEIADES are a herd of caribou.

There were once three young men who spent most of the time hunting with their two dogs. They lived with an old woman, their grandmother; and when they came in from the hunt, they used to give the old woman a little of the game, caribou-liver, and other good bits. One day, after hunting all day and killing nothing, they came home to the old woman, and, taking some rotten wood, gave it to her and said, "Here, Grandmother, here is some caribou-liver for you."

The old woman, who was blind, took it and tried to eat it,

and when she discovered the trick was very angry. So the next day, when the young men started out, the old woman took a bear's foot and heated it in the fire and danced about the camp and sang her song; and in this way by her magic she prevented them from coming back, and turned them into stars.

After this the young men lived in the sky. One day, while hunting, they found the tracks of a great moose and followed them for several days. And as they were tracking the moose they looked down and saw the earth, and the eldest brother decided to try to get back to the earth. So he told his brothers to cover themselves with their blankets and not to look. Then he started, but when he was only partway down, the youngest brother looked through a hole in his blanket. So his brother could go no farther. They have lived, all three, in the sky ever since, and can be seen to this day, as well as the moose and the dogs. The Morning Star is the old woman with a torch, looking for the young men, her grandsons.

(Reprinted from Farrand, Traditions of the Chilcotin Indians, 31.)

?: Great Bear

Like the tribes of the eastern Arctic, the Chilcotin performed an elaborate ceremony upon killing a bear. They sang songs, marked the skull with charcoal and set it in a tree, feasted, and performed other rituals. In this story the hunter does not kill the bear but instead gains valuable information about how to increase his skill in hunting. The story does not identify the bear in the sky.

Chilcotin narrators, like their counter-parts in other tribes, shared their tales in the wintertime.

Once a man went out with his two dogs to hunt. It was in the autumn and there was a little snow on the ground. At night he camped.

As he lay on the ground under the trees, Great Bear appeared in the sky. Then the hunter started on, because he knew it was nearly morning. When he had gone but a little way, however, his dogs started a bear.

The bear ran fast and the dogs followed. The man followed both as rapidly as he could, and soon he came to a man sitting on a log. At once the hunter knew this was the bear. The man wore a blanket made of many different kinds of skins.

When the hunter came up, Bear said, "You thought last night I was slow in coming up, but my trail in the Sky Land is very hard and rough. Sun has the same trouble. He travels rapidly at first when the trail is smooth, but in the middle of the day the trail is rough and he travels more slowly. Then his trail grows easier again, and he travels more rapidly to the going-down place at the edge of the earth."

Great Bear then told the young man to pull out from the blanket that he wore the skins of whatever animals he wished most to kill. The man took the skins of marten and fisher. Great Bear told him whenever he went out to hunt, to put the skin of whatever animal he wished to kill in his pouch, and then he would easily kill as many as he wished.

Then Great Bear went back into the Sky Land.

(Reprinted from Judson, *Myths and Legends of British North America*, 79–80.)

Tahltan

The Tahltan live along the upper section of the Stikine River in what is now northwest British Columbia. These Native peoples traded with and borrowed some cultural elements from tribes in the nearby Pacific Northwest. Coastal trade goods from the Northwest became important in Tahltan life as individuals accumulated more material possessions. The Tahltan tell stories about Raven, the culture hero of the Pacific Northwest, and developed a structured class system with chieftains, titles, and inherited rights similar to the class system of the coastal Tlingit, their neighbors to the west. The Tahltan lived in independent bands.

The yearly cycle consisted of fishing, feasting, and visiting in the summer; trading with the Tlingit in early fall; and breaking up into two-family groups for hunting in winter. People traveled by snowshoe in the winter and on foot in the summer.

The Tahltan hunted big game including caribou, moose, buffalo, bear, mountain sheep, and mountain goat, as well as smaller game such as beaver, marmot, and muskrat. They believed that Game Mother was in charge of the animals and that if hunters did not show her proper respect she would call back the game. In past times, many other Native peoples also believed that a supernatural being controlled the availability of animals and could release them at his or her pleasure.

CROW: Raven Carrying the Sun

The magical figure Big-Raven, born in the land of the Tlingit, travels around the Tahltan countryside helping the people by giving them knowledge. When he completes his work, he flies toward the setting sun and disappears. The Tahltan describe the classical constellation CROW as a raven carrying the sun.

BIG DIPPER: The Grandfather Stars

This story illustrates how important the BIG DIPPER is in northern latitudes. Although the moon may wax and wane, if the BIG DIPPER disappears, catastrophe will follow.

> Once the [Dipper stars] called down to the people, saying, "My grandchildren, I will tell you something. Watch me, and as long as you see me going around, everything will be well with you, and you need not be afraid. But if I get lost, light will nevermore come to you and all of you will die. It is nothing if the moon is lost, for it will not be for long, but if I am lost, I can nevermore come back."
>
> For this reason the Tahltan watch the Grandfather Stars whenever there is an eclipse, and if they see him going as usual, they say, "Everything is well."
>
> (Reprinted from Teit, Tahltan Tales, 228–229.)

Milky Way: Snowshoers in the Sky

This Tahltan tale says that people and their snowshoe tracks are the Milky Way. (The Alaskan Inuit to the north, in contrast, say that the Milky Way is Raven's snowshoe tracks.)

> Many people were traveling towards the sky on snowshoes. They had nearly reached the sky. It is said that the people were on their way to the country of the dead in the sky. Someone transformed the people and their trail into stars, and this is now the Milky Way.
>
> (Reprinted from Teit, Tahltan Tales, 229.)

Tutchone

The historical hunting lands of the Tutchone lay in the Yukon Plateau between the Saint Elias Mountains on the southwest and the Selwyn Mountains on the northeast. Contemporary Tutchone still live in this region. Like other hunter-gatherers, these Native people moved across the land in an annual cycle of hunting and fishing. They traveled primarily by foot and snowshoe, although the Northern Tutchone used birchbark canoes on the rivers that flow across the plateau.

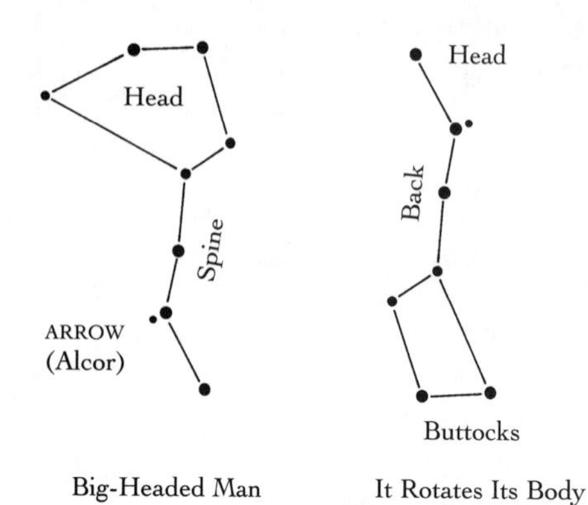

Fig. 4.2. Two Subarctic constellations using the stars of the classical BIG DIPPER: *left,* Big-Headed Man (Tutchone); *right,* It Rotates Its Body (Koyukon).

Stars and Constellations

A loon once cured a blind man, then flew into the sky; its flight path is the Milky Way.

The Southern Tutchone say that there is a big-headed man in the sky. The man's big head is the BIG DIPPER's bowl and a star in front of the bowl. His spine is the handle of the BIG DIPPER. The arrow is the small star Alcor, which is next to Mizar in the handle.

The Big-Headed Man has supernatural powers: He can take rabbits from snares by chanting the magical phrase, "*Hu, Hu.*" One day a hunter shoots an arrow into Big-Headed Man's spine. Before he dies, Big-Headed Man says that he will reappear in the northern sky at night, and so he does.

(Adapted from McClellan, *My Old People Say: An Ethnographic Survey of Southern Yukon Territory*, 78.)

Koyukon

The Koyukon, who live in the Yukon River valley in interior Alaska, also call the BIG DIPPER a man. According to legend, this man argues fiercely with Raven and then leaves for the sky, where he becomes *Nosikghaltaala* (It Rotates Its Body), a "measurer of time." The tip of the handle is his head, the rest of the handle is his back, and the two pointer stars of the bowl are his buttocks.

Like many other tribes, the Koyukon used the stars to mark the passage of the hours during the night. The BIG DIPPER serves as the primary clock, making it the "measurer of time." The Koyukon also marked the movement of Altair, CASSIOPEIA, PLEIADES, and ORION's belt to keep track of time.

CONSTELLATIONS OF THE SUBARCTIC

Culture Group	Star or Constellation	Classical Equivalent
East Cree	Fisher Stars	BIG DIPPER
East Cree	Great Star, Guide of the People	North Star
Eastern Subarctic	Ghost Road	Milky Way
Eastern Subarctic	Three Chiefs	ORION's belt
Chilcotin	Three Hunters	ORION's belt?
Chilcotin	Old Woman with Torch	Morning Star
Chilcotin	Great Bear	Unidentified
Carrier	Herd of Caribou	PLEIADES
Tahltan	Grandfather Stars	BIG DIPPER
Tahltan	Raven Carrying the Sun	CROW
Tahltan	Snowshoers in the Sky	Milky Way
Tutchone	Big-Headed Man	BIG DIPPER
Tutchone	Flight of the Loon	Milky Way
Koyukon	It Rotates Its Body	BIG DIPPER

Chapter five

Raven's Snowshoe Tracks

Stars of the Arctic

The Arctic is a land of extremes. The winter sky is dark and the summer sky is light for months at a time. Dwarfed trees hug the ground for protection and warmth. In vast areas the ground is permanently frozen below the surface. Plants have adapted to the harsh conditions by reproducing swiftly in the short summer. Many animals migrate south to warmer climates when the wind turns cold. It is a tribute to human adaptability and tenacity that people have developed skills for living year-round in this demanding region.

The Inuit (Eskimo), a group of culturally and linguistically similar peoples, have long inhabited the Arctic region that stretches from eastern Siberia across the vast expanse of Canada to Greenland. The native language of North Alaskan, Canadian, and Greenland Inuit—the groups included here—is Inuit-Inupiaq and related dialects in the Inuit-Aleut language family.

In the past, Inuit life revolved around the hunting of marine mammals (seal, walrus, and whale) for those who lived near the coast, as well as caribou, waterfowl, ptarmigan, fish, and other animals, depending on availability.

Although the Inuit have recently developed social and political organizations, in the past, individuals functioned within a small cooperative hunting group that included the nuclear family and extended kin. People considered themselves members of their kinship group rather than members of a larger association or culture group.

Although recorded Inuit constellations are few, they do reflect the importance of hunting in Inuit life. The Alaskan Inuit identify a constellation called Sharing-Out of Food, which they see as dogs chasing a bear, and Three Lost Hunters. The Canadian Inuit tell a story about hunters, their sledge, and a bear they are hunting. The Polar Inuit describe a bear and dogs in the sky, as well as a seal hunter.

Alaskan Inuit

Half a dozen communities from Nuvuk (Point Barrow) to Tikiġaq (Point Hope on the Bering Strait) traditionally hunted bowhead and beluga whales and narwhal; some communities continue to engage in subsistence whaling today. Each village had one or more

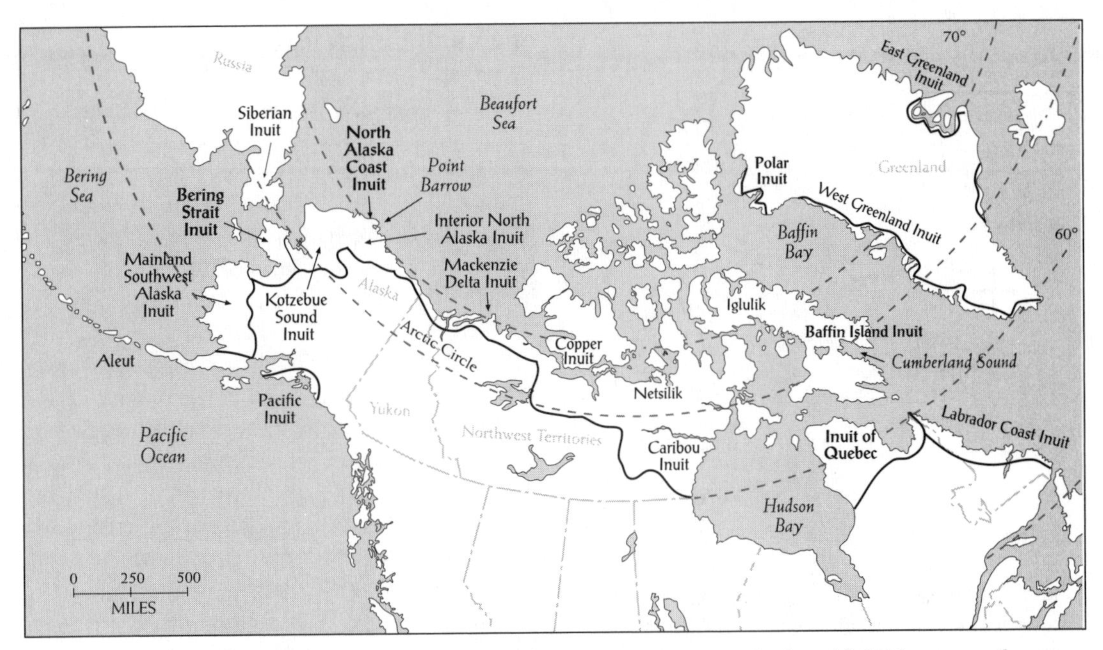

Fig. 5.1. Tribes of the Arctic culture area and their approximate territories in the mid-1800s; some tribes are shown at a somewhat earlier time, others at a somewhat later time. The tribes indicated in **boldface type** are included in this chapter. Most tribes that inhabited this area are shown.

whaling crews of highly trained hunters. The *kashim* (men's social house) was generally affiliated with these crews. After a kill, whale meat was divided according to how much effort each person or crew contributed, although the boat owner was obliged to give generously to everyone in the community.

The Inuit along the northern coast made their homes of blocks of sod, not snow, built around an excavation. They constructed a passageway entrance that began at ground level, several storage rooms, and a main room lined with driftwood. The domed roof had a skylight made of animal gut. Each structure housed eight to twelve people, a nuclear or extended family.

The Inuit in the Bering Strait area also built homes partly underground but made of turf and wood. During the summer they used wood structures built above ground, or temporary shelter consisting of tents of caribou, seal, or walrus skin.

At Point Barrow there are seventy-two days of total light between June and August and a corresponding absence of sun between December and February. At least through the late 1800s, the seal-netters of Point Barrow used, in the dark season, the position of the bright star Arcturus near the BIG DIPPER to judge when to stop netting. (At seventy degrees north latitude, Arcturus is visible almost all day on January 1. In contrast, on the same day at forty degrees north latitude—the latitude of Philadelphia and Denver—Arcturus is visible only from 1 A.M. until dawn.)

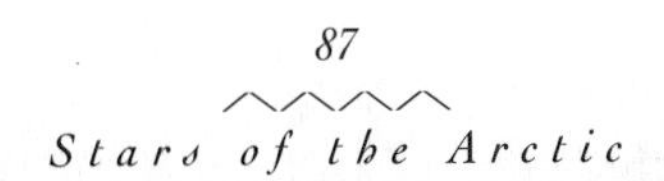

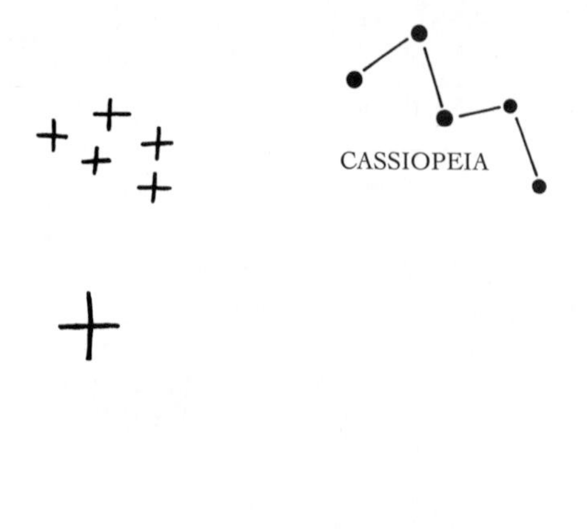

Fig. 5.2. In the early 1900s, upon seeing an Inuit drawing of a flock of birds (*left*), Zelia Nuttall, an ethnologist, wondered if the drawing might not represent a flock of birds wheeling around the North Star (*right*), or the classical constellation CASSIOPEIA. Many native groups used crosses to represent stars in their drawings. The Polar Inuit of Greenland call the classical constellation CASSIOPEIA Stones Supporting a Lamp. (Illustration from Nuttall, Fundamental Principles of Old and New World Civilizations, 50.)

Stars and Constellations

According to inhabitants of the Bering Strait, Raven walked across the sky in his snowshoes during one of his journeys to create the people of the Earth. His snowshoe tracks are the Milky Way. Moon Dog is Sirius; when the star is near the moon there are high winds. The Great Stretchers, posts on which rawhide lines can be stretched, are the three stars of ORION's belt. The constellation Little Foxes, a litter of fox cubs, is the PLEIADES.

Raven Boy, Raven's brother, is responsible for creating the Morning Star. Raven Boy stole the sun from Raven (who had taken it away from the people) and put it back in the sky. "While he was standing close by the edge of the earth, just before sunrise, Raven Boy stuck into the sky a bunch of the glowing grass that he held in his hand, and it has stayed there ever since, forming the brilliant Morning Star."

The people of Point Barrow identify *Pa shukh lurin* (Sharing-Out of Food) as Aldebaran and PLEIADES. Aldebaran is the polar bear, and the stars of the PLEIADES are hunters; "sharing-out" refers to the distribution of meat when the hunt is successful. The three stars in ORION's belt are three hunters who become lost in the winter and are covered with snow. They at last discover an opening and find themselves in the sky as stars. There is also an unidentified constellation, called House-Building, in which several people are building a winter hut. The NORTHERN CROWN, a bright semicircle that is high in the sky in the far north, would seem a likely candidate for House-Building, because the semicircle could easily be interpreted as a round winter hut with an entrance.

Canadian Inuit

The Inuit of Baffin Island hunted and fished in a seasonal cycle: seals through holes in the ice in winter; seals, beluga whale, and walrus at floe edge in spring; fish in early summer; caribou in late summer; fish in early fall; and walrus and whales in late fall. In summer people lived in tents; in winter they lived in snow houses.

ORION: Three Hunters and a Bear

The tale told here is from the Inuit of Cumberland Sound and the east coast of Baffin Island. A sledge is a sled with low runners used to carry loads over ice or packed snow.

Three men went bear hunting with a sledge and took a young boy with them. When they approached the edge of the floe they saw a bear and went in pursuit. Though the dogs ran fast, they could not get nearer, and all of a sudden they observed that the bear was lifted up and their sledge followed. At this moment the boy lost one of his mittens and in the attempt to pick it up fell from the sledge. There he saw the men ascending higher and higher, finally being transformed into stars. The bear became the star *Nanuqdjung* (Betelgeuse); the pursuers, *Udleqdjun* (ORION's belt); and the sledge, *Kamutiqdjung* (ORION's sword). The men continue the pursuit up to this day. The boy, however, returned to the village and told how the men were lost.

(Reprinted from Boas, *The Central Eskimo*, 228.)

Origin of the Sun, Moon, and Stars

The Polar Inuit in northwestern Greenland and the Baffin Island Inuit both tell stories similar to the following one, which is from the Inuit of Quebec. The Inuit consider the sun to be female and the moon to be male; in almost all other Native American cultures the opposite is true.

Incest is a common and widespread theme in Native American myth. Sometimes it occurs during the time of creation. In a California myth, for example, Condor and his sister escape a great flood that kills all living things on Earth except for a few animals. Condor and his sister marry, and their offspring begin the human race.

At other times a trickster is the perpetrator of incest, and the act is viewed with great revulsion. In the following narration, the female sun is disgusted when she finds that it is her brother who has been coming to her in the darkness, and he races into the night sky to escape her wrath. The sparks from the girl's torch become the stars in the sky; the soot forms the marks on the moon.

At a time when darkness covered the earth, a girl was nightly visited by someone whose identity she could not discover. She determined to find out who it could be. She mixed some soot with oil and painted her breast with it. The next time she discovered, to her horror, that her brother had a black circle of soot around his mouth. She upbraided him and he denied it. The father and mother were very angry and scolded the pair so severely that the son fled from their presence. The daughter seized a brand from the fire and pursued him. He ran to the sky to avoid her but she flew after him. The man changed into the moon and the girl who bore the torch became the sun. The sparks that flew from the brand became the stars. The sun is constantly pursuing the moon, which keeps in the darkness to avoid being discovered. When an eclipse occurs they are supposed to meet.

(Reprinted from Turner, *Ethnology of the Ungava District, Hudson Bay Territory,* 266.)

Polar Inuit of Greenland

Though slightly out of the range of this book, the tales of the Polar Inuit of Smith Sound, Greenland, are included here because they supplement the relatively few recorded stories of the North Alaskan and Canadian Inuit.

According to the Polar Inuit at Smith Sound, soil and stones fell from the sky a long, long time ago, creating the Earth. Humans came forth, and their numbers grew because they did not know how to die. An old woman said that she would prefer to have death if it meant that there could be light, for until that time there was only darkness. It happened as she wished: Death came to the Earth, and so did light in the form of the sun, moon, and stars. When people die, they go to the heavens and become stars.

PLEIADES: They Who Are Like a Flock of Barking Dogs After a Bear

Knud Rasmussen (1879–1933) was born in Greenland of Inuit ancestry on his mother's side. He became an Arctic explorer, an ethnologist, and an authority on the native people of Greenland. A Polar Inuit named Aisivak told the following story to Rasmussen. The narration describes the creation of PLEIADES.

A woman who had had a miscarriage had run away from her family. As she ran, she came to a house. In the passage lay the skins of bears. She went in. The inhabitants turned out to be bears in human shape.

But she stayed with them. One big bear caught seals for them. He pulled on his skin, went out, and remained away some time, but always brought something home. One day the woman who had run away took a fancy to see her relations and wanted to go home, and then the bear spoke to her.

"Do not talk about us when you get back to men," he said to her. He was afraid that his two young ones might be killed by men.

So the woman went home, and a great desire to tell came over her; and one day, as she sat caressing her husband, she whispered in his ear, "I have seen bears!"

Many sledges drove out, and when the bear saw them coming towards his house he had great compassion on his young ones and bit them to death. He did not wish them to fall into the power of men. Then he rushed out to look for the woman who had deceived him, broke into the house where she was, and bit her to death. When he came out again the dogs closed up in a circle around him and rushed upon him. The bear defended himself, and suddenly they all became luminous and rose up into the sky as stars. And those are what they call *Qilugtûssat* (They Who Are Like a Flock of Barking Dogs After a Bear).

Since then men have been cautious about bears, for bears hear what men say.

(Reprinted from Rasmussen, *The People of the Polar North*, 176.)

Venus: He Who Stands and Listens

Another Smith Sound resident, Maisanguaq, told Rasmussen this story, which is set in winter, when hunters sought seals at breathing holes. Seals, being mammals, must periodically come to the water's surface to breathe, and in the winter they do so at natural breaks in the ice. Once a hunter's dog located a seal hole, the hunter cleared off the snow, checked it to see if seals still used the site, re-covered the opening, and made a spear hole to set the angle for the harpoon. He then placed a downy feather or similar sensitive material by the hole so that when the seal came to breathe, the indicator would quiver. The hunter would then thrust the spear into the hole, killing the seal. While he was waiting for the seal to appear, it was very important that the hunter stand quietly and not scare away his prey.

There was once an old man who stood out on the ice and waited for seals to come to the breathing-holes to breathe. But close to him, on the shore, a large troop of children were playing in a cleft of the fjeld [a barren plateau]; and time after time they frightened the seals away from him just as he was about to harpoon them.

At last the old man became furious with them for disturbing his seal-catching and shouted, "Close, cleft, over those who frighten my catch away!" And immediately the cleft closed in over the playing children. One of them, who was carrying a little child, got the tail of her fur coat cut to bits.

Then they all began to scream inside the cleft of the rock because they could not get out. And no one could take food to them down there, but they poured a little water down to them through a tiny opening in the fissure. And they licked it up from the side of the rock. At last they all died of hunger.

People then attacked the old man who had made the rock close over the children by his magic. He started off at a run, and the others ran after him. All at once he became luminous and

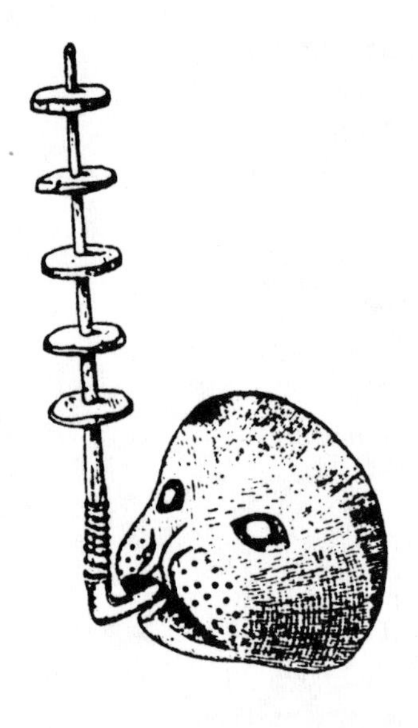
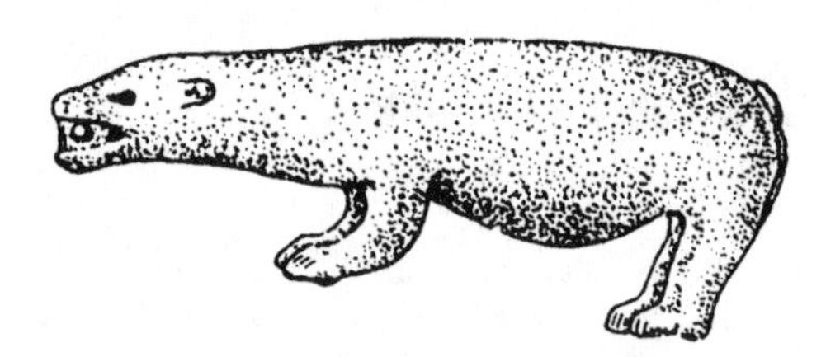

Fig. 5.3. This small mask *(left)* from the Bering Strait represents a seal with rising air bubbles. In a Polar Inuit tale, a seal hunter turns into He Who Stands and Listens (Venus). Bears figure frequently in Inuit art. The toy bear illustrated (*right*) is also from the Bering Strait. The Canadian Inuit tell of a bear and three hunters with a sledge who become stars in the classical constellation ORION. A Polar Inuit myth portrays similar events leading to the creation of the PLEIADES. (Illustrations from Nelson, *The Eskimo About Bering Strait,* 414, 346.)

shot up to the sky, and now he sits up there as a great star. We see it in the west when the light begins to return after the long Dark; but very low down—it never comes up very high. We call it *Nâlagssartoq* (He Who Stands and Listens), perhaps because the old man stood out on the ice and listened for the seals to come up to breathe.

(Reprinted from Rasmussen,
The People of the Polar North, 176–177.)

Other Stars and Constellations

The Polar Inuit identify several other constellations: Giant Caribou (BIG DIPPER), Stones Supporting a Lamp (CASSIOPEIA), and Three Steps Cut in a Snowbank, which lead from Earth to the sky (belt of ORION).

CONSTELLATIONS OF THE ARCTIC

Culture Group	Star or Constellation	Classical Equivalent
Alaskan Inuit	Raven's Snowshoe Tracks	Milky Way
Alaskan Inuit	Bunch of Glowing Grass	Morning Star
Alaskan Inuit	The Great Stretchers	ORION's belt
Alaskan Inuit	Three Lost Hunters	ORION's belt
Alaskan Inuit	Little Foxes	PLEIADES
Alaskan Inuit	Sharing-Out of Food	PLEIADES, Aldebaran
Alaskan Inuit	Moon Dog	Sirius
Alaskan Inuit	House-Building	Unidentified
Canadian Inuit	Bear	Betelgeuse
Canadian Inuit	Hunters	ORION's belt
Canadian Inuit	Sledge	ORION's sword
Polar Inuit	Giant Caribou	BIG DIPPER
Polar Inuit	Stones Supporting a Lamp	CASSIOPEIA
Polar Inuit	Three Steps Cut in a Snowbank	ORION's belt
Polar Inuit	They Who Are Like a Flock of Barking Dogs After a Bear	PLEIADES
Polar Inuit	He Who Stands and Listens	Venus

Chapter six

The Harpooner-of-Heaven

Stars of the Pacific Northwest

The Native Americans of the Pacific Northwest live in a distinct geographic area, a narrow band of temperate rain forest largely bordered by an inland mountain range, extending from southern Alaska to southern Oregon. This region has a mild climate of cool summers and wet winters. In the past, these Native people did not grow food but relied on plentiful natural resources—fish, sea mammals, waterfowl, seabirds, shellfish, land mammals, berries, and greens.

Of the groups for which there are recorded constellations and star myths, the Tlingit, Tsimshian, and Haida live in the northern half of the culture area; the Kwakiutl, Central Coast Salish, and Southern Coast Salish live in the area from Vancouver Island to Puget Sound; the Makah, Quileute, and Quinault live along the Pacific Coast in what is now Washington State; and the Cathlamet live along the lower Columbia River.

The people of the Northwest were hunters and gatherers, but abundant and concentrated food sources allowed them to build permanent communities, amass personal belongings, and develop a complex social organization not usually found in hunter-gatherer societies. Hunter-gatherers in the Great Basin, for example, were nomadic. They lived in small groups, built temporary shelters, and moved from place to place as food became available throughout the year. In contrast, farmers—including those in the Southeast and the Southwest—were able to build homes in permanent communities because they could secure a major portion of their food from a relatively small area. Native people in the Northwest essentially "farmed" the sea and coastal lands and so were able to develop a substantial material culture.

Together, the Native people of the Northwest spoke at least forty-five languages, making this the most linguistically complex area outside of California. Despite the linguistic diversification, there are common cultural elements. Native Americans of the Pacific Northwest became master woodworkers, carving totem poles; dugout canoes; wooden masks, boxes, other utensils; and ceremonial pieces. They also made twined baskets, woolen and plant-fiber clothing, basketry hats, and decorative copper shields called *coppers*. They traveled by canoe, making many different types to fit local needs and navigational challenges. Contemporary Native peoples of the Pacific Northwest still

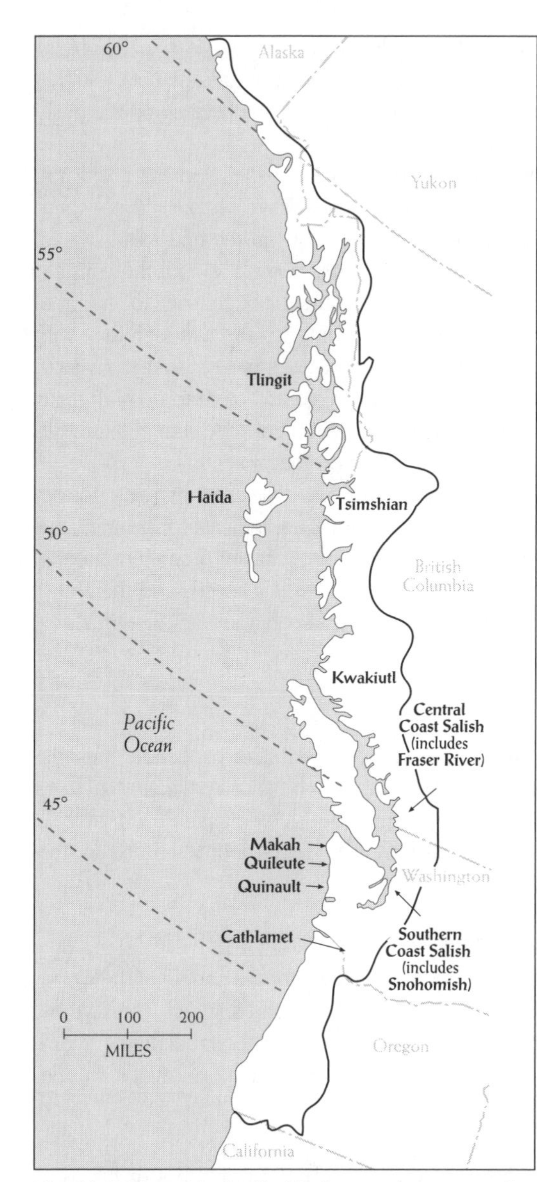

Fig. 6.1. Tribes of the Pacific Northwest culture area and their approximate territories in the early 1800s; the Kwakiutl are shown as of the mid-1800s. The tribes indicated in **boldface type** are included in this chapter. Not all tribes that inhabited this area are shown. (After Suttles, *Handbook of North American Indians* Vol. 7, 10–11; and Waldman, *Encyclopedia of Native American Tribes*, 31.)

live in their historical homelands, and many have thriving cultures that include the performance of ceremonial dances and the creation of beautifully executed artwork.

Most communities were divided into the wealthy, their followers, and—in most areas—slaves, who could be acquired in raids of neighboring tribes or through purchase. There was an emphasis on accumulating material goods such as those described above. Many tribes had secret societies to which the rich and powerful belonged.

To celebrate special occasions like honoring the dead or initiating people into societies, wealthy individuals put on feasts with theatrical dances that reenacted the time when their ancestors met supernatural spirits. It was in this legendary time that crests signifying lineage and clans were presented to the ancestors. From the Tlingit south to the Puget Sound area—and later south to the Columbia River—feasts included the distribution of lavish gifts to those attending. The giveaway, called a *potlatch,* demonstrated prosperity, for the wealthier the host, the bigger and more valuable the gifts. The guests were generally expected to reciprocate at a later time.

Although the foggy shores and cloudy skies of the Pacific Northwest might not seem the best place to develop sky lore, there were in fact elders whose job it was to make observations of the sky and provide information about the movement of stars, planets, and the moon. One Tlingit clan history describes elders sitting outside and checking on the position of various stars, Venus, the Milky Way, the moon, and the sun.

Much celestial information, however, was not recorded, either because it did not seem important to Euro-Americans at the time, or because Native people did not want to provide details. James Swan, a teacher on the Pacific Coast in the mid-1800s, wrote that the Makah were reluctant to show him constellations. "Most, if not all the constellations have names, . . . but I have never had any of them pointed out to me; [the Makah] seemed to have a superstitious repugnance to doing so, and although they will at times talk about the stars, they generally prefer cloudy weather for such conversations." The Makah's reluctance to discuss the stars may have had more to do with their aversion to sharing sacred beliefs with an outsider than with mere superstition; the result was that Swan was unable to gather much information on the topic.

It is no surprise that many of the recorded Northwest constellations relate to the ocean. The stories describe fish, marine mammals, canoes, implements for fishing, and halibut fishers in the sky. The identified constellations represent the prominent groups of stars—the BIG DIPPER, the LITTLE DIPPER, ORION, the PLEIADES, CASSIOPEIA, and the Milky Way. Other constellations are mentioned in tantalizing asides but are not identified.

Tlingit

The Tlingit, who live along the fjords and islands of the Alaskan panhandle and the coast of British Columbia, were historically divided into two major ceremonial groups, the Raven moiety and the Wolf moiety. There were about thirty clans in each moiety, and various lineages or house groups in each clan. Thus, each individual, except for slaves, had a distinct lineage.

Origin of the Sun, Moon, and Stars

In this myth, Raven, the Tlingit culture hero, secures daylight for the people of the world and in the process causes the creation of the animals of the land and sea. The myth is told with variations by several tribes in the Pacific Northwest. The grandfather's "wonderful boxes" are likely similar to the carefully crafted wooden boxes made by the Tlingit and other Northwest woodworkers.

Long ago, the world was entirely dark. A chief kept the sun [or sunlight], the moon, and the stars in three wooden boxes. To secure sunlight for the people Raven turned himself into a hemlock needle. The chief's daughter ate the needle and then gave birth to a baby boy (Raven). The boy pleaded for his grandfather's wonderful boxes. When his grandfather relented, the boy opened the boxes with the moon and stars and threw their contents into the sky. Then he turned himself back into a raven and took the box containing the sun to a place called Dry Bay. When Raven opened the box, all the people were afraid of the sunlight and they ran from him. People wearing seal skins ran into the water and became seals while people wearing the skins of various other animals became those animals.

(Adapted from de Laguna, *Under Mount Saint Elias: The History and Culture of the Yakutat Tlingit,* 796.)

Fig. 6.2. Raven is the culture hero of the Tlingit and other Pacific Northwest tribes. In Tlingit myth, Raven released the sun, the moon, and the stars so that they could shine in the sky. He also created the constellations Sculpin (PLEIADES) and Halibut Fishers in a Canoe (classical constellation unidentified). The double raven in the illustration is of Haida design. (Illustration from Mallery, *Picture-Writing of the American Indians,* fig. 523.)

Various Stars

The Tlingit say that the sky is a giant vault above a flat world, and that spirits called *yēk* live in between. The stars are houses or towns, and the light that shows from them is the reflection of the sea.

NA´s!gînAX-qa (Three Men in a Line) are three stars in a row, probably ORION's belt. *Kēq!A´cAGULÎ´* (Morning Round Thing) is Venus, the Morning Star. *K!uxdî´sî* (Marten Month or Marten Moon) is the Evening Star, perhaps Jupiter; the marten, a member of the weasel family that has the ability to climb trees, is about the size of a cat. The Milky Way is *Łq!ayā´k! q!ō´siyite* (*Łq!ayā´k!*'s Tracks), the tracks that the Tlingit hero *Łq!ayā´k!* left while chasing a monster across the sky.

BIG DIPPER is *YAxtê´*. In earlier times, the Tlingit used the handle of the BIG DIPPER as a celestial clock, getting up in the dark when the handle reached a certain position.

PLEIADES: *Wēq!,* Sculpin

The mother of Katishan (the chief of a Tlingit community in Wrangell, southern Alaska), related this myth to John Swanton in 1904 and then explained its connection to everyday life: "So nowadays, when a person wants people to think he knows a great deal and says, 'I am very old,' they will answer, 'If Sculpin could not make Raven believe he was so old and knew so much, neither can you make us believe it of you.'"

A sculpin is a spiny but edible large-headed fish.

As Raven was traveling along he saw a sculpin on the beach looking at him and hid from it to see what it would do. Then he saw it swim out on the surface of the ocean and go down out of sight some distance off. After that he opened the door of the sea, went to the house of the sculpin, which was under a large rock, and said to it, "My younger brother, this is you, is it?"

"I am not your younger brother."

"Oh! yes, you are my younger brother. We were once coming down Nass River in a canoe with our father and had just reached its mouth when you fell overboard and sank forever."

Then the Sculpin said, "I cannot be your younger brother for I am a very old person."

Said Raven, "I want you to be next to me. There will be many sculpins, but you shall be the principal one." So he placed the sculpin (*wēq!*) in the sky where it may still be seen.

(Reprinted from Swanton, *Tlingit Myths and Texts,* 107.)

?: *DANA´qᵘs!ike,* the Halibut Fishers

Katishan's mother also told this story, which moralizes about why children should not be idle. "When a child was lazy and disobedient, they told him how the halibut fishermen got up into the sky for their laziness. Therefore the children were afraid of being lazy," she said. Katishan's mother did not identify the stars of The Halibut Fishers.

The Tlingit used hook and line for catching halibut. Carvers made the upper arm of the hook out of lightweight cedar and lashed on a bone barb to hold the octopus-meat bait. They carved the lower arm, which was alder, into an animal or human figure to attract the fish. Fishermen then attached the hook, which was about a foot long, to a long bottom line. Because halibut are large, sometimes up to one hundred pounds, fishermen needed skill to haul the hooked fish into their canoes.

It is interesting to note that the Tsimshian and Makah also name halibut-related constellations, though their classical counterparts are not known. The halibut's large size and its obvious contribution to the larder may have prompted Native peoples to give it an important place in the sky.

Raven saw a canoe out after halibut and said, "Come ashore and take me across," but they paid no attention to him. Then he said, "If you do not I will put you up in the sky also [in addition to the sculpin]. I will make an example of you, too." Then he held his walking stick out toward the canoe and they found themselves going up into the sky. That is what you can see in the sky now. It is called *DANA´qᵘs!ike* (the Halibut Fishers).

(Reprinted from Swanton, *Tlingit Myths and Texts,* 107.)

Tsimshian and Haida

The Tsimshian lived on and around the Skeena River in what is now British Columbia, and today they live in reserves in British Columbia and southern Alaska. They call themselves *Ts!EM-sia´n* (Inside of the Skeena River). In Tsimshian society, the "house" is a matrilineal descent group, that is, descendants of a mother's mother. In the past, these descendants—together with relatives by marriage, children of other descent groups, and slaves—lived together in one or sometimes two large buildings. Each descent group owned a large structure in the winter village; fishing, hunting, and gathering territories; and certain crests and songs. Crests featured animals, mythological beings, and natural phenomena. There was a chief for each house, and each chief had an established rank when all of the chiefs of the village gathered.

The Haida, for whom there is but one recorded star story, live to the west of the Tsimshian, on the Queen Charlotte Islands.

Stars and Constellations

The Tsimshian believe that there was initially only darkness. A chief's elder son became the moon, his younger son became the sun, and his daughter became the fog. The stars are the sparks that fly out of the sun's mouth when he is asleep.

In another myth, the sun is described as a chief and the stars as his tribe. The Morning Star, the sun's daughter, lives with him in his house, as do the unidentified constellations Kite, Halibut Fishing-Line, Stern-Board in the Canoe, Old Bark Box, and Dipper. Given the simplicity of the image, it would be no surprise if the Dipper were the BIG DIPPER. Tsimshian workers made wooden dippers along with other household items.

PLEIADES: Seven Brothers in a Boat

The Haida tell of seven brothers who become the PLEIADES.

> Seven giant brothers, while fishing one day, harpooned a great monster. The whale dragged their vessel swiftly over the ocean toward a whirlpool, where death awaited them. But as they drew near, the rope broke and the whale sank into the ocean. The boat was traveling so fast that the brothers sailed into the sky, where they became the Pleiades.
>
> (Reprinted from Hamilton, Stellar Legends of American Indians, 48.)

Kwakiutl

The Kwakiutl include many tribal groups in the area around Queen Charlotte Strait at the northwestern end of Vancouver Island. The Kwakiutl were and are outstanding wood carvers who make beautiful masks, totem poles, and wooden hats. Historically, they wore the masks, which were painted and decorated with feathers and hair, in various ceremonies and when telling stories. Most Kwakiutl myths describe the time when ancestors met supernatural spirits and crests and ceremonies were created.

Stars and Constellations

For the Kwakiutl, ORION is Harpooner-of-Heaven, or Paddler. He is a hunter in the sky who paddles a canoe and owns the fog. He has a box that, when opened, brings forth fog. Harpooner-of-Heaven hunts an otter that escapes to the sky and becomes the PLEIADES. At one point, an ancestor climbs into the canoe of the hunter and visits the Evening Sky.

Traditionally, the Kwakiutl celebrated two major festivals, the Cedar Bark Dance and the Weasel Dance. In the first ceremony, one person portrays the Cannibal Spirit. This man-eating spirit, similar to the devil in Western culture, is the leader of animals that have a taste for human flesh. In a song that accompanies the Cannibal Dance, the Milky Way is referred to as the pole of the cannibal.

Central Coast and Southern Coast Salish, Including the Snohomish

Dozens of small tribal groups broadly called the Central Coast and Southern Coast Salish live in the coastal Washington area around the Strait of Georgia, Juan de Fuca Strait, and Puget Sound. Only a few star stories have been recorded from this region.

Stars and Constellations

One Central Coast Salish group, identified only as living along the lower Fraser River, described two constellations in the early 1900s. The first is a group of children who are crying because their parents have left them; they are placed in the sky where they become the PLEIADES. In the second, a man and a dog have chased an elk, and the elk flees to the sky, where it becomes the BIG DIPPER.

The theme of the elk and the BIG DIPPER reoccurs in a tale told by the Snohomish, a tribe of the Southern Coast Salish. The Snohomish lived along the Snohomish River in northwestern Washington and today live on the Tulalip Reservation, also in Washington.

BIG DIPPER: Three Hunters and an Elk
ORION's belt and sword?: Two Canoes and a Fish

The Snohomish call the Creator *Dohkwibuhch* (Changer). Changer created the world and performed many other feats of importance, including distributing languages to the Native Americans of the continent. He set the height of the sky imperfectly, however, and the people in the Northwest decide to change it.

When they raise the sky, they trap an elk, hunters, a dog, two canoes, and a fish in the Sky World, and these all become constellations. The story specifies that the elk, hunters, and dog become the BIG DIPPER but doesn't describe the fate of the canoes and fish. The Wasco, a Plateau tribe located inland on the Columbia River, tell a myth about two canoes and a fish that become the belt and sword of ORION. Given the similarities in detail—not just two canoes, but two canoes and a fish—and the proximity of the tribes, it seems likely that the Snohomish canoes also become stars in ORION.

When the Changer made the world, he started in the east and worked his way west, making groups of people and giving each a language. By the time he got to Puget Sound he decided to stop, and he dispersed all of the remaining languages in that area. That's why the people of that part of the country speak so many different languages.

The Changer made the sky too low. People bumped their heads on it and sometimes ventured into the Sky World. Everyone wanted to raise the sky, but they spoke different languages and lived in different areas. How could they solve the problem? Some wise men got together and decided that everyone would push at a signal, *Ya-hoh*, which in all of the languages means "lift together."

Everyone in the area, people and animals alike, made giant poles from fir trees with which to push up the sky. At the signal *Ya-hoh,* they all pushed as hard as they could. Each time the wise man shouted *Ya-hoh,* they pushed again. Eventually, they pushed the sky to its present position, where no one can bump into it and no one can climb into the Sky World.

There were several people and animals from the earth, however, who did not know about the scheme to raise the sky. Three hunters who had been tracking an elk in the Sky World were caught there. The elk became the stars of the bowl of the Big Dipper, the hunters became the stars of the handle, and their little dog became the tiny star (Alcor) next to the middle hunter.

Two canoes, each with three men, were also trapped in the sky, as was a little fish.

(Adapted from Clark, *Indian Legends of the Pacific Northwest*, 148–149 and Shelton, *The Story of the Totem Pole*, 11–12.)

Milky Way: A River in the Sky

Many tribes across North America narrate myths about how mythical people or animals stole fire for humans to use. The following Snohomish tale grafts a plot about a canoe maker onto this widespread myth.

A Snohomish canoe maker worked from early morning until late at night. The Sky Chief, who lived on a river (the Milky Way) became irritated by the continual hammering, so he told four of his men to steal the workman, which they did.

The Snohomish people set out to find the man, who was their only canoe maker. They constructed a chain of arrows that they could climb into the sky. Once there, they freed the canoe maker and stole fire as well. Then they retreated down the ladder of arrows.

Four shamans—Grizzly Bear, Black Bear, Cougar, and Wild Cat—began to dance. They broke the chain and make everyone fall, but no one was hurt. The snake and the lizard were left behind and could not return until spring.

While in the sky, the canoe maker had promised the Sky Chief not to make canoes early in the morning or late at night. All of the tribes around this place also abide by these regulations, for the Sky Chief has said that he will attack the earth people if they disturb him again.

(Adapted from Haeberlin, Mythology of Puget Sound, 412.)

Makah, Quileute and Quinault

The Makah (Cape Dwellers) were whaling people who lived on Cape Flattery, on the northwest tip of what is now Washington State. They and several nearby groups hunted gray and humpback whales. Whaling was a prestigious occupation that demanded knowledge of particular ceremonies as well as skill in navigation and boat handling. During the summer the Makah gathered food; in the winter they worked on crafts and held ceremonies. The Makah still live on Cape Flattery.

The Quileute, also a whaling tribe, lived south of the Makah, along the Quillayute River and its tributaries. This group had five societies, one each for warriors, fishermen, hunters, whale hunters, and weather men. Initiations were held during the winter months. The Quileute now live on two reservations in Washington state.

The Quinault lived south of the Quileute in about twenty villages, each with its own chief. Like other groups in the Northwest area, the Quinault lived in gable-roofed houses made of cedar planks. They had some social stratification into nobility and commoners but owned few slaves. The descendants of the Quinault live on the Quinault Reservation in Washington State.

Stars and Constellations

The Makah say that the stars are the spirits of Indians from the mythological age and animals of every kind, including mammals, birds, and fish. Comets and meteors are spirits of departed chiefs. In times past, the Makah were reluctant to talk with Euro-Americans about the stars and mentioned but did not identify several constellations, including Whale, Halibut, Skate, and Shark.

Many Northwest groups tell myths of how animals venture into the sky. In one Quileute story, several animals go there and are trapped. Skate becomes the LITTLE DIPPER, a bear skin becomes the BIG DIPPER, and Beaver becomes an unidentified set of stars "in the shape of a pancake turner." The brightest star in the sky (Venus?) is the single eye of a man who used to live in the ocean but who is no longer there.

In a Quinault tale, all the animals venturing into the sky return except Skate (BIG DIPPER) and Fisher (unidentified).

CASSIOPEIA: The Giant Elk Skin

Although the Quileute relied heavily on the ocean for food, they did hunt deer and elk with bow and arrow, deadfalls, and snares. This elk-hunting story was recorded in the early 1900s.

Fig. 6.3. To the Makah, a whaling tribe that hunted gray and humpback whales, a whale swims through the night sky as a group of stars. In a Haida myth, seven fishermen harpoon a great whale and are about to disappear in a whirlpool when the line breaks and they are thrown into the sky, becoming the PLEIADES. This Makah design shows a thunderbird, a whale, and two other animals. (Illustration from Swan, The Indians of Cape Flattery, fig. 1.)

It was in the long ago. There were five brothers living here at La Push. One day four of them went up the river hunting in their canoes. The younger brother, who was named Tuscobuk, stayed at home. When they got up to Salalet near Forks Prairie, they left the river in search of game. When they came to the Prairie, they saw a lone man walking about with his bow and arrows. He came to meet the brothers. When he came to them, he asked, "Where are you going?"

They told him, "We are hunting for elk."

He said, "I know where there are lots of elk. If you will all get behind that tree there, I will drive the elk down the trail that goes by there. Then you can shoot them." He started to walk away to round up the elk. Then he stopped and looked at their bows and arrows. He said, "I have better arrows than you have. Let's trade." The brothers traded with the stranger. He was working power on them, so they did not know what they were doing. The stranger's arrows were made of salal bush stems and the points of salal leaves. They were of no use whatsoever. After they had exchanged arrows, the stranger went off to find the elk.

Soon a big elk came down the trail, as the stranger had said. The brothers made ready to shoot it. They drew their bows. The arrows hit the elk. Every arrow doubled up against the elk's flanks. Then the elk charged upon them and killed them all. The elk then turned himself back into a man. That was the man of the Prairies that had met the brothers. This was his way of killing people.

When the four brothers did not come back from the hunting trip, Tuscobuk went to look for them. He went in his canoe to the Prairie. At the Salalet landing he saw his brothers' canoe. "They have gone hunting from here," he thought. He climbed a hill and entered a grassy place. As soon as he went there, he saw a large man walking back and forth across the Prairie. When the man saw Tuscobuk, he came to him. Tuscobuk inquired, "Have you seen my brothers who were hunting elk?"

Man of the Prairies answered, "No," although it was he who had killed them. They talked quite a while. Man of the Prairies said to Tuscobuk, "I know where there is a band of elk. If you will hide behind that tree over there, I will drive the elk down the trail. Then you can kill at least one of them."

Tuscobuk said, "All right." He was hungry.

Man of the Prairies prepared to start to drive up the elk. He was just leaving when he stopped. He said to Tuscobuk, "Let me see your arrows," the same manner as he had spoken to the four brothers. After he looked at the arrows carefully, he said, "These arrows are no good. I have better arrows. Do you want to trade?"

Tuscobuk was a medicine man. He was not so easily fooled as his brothers had been. He sensed that something was wrong. His power was as great as the other man's. He could easily see that the arrows that the man wished to trade him were worthless. He said, "No, I won't trade."

Man of the Prairies went to drive down the elk anyway. He had not been gone long when the elk with the long horns that had killed the four brothers came down the trail looking for the man behind the tree. The arrows had made Tuscobuk suspicious. He had hidden himself behind another tree. When the elk was in easy range, he began to shoot at it with his own yew-wood arrows. He shot it four times and each arrow struck a vital spot. The elk saw where he was hidden. He ran and jumped upon Tuscobuk. But the elk had been weakened by the arrow wounds. Tuscobuk fought with the elk. He cut his throat with his clamshell knife. The elk died there. He was a monster of an elk as he lay stretched there on the ground.

Tuscobuk skinned him. When he went to stretch the skin, he was surprised. He saw that the skin was bigger than the prairie. So Tuscobuk threw the skin up in the sky. You can see it any clear night. It is Cassiopeia. The stars indicate the places in the hide where Tuscobuk drove the stakes when he was stretching it. The handle is the elk's tail.

(Reprinted from Reagan and Walters, Tales from the Hoh and Quileute, 325–326.)

Cathlamet

The Cathlamet are one of the four divisions of the Chinookans of the lower Columbia River. Because they lived on a river instead of the ocean, the Cathlamet did not hunt whales but relied on fish, land mammals, and wild foods.

In the early 1890s, Charles Cultee narrated the following story, in which many animals become stars, for the ethnologist Franz Boas. As in other tales that take place in the "early times" or mythological era, the "people" then are animals who talk, interact, and solve problems. These people are able to find their way into the sky world, and they can perform miraculous feats. When Beaver is killed, for example, he gets up shortly thereafter and runs outside.

This myth is reminiscent of the eastern Subarctic tale of securing summer. There, the animals want to leave Otter behind because he laughs too much; here, the animals want to exclude Skate because they think he will be too easy to hit with an arrow. In both stories, the attackers sabotage the equipment or clothing of the enemy. At the end of both narrations, one or more animals turn into stars. In the Subarctic, Fisher goes to the sky, whereas here Woodpecker, Fisher, Skate, Elk, and Deer are trapped in Sky Land.

There were five Southwest winds. The people were poor all the year round. Their canoes and their houses were broken. The houses were blown down. Then Bluejay said, "What do you think? We will sing to bring the sky down." He continued to say so for five years. Then their chief said, "Quick! Call the people." All the people were called. Then they sang, sang, and sang, but the sky did not move. They all sang, but the sky did not move. Last of all the Snowbird (?) sang. Then the sky began to tilt. Finally it tilted so that it touched the earth. Then it was fastened to the earth and all the people went up. They arrived in the sky. Bluejay said, "Skate, you had better go home. You are too wide. They will hit you and you will be killed. Quick! Go home."

Skate said, "Shoot at me; afterward I will shoot at you." Skate stood up. Bluejay took his bow and shot at him. But Skate turned sideways and Bluejay missed him. Then he told Bluejay, "Now I shall shoot at you." Bluejay stood up. Skate said, "Raise your foot before your body; if I should hit your body, you would die." Bluejay held up his foot. Then the Skate shot him right in the middle of his foot. He fell down crying.

Now the people had arrived in the sky. It was cold. When it got dark, they said to Beaver, "Quick! Go and fetch the fire." Beaver went up to the town. Then he swam about in the water. A sky person saw him and said, "A beaver is swimming about." Then the man ran down to the water, struck Beaver, and killed him at once. He hauled him to the house and said, "What shall we do with Beaver?" "We shall singe him," his companions said. They placed him over the fire and the sparks caught in his fur. Then Beaver arose and ran outside. He swam away from the shore, carrying the fire. Soon he arrived at the place where his relatives were staying and brought them the fire. The people made a fire.

Then they said to Skunk, "Go and examine the house, and try to find a hole where we can enter in the night." Skunk went and laughed, running about under the houses. Then an old man said, "Behold! There is a skunk. Never before has a skunk been here, and now we hear it. Search for it. Kill it." They looked for the Skunk. Then Skunk ran home because it became afraid.

They told Robin, "Quick! Go and look at the house. See if there is a hole where we can enter at night." Robin went and entered a small house. There were two old women. He warmed himself and remained there.

Then they said to Mouse and Rat, "Quick! Go and look for Robin." Mouse and Rat went. They entered the last house. Then they cut the bowstrings and the strings of the coats of the women. They did so in all the houses. They cut all the bowstrings. Then they went home. They said, "We cut all their bowstrings."

Robin had disappeared, and they said, "Perhaps they have killed him." Then they attacked the town. After a while Robin went home. His belly was burnt red by the fire. Then these people in the sky were killed. They tried to span their bows, but they had no strings. The women intended to put on their coats and to run away, but the strings were cut. They stayed there and they were killed. The Eagle took the eldest Southwest wind by its head; the Owl took another one, the Golden Eagle a third one, the Turkey the fourth one, and the Chicken-hawk took the youngest one by its head. After a while the four elder ones were killed. Then the youngest one escaped from Chicken-hawk. The one that Turkey held would have escaped, if they had not helped him. Only the youngest Southwest wind escaped from them.

Then the earth people went home. Bluejay went down to the earth first. His foot was sore. Then the others descended. Skate was still above when Bluejay cut the rope and the sky sprang back. That is why some of the people are still above. They became stars. There are all kinds of things in the sky—Woodpecker, Fisher, Skate, Elk, and Deer. Many things are there.

(Reprinted from Boas, *Kathlamet Texts*, 67–71.)

CONSTELLATIONS OF THE PACIFIC NORTHWEST

Culture Group	Star or Constellation	Classical Equivalent
Tlingit	Yaxtê´	BIG DIPPER
Tlingit	Marten Month	Evening Star (Jupiter?)
Tlingit	Łq!ayā´k!´s Tracks	Milky Way
Tlingit	Three Men in a Line	ORION?
Tlingit	Sculpin	PLEIADES
Tlingit	Morning Round Thing	Venus
Tlingit	Halibut Fishers	Unidentified
Tsimshian	Dipper	BIG DIPPER?
Tsimshian	Sun's Daughter	Morning Star
Tsimshian	Kite	Unidentified
Tsimshian	Halibut Fishing-line	Unidentified
Tsimshian	Stern-board in the Canoe	Unidentified
Tsimshian	Old Bark Box	Unidentified
Haida	Seven Brothers in a Boat	PLEIADES
Kwakiutl	Cannibal Pole	Milky Way
Kwakiutl	Harpooner-of-Heaven; Paddler	ORION
Kwakiutl	Otter	PLEIADES
Fraser River	Elk	BIG DIPPER
Fraser River	Crying Children	PLEIADES
Snohomish	Three Hunters and an Elk	BIG DIPPER
Snohomish	River	Milky Way
Snohomish	Two Canoes and a Fish	ORION?
Makah	Halibut	Unidentified
Makah	Shark	Unidentified
Makah	Skate	Unidentified
Makah	Whale	Unidentified
Quileute	Bear Skin	BIG DIPPER
Quileute	Elk Skin	CASSIOPEIA
Quileute	Skate	LITTLE DIPPER
Quileute	Eye	Venus?
Quileute	Beaver	Unidentified

CONSTELLATIONS OF THE PACIFIC NORTHWEST (*cont.*)

Culture Group	Star or Constellation	Classical Equivalent
Quinault	Skate	BIG DIPPER
Quinault	Fisher	Unidentified
Cathlamet	Deer	Unidentified
Cathlamet	Elk	Unidentified
Cathlamet	Fisher	Unidentified
Cathlamet	Skate	Unidentified
Cathlamet	Woodpecker	Unidentified

Chapter Seven

Bringing the Day

Stars of the Plateau

The Plateau culture area encompasses the region that stretches from the Cascade Mountains to the Rockies and from the Fraser River in British Columbia to the Great Basin. This region includes southeastern British Columbia, eastern Washington, northern Idaho, sections of Montana and Oregon, and the northeastern tip of California.

The Plateau lies in the rain shadow of the Cascade Mountains in the United States and the Coastal Range in Canada. The land receives little precipitation, and much of the area is covered with sagebrush and grasses. Villages developed along the rivers (including the Willamette, Snake, Umatilla, Deschutes, Okanagan, Thompson, Columbia, and Fraser), which offered salmon, trout and sturgeon. Plateau tribes relied on fishing and gathering, supplementing these food resources with game.

In this region, the primary social unit was the family, and the next larger group was the band. Each band had a chief, but a council of elders made major decisions. In general, the tribes of the Plateau did not develop a tradition of hereditary nobility, classes with special privileges, clans, or secret societies, as did their neighbors in the Pacific Northwest.

A significant portion of the star information of Plateau tribes comes from the work of James Teit, who settled near the Thompson tribe (also called the Ntlakyapamuk) after emigrating from the Shetland Islands. Teit was not trained as an ethnologist, but he showed a lively interest in Native people. His first wife was a Thompson. When he carried out ethnological studies he stayed not at hotels but with the people themselves. Although he submitted his material to a noted ethnologist, Franz Boas, for editing, the range of information and the detail with which it is presented is clearly his own.

Teit preserved information about the Thompson and Coeur d'Alene, two tribes that spoke similar dialects of the Salishan language family. Because Teit had worked with the Thompson, he was able to ask the Coeur d'Alene about specific constellations and to compare the star beliefs of the two groups, in which he found many similarities despite the distance separating them. (Of the other tribes considered in this chapter, the Klamath spoke a dialect in the Penutian language family; the Nez Perce a dialect of the Sahaptian language family; and the Kootenai Kutenian, a language in its own language family.)

Some star images appear to be shared on a wider basis than dialect or language family. The Kootenai, Thompson, and Coeur d'Alene believe that most stars are people from Earth who have been transformed. Four tribes—the Thompson, Coeur d'Alene, Flathead, and Wasco—tell stories about canoes that become stars in ORION or other constellations. The Thompson, Coeur d'Alene, and Wasco say that a grizzly bear and hunters or brothers become the BIG DIPPER, and that Cluster, a bunch of stars, is PLEIADES.

Despite the efforts of Teit and other ethnologists, there are frustrating gaps in the recorded star information of this region. An ethnologist who worked with Plateau tribes in the early 1940s noted that the Lillooet and Carrier have "many constellations," and that various tribes use constellations to indicate directions and time, but he did not name or describe any of these star groups.

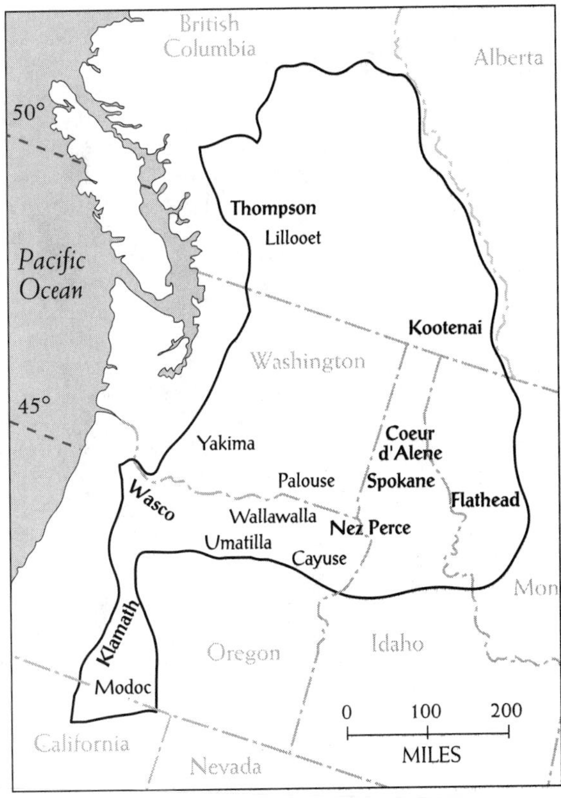

Fig. 7.1. Tribes of the Plateau culture area and their approximate territories in the 1800s. The tribes indicated in **boldface type** are included in this chapter. Not all tribes that inhabited this area are shown. (After Waldman, *Encyclopedia of Native American Tribes*, 194.)

Kootenai (Kootenay)

The Kootenai (Kootenay in Canada), who call themselves *San'ka*, lived in the area that is now northwestern Montana, northern Idaho, northeastern Washington, and southern British Columbia. Today these Native people live on reservations in Idaho, Montana, and British Columbia.

The Kootenai used tipis in the summer and either semisubterranean dwellings or longhouses in the winter. The Upper Kootenai depended on bison, which they hunted on the Great Plains, and the Lower Kootenai relied on fish; both groups also hunted other game.

Stars and Constellations

Guwitl itlnohos (Big Star) refers to both Morning Star and Evening Star. *Aqkemais qaetltsin* (Trail of the Dog) is the Milky Way. *Tlautla* (Grizzly Bear) is the North Star. *Aqkitl kanka* (untranslated) is the PLEIADES.

Like other culture groups, the Kootenai believe that a time long ago animals talked just as people do now. The following myth about animals going into the Sky World also appears in tribes in the Pacific Northwest. According to the Kootenai:

> Coyote, Grizzly, Raven, two hawks and other animals decided to visit the sky world and climbed a chain of arrows to get there. But the chain was no longer in place when they tried to return. Some of the animals used feathers from Thunderbird to fly back to earth while others jumped down. The Sky People killed the animals who remained and turned them into stars.
>
> (Adapted from Clark, *Indian Legends of Canada*, 32–33.)

Thompson (Ntlakyapamuk)

Fig. 7.2 A Thompson boy painted this apron and wore it during his puberty ceremony. The boy, represented in the center, wears feathers and an apron. Two moons and six stars symbolize his nightly travels. Game animals and mountains appear below him. He seeks a lizard, which he hopes will be his guardian spirit, on one special mountain shown at the bottom of the apron. (Illustration from Teit, The Thompson Indians of British Columbia, 380.)

The Thompson, who call themselves Ntlakyapamuk, made their winter homes in the valleys of the Fraser, Thompson, and Nicola Rivers in interior British Columbia. During other times of the year they dispersed to hunting and fishing areas in the mountains. They now live on reserves in these areas.

These Native Americans lived in semisubterranean structures from December to February or early March and at other times used lodges made of poles covered with bark or mats. Roots, berries, salmon, and deer were the most important foods in their diet. Men, who did the hunting, also secured elk, mountain sheep, mountain goat, marmot, beaver, rabbit, squirrel, grouse, and waterfowl. Women collected roots, berries, and other wild foods, which together made up a vital portion of the food supply.

The Thompson showed respect for the spirits of the animals upon which they subsisted. A bear hunter, for example, took a sweat bath before hunting and, if he was successful, placed the bear skull in a tree to honor the animal's spirit. (Tribes in the Subarctic also followed this custom.) Thompson women refrained from eating before gathering roots or robbing the caches of squirrels.

Because all plants, animals, and even some objects that the Western worldview considers inanimate were people during the "early times" of the world, these entities continue to have souls. Many tribes differentiate between the early times, when the world was created and animals talked, and historical time, in which events that can actually be recalled by an individual or through the oral tradition take place. Ethnologists call these early times, which are steeped in myth, the mythological age.

Shamans enlisted the aid of guardian spirits to perform supernatural feats and cure the sick. Favorite spirits included the sun, Milky Way, PLEIADES, Morning Star, and other stars. Men, women, hunters, and even gamblers and runners might also secure the aid of guardian spirits.

The Thompson believe that various beings with supernatural powers transformed the landscape into the mountains and valleys that surround them. Old Coyote is the most powerful transformer, and his mission is to put the world in order. There are also three brothers collectively named Qwa´qtqwEtl/ who travel across the countryside performing miracles.

At puberty, a boy went out by himself to fast and pray, hoping that a guardian spirit would come to him in a vision. Boys as well as men recorded their dreams on rocks or articles of clothing, and some of these drawings contain stars.

Stars and Constellations

Although the stars are generally seen as transformed people, in one myth they are portrayed as the roots of plants in the upper world. Coyote's son climbs into sky country and walks around a broad plateau. He sees large wild potato plants (a species of spring beauty or *Claytonia* whose enlarged stem is edible) and pulls one up. Below lies the world, and he realizes that the roots are the stars.

The Thompson stars Bunch or Cluster (PLEIADES) are friends of the Moon and visit him at his house. A star that rises after Bunch (Aldebaran?) is a dog following Bunch's trail. The Morning Star is Bright Face or Bringing in the Daybreak.

Three hunters (the handle of the BIG DIPPER) follow a grizzly bear (the bowl) in the sky. The first brother, who is quick and brave, is nearest the bear. The second leads a small dog, so he is slower, and the third is not as brave, so he lags behind. The Lower Thompson believe that the three hunters are the three transformer brothers, Qwa´qtqwEtł. When these brothers finish their work on the Earth, they go into the sky in pursuit of a bear.

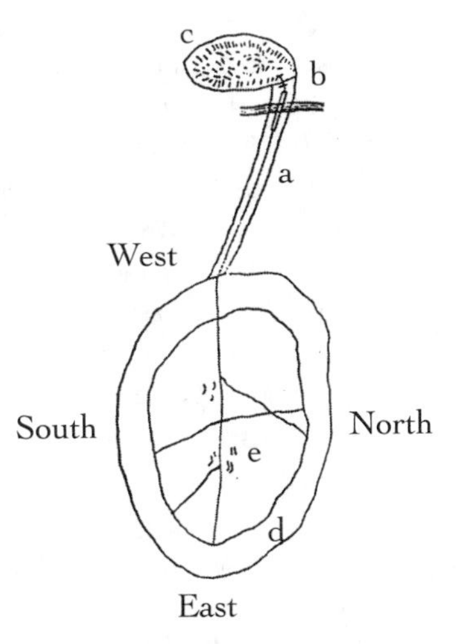

Fig. 7.3. A Thompson sketch of the world, showing (a) trail from Earth to land of ghosts (Milky Way), (b) river with log bridge, (c) land of the ghosts, (d) lake surrounding Earth, (e) rivers and villages. The trail to the land of the ghosts lies in the west. (Illustration from Teit, The Thompson Indians of British Columbia, 343.)

The Milky Way is variously called Tracks of the Dead, Trail of the Stars, That Which has been Emptied on the Trail of Stars, and Gray Trail. A Thompson map of the world shows a trail (the Milky Way) leading from the Earth across a river to the land of the ghosts. This ghost land is a beautiful country with flowers, ripe fruit, and warm air.

The Thompson made cedar dugout canoes as well as bark canoes, calling the latter into service on lakes in which they fished and on steep mountain streams. Canoe builders often carved the prows, sterns, and gunwales of the canoes and decorated the craft with elk teeth. The Thompson say that Bark Canoe (ORION) makes its way across the sky.

Women Engaged in Roasting Roots (also called Cooking in an Earth Oven) is likely the same as the Coeur d'Alene figure Women Cooking Camas Roots, thought to be stars in the CHARIOTEER. (See page 114 for more information on this constellation.)

Foot racing was a popular pastime among the Thompson and provided an occasion for betting. There were periods when most of the men were in training for these events. Some races were intertribal competitions. Two unidentified stars are called Runners; men (or, in one story, Elk and Antelope) are transformed into stars as they are racing. Three other unidentified stars, Following Each Other, are likely the same as the Coeur d'Alene Three Persons Running a Race.

The Thompson name several other unidentified constellations: Swan or Lake with a Swan on It; Canoe, composed of men who are pursuing the swan (this constellation follows Swan into the sky); Fishermen Fishing with Hook and Line; Weasel's Tracks; and Arrows Slung on the Body, that is, a hunter carrying his bows and arrows.

Coeur d'Alene (Skitswish)

The Coeur d'Alene lived along the headwaters of the Spokane River in what is now northern Idaho and eastern Washington State; today they live on a reservation in northern Idaho. They used conical lodges or tents made of poles covered with mats—later, buffalo skins—but did not build lodges set into the earth.

This mountainous, heavily forested area provided fish and game in equal measure. Hunters sought deer, elk, bear, beaver, smaller mammals, and birds; in neighboring Flathead or Kalispel (also called Pend d'Oreille) country they hunted bison. Women collected roots, berries and nuts to supplement the meat.

PLEIADES: *Qô'ẓqôzt,* Cluster
Aldebaran?: Badger

The Coeur d'Alene place PLEIADES and nearby stars in a constellation that represents a kidnapping. Badger, a large red star (probably Aldebaran), steals Coyote's favorite child, a small star nearby. Coyote chases after them, but just as he is about to catch up, Badger and the child are transformed into stars. People standing nearby, together called Cluster (PLEIADES), watch the chase and also become stars.

CHARIOTEER (?): *Sgwelkai'len,* Women Cooking Camas Roots

Camas (*Camassia quamash*) has an edible, starchy, onionlike root that was an important source of food for the Coeur d'Alene and other Plateau tribes. Camas is a member of the lily family that grows in moist meadows in the Pacific states and provinces. The Coeur d'Alene held a "first fruit" or first harvest ceremony when the roots were ready to be collected. The chief invoked a prayer, then the rest of the band danced and dug. One of the summer months, *yaltsk*, is defined as the time in which people dig camas.

Women boiled these roots or roasted them in an earth oven. It was taboo for men to approach the oven because, it was thought, if men came too close the roots would not cook properly. Among the Thompson, this prohibition applied more generally to other roots, too. Coeur d'Alene women dug a pit, lined it with rocks, laid down layers of grass, roots, bark, and earth, and then built a fire on top. They kept the fire going for as long as two days.

The constellation Women Cooking Camas Roots is thought to be stars in CHARIOTEER.

One day a group of women were cooking camas when Skunk approached the pit, thinking he would ruin the food. The women sat in a tight circle, though, and he could not get in. He sat to the side, and both he and the women were transformed into stars.

(Adapted from Teit, *The Salishan Tribes of the Western Plateaus*, 179.)

ORION: *Etsko'.lko.l*, the Canoe

The Coeur d'Alene made cedar-bark rather than dugout canoes. Canoe builders stripped the bark in one large piece when the cedar sap ran in May, June, and July and constructed most boats during this time. The Coeur d'Alene star group, the Canoe, is likely the same constellation as the Thompson Bark Canoe (ORION).

> Five men were building a bark canoe. One man stood at the bow and one at the stern. Two men were on one side, and one on the other. In this position they became stars.
>
> (Adapted from Teit, *The Salishan Tribes of the Western Plateaus*, 179.)

?: Bird Flying over a Lake

This unidentified Coeur d'Alene constellation is probably the same as the Thompson Swan (or Lake with a Swan on It), also unidentified. Perhaps coincidentally, there is also a classical constellation, the SWAN, that depicts a large bird with a long neck and outstretched wings.

> Some men were hunting and shot a bird called *t'äq'tul* (snow goose?). The bird spread its wings and was transformed, with the lake, into stars.
>
> (Adapted from Teit, *The Salishan Tribes of the Western Plateaus*, 179.)

BIG DIPPER: Three Brothers and a Grizzly

This Coeur d'Alene tale is similar to the Thompson story of the BIG DIPPER. Teit, who recorded the tale, notes that some Coeur d'Alene say that the stars of the BIG DIPPER'S bowl are the three brothers and the bear, while others say that these stars are the four feet of the grizzly and the stars of the handle are the brothers. The star nearest the bowl is the youngest brother.

> There were three brothers who were hunters. Their brother-in-law was Grizzly Bear. Although the youngest brother liked Grizzly Bear, the other two hated him and decided to kill him.
>
> Grizzly Bear went out hunting, and the youngest brother followed him to give warning. Just then the two older brothers caught up and were about to shoot. The youngest cried, "Brother-in-law, watch out!"
>
> In this position, they were all transformed into stars.
>
> (Adapted from Teit, *The Salishan Tribes of the Western Plateaus*, 178–179.)

Other Stars and Constellations

Like the Thompson, the Coeur d'Alene enjoyed footraces, and they recognize three stars in a row as three people in a race.

Since the two obvious examples of three stars are the belt of ORION and the head of the EAGLE—and ORION is identified as a bark canoe—a logical candidate for Three Persons Running a Race is Altair and two other stars that form the EAGLE's head. (See illustration of EAGLE on page 17.)

Bringing the Day is the Morning Star, while Dusty Road is the Milky Way.

Flathead, Nez Perce, and Spokane

The Flathead lived in and around the Bitterroot Valley in Montana. After gaining the use of horses, this group became seminomadic, hunting bison and adopting other elements of Plains culture. The Nez Perce lived in fishing villages in the high country of eastern Oregon and Washington, and in Idaho. They built large lodges that held several families and became skillful traders. Many Nez Perce now live on the Nez Perce Reservation in Idaho. The Spokane lived in what is now eastern Washington, fishing on the Spokane River and hunting bison to the east. They now live on the Spokane Reservation in Washington and with other tribes on reservations in Washington and Idaho.

These three tribes are considered very briefly here because only one star story has been recorded from each.

ORION?: Five Canoe Builders

Although this Flathead narration does not identify the canoe builders as ORION, several other Plateau stories do. (The Snohomish, a Pacific Northwest tribe, also tell a tale about a canoe builder.) This story appears to describe the time prior to acquisition of the horse, when canoes were more important in food gathering.

Although there are many stars in the sky now, a long time ago when the earth was new there were only a few. The stars that we see are campfires that the Creator has made as well as people and animals from this world that have been transformed.

Once five friends lived on the shore of Flathead Lake. These friends always hunted, fished, and worked together. Their canoes, however, were old and decrepit. "Let's build a canoe that is big enough for all of us," proposed the oldest friend. The others agreed, and they set to work. They planned to use the canoe to go fishing in a shallow bay where fish were plentiful.

Bluejay, who usually jabbered away, listened carefully to the builders. She knew that a huge storm was approaching and saw that the builders would be killed when they ventured out in their new canoe. She went to Old Man Coyote and asked for his help.

"I cannot control the water," he answered, "but I can lift them into the sky." Old Man Coyote placed the five friends and their canoe in the Sky World, where they remain to this day, working together as they did on the earth.

(Adapted from Clark, *Indian Legends from the Northern Rockies*, 104–105.)

PLEIADES: Eyes-in-Different-Colors and Her Sisters

This Nez Perce tale explains why only six of the seven sisters forming PLEIADES are visible.

Long ago, the sun, moon, and stars were people. The sun governed the day, the moon governed the night, and both the sun and the moon governed the stars.

At this time there were seven sisters who were stars. Each sister loved something or someone in nature, but none talked about her love. Each believed that if she shared these personal feelings, she would vanish from the sky.

Eyes-in-Different-Colors, the next to youngest sister, loved a mortal, but he died. She mourned him and told her sisters of her sadness. They made fun of her and she was stricken with guilt and grief.

Finally, she hid herself behind the sky's veil so that neither her sisters nor people on the earth can see her. For this reason, we can only see six of the seven star sisters in the sky.

(Adapted from Clark, *Indian Legends of the Pacific Northwest*, 155–156.)

Aldebaran: Coyote's Eyeball

For the Spokane as for other tribes, Coyote is a culture hero and trickster. This brief tale describes how Coyote's Eyeball became the bright red star Aldebaran.

Coyote is a trickster who enjoys performing in front of girls. One day he took out his eyeballs and juggled them. One eyeball went so high that it lodged in the sky and is still there.

(Adapted from Lawrence Hall of Science, Native American constellations.)

Wasco

The Wasco, fishing people who lived along the Columbia River in Oregon, played an important role as traders between tribes on the coast and those in the interior. Some ethnologists consider the Wasco to be more similar to the Pacific Northwest tribes than to the Plateau tribes and classify them with the Northwest.

In winter, the Wasco resided in lodges set partially underground and with cedar-

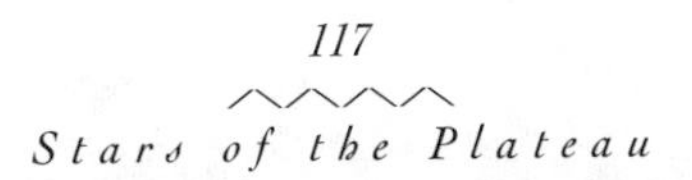

bark roofs, and in summer they used structures made of poles. They also built ceremonial sweat lodges. Most of their food came from the river, and they ate heartily of the salmon, sturgeon, and eels that lived there. Today the Wasco live with other tribes on a reservation in Oregon.

BIG DIPPER: Five Wolf Brothers and Two Grizzlies
Milky Way: Big White Road
PLEIADES: Bunch of Stars
?: Knife

Lucullus McWhorter recorded the following Wasco legend about stars in the BIG DIPPER in 1921. The Grizzlies are the two Pointer Stars, and the five Wolf Brothers are the bowl's two other stars and the stars of the handle. The oldest brother carries a dog, the small star Alcor. McWhorter notes that a Yakima-Klikitat version of the story, also from the Plateau area, identifies the two Pointer Stars as a black bear and a grizzly bear; the youngest wolf fights the grizzly bear and wins, which is why the wolf has been the master of the grizzly ever since.

The Big White Road mentioned in the story is, of course, the Milky Way. The "bunch of stars" is PLEIADES, a group of stars that, McWhorter writes, gives luck to gamblers. The Knife is not identified.

The western meadowlark, which is mentioned in this story, sings a flutelike song. It lives in grasslands, pastures, meadows and prairie and is easily identified by its bright yellow breast sporting a black V. In Plateau and Pacific Northwest tales, a chain of arrows is a typical device for climbing to the sky.

Five Wolf Brothers lived on the earth and hunted deer, elk, and other game. Coyote lived with them, sharing their food. One evening, Coyote heard the brothers talking about something that they had seen in the sky. He asked each brother what they had seen, but the wolves did not want to tell him. Finally, the brothers acquiesced.

"We saw two animals up there, but they are so high that we cannot get to them," they said. At this time, there were no stars in the sky.

"Oh, we can get there!" replied Coyote. He took his arrows and began shooting them. The first arrow stuck in the sky, the second hit the shaft of the first arrow, and so on until there was a chain of arrows into the sky land. Coyote had notched the arrows so that they would be easy to hold on to. He led the way and the brothers followed. The oldest brother carried a dog. It took them many days to reach the sky. When they did, they found themselves near the Grizzly Bears.

"Be careful!" cried Coyote. "They will rip you to bits!"

But the two youngest Wolf Brothers were not afraid and went right up to the Grizzlies. The oldest brother, with the dog, hung back because he was fearful.

Coyote looked at the Wolves and Grizzlies standing there, and he liked what he saw. He thought, "I'll make them into a picture that the people on earth will see." He left the Wolf Brothers in the sky, taking down the ladder arrow by arrow as he descended.

When he got to the ground he again admired his handiwork. But he was apprehensive that people would not realize that he, Coyote, had made the figure in the sky. He told the bird Whoch-whoch (Meadowlark) to sing this story to the inhabitants of the earth.

One night Coyote looked up at the sky and saw that there were many stars there, and they were multiplying rapidly. "There are too many stars! If they grow too thick, they will fall down and the earth will be covered with frost," he told Whoch-whoch.

Coyote climbed back up in the sky and gathered the stars together. He put the Big White Road across the sky, then he put up stars in the shape of a knife. He arranged some into a bunch, and then into other shapes. He told the stars that they must stay together. "If you must move, go like the lightning. And do not multiply too quickly," he warned them.

(Adapted from McWhorter, The Legend of the Great Dipper.)

ORION: The Celestial Canoemen

McWhorter recounts another story, a fragment of a much longer myth about Cold Wind brothers and Chinook Wind brothers. The story interprets the positions of the stars in the sky and explains an unusual rite practiced by a nearby tribe. According to McWhorter, a being called Speel-yí ate a salmon that had spawned and died. Normally, these dead salmon are not taken for food. Speel-yí decreed that others should do as he did.

McWhorter identified ORION's belt as the canoe of Cold Wind brothers and the sword as the canoe of Chinook Wind brothers. The two canoes in the sky are racing toward a dead salmon to secure the fish for ceremonial consumption. The salmon is a small star between the canoes. Although McWhorter specifies that the star is closer to Chinook Wind (ORION's sword), the obvious candidate for the salmon in fact lies slightly nearer Cold Wind (ORION's belt).

The tale also indirectly explains the balance between warm and cold weather in the region. The warm, moist wind that comes from the southwest, called a chinook wind, prevails against the cold winds from the north. In the sky, Chinook Wind's canoe is positioned in the south and Cold Wind's canoe is positioned in the north, thus reflecting their

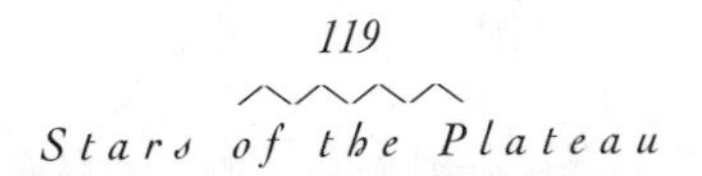

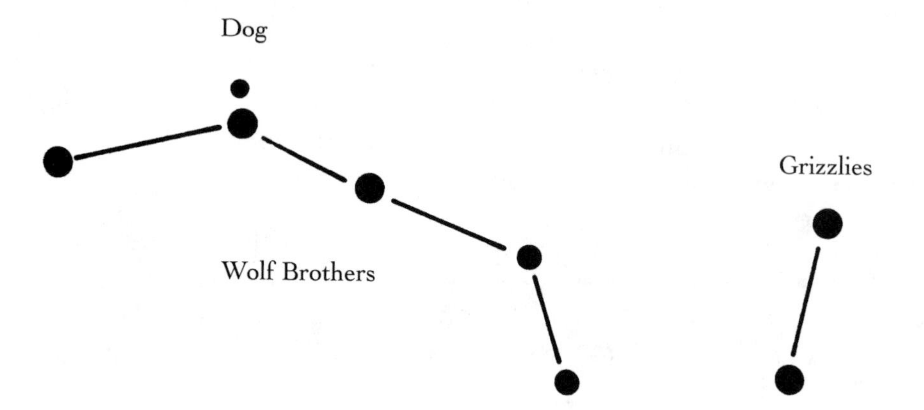

Fig. 7.4. In the Wasco constellation Wolf Brothers and Grizzlies (the BIG DIPPER), one of the brothers carries a dog (Alcor).

respective situations in nature. Chinook Wind's capture of the salmon and victory in the wrestling match show that he is stronger than Cold Wind and that Cold Wind can dominate no longer.

Cold Wind always arrived too late to catch his own fish. The grandfather of Chinook Wind caught many salmon, but Cold Wind took them away from the old man. After this had gone on for some time, Chinook Wind arrived from the Hood River and concealed himself in his grandfather's lodge. The next time that Cold Wind arrived and demanded the salmon as usual, the grandson stepped out and said, "You cannot have it."

Cold Wind demanded that they wrestle for the fish and, though he was younger, Chinook Wind agreed. Chinook Wind's aunt had taught him to grow up strong, and Chinook Wind had been practicing since childhood.

The time for the match arrived, and the two began to wrestle. Chinook Wind's grandmother threw salmon oil on the ice so that Chinook Wind could stand, while Cold Wind's sister pitched water on the ice so that Chinook Wind would slip, but Chinook Wind was more powerful than Cold Wind and won the match. After that time Cold Wind was never as strong and has not been able to take the grandfather's fish.

In the sky, the canoes of Chinook Wind brothers and Cold Wind brothers are headed for a salmon, but the canoe of the

Chinook Wind brothers is closer. Cold Wind brothers cannot get there fast enough; Cold Wind brothers will never get the salmon floating in the river.

(Adapted from McWhorter, The Celestial Canoemen.)

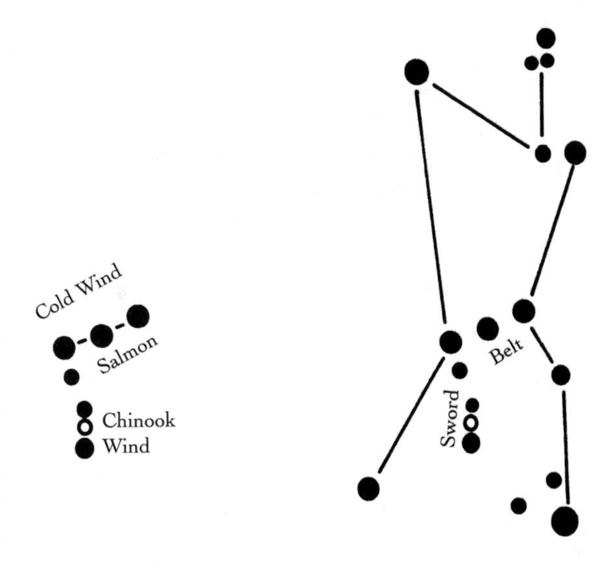

Fig. 7.5. The Wasco constellation at left shows the canoe of Cold Wind brothers (ORION's belt) and the canoe of Chinook Wind brothers (ORION's sword) racing toward a salmon.

Klamath

The Klamath, who call themselves *Maklaks* (Men, People, and/or Community) lived in southern Oregon near the California border, where their descendants still live. The Klamath used semisubterranean earth lodges in the winter and moved to other structures in warmer weather. The annual cycle centered on gathering food. They caught and dried great quantities of migrating fish; used the roots of camas, arrowroot, cattail, tule, and other plants; collected the eggs of waterfowl; and to a limited extent hunted antelope, deer, mountain sheep, and other game. Winter villages were located in areas where fish could be secured most of the year.

The Klamath and the nearby Modoc were wedged between the Pacific Northwest, California, and Great Basin cultures. One might expect Klamath constellations to be similar to those of nearby cultures—Northern Paiute (Great Basin), Shasta (California), or tribes of the Pacific Northwest. The identification of the BIG DIPPER as a group of loons, however, is found nowhere else on the continent. The Klamath do share with nearby tribes (Modoc, Shasta, Atsugewi) the belief that a single star close to the moon is a sign of evil.

Ethnographers gathered most of the following information in 1925 and 1926 from several Klamath; the older people who were interviewed had lived the traditional life of hunting and gathering when they were young.

Stars and Constellations

Ḳo'ḳiuḳs (The Divers, or Loons) are the BIG DIPPER. The people in this group, like loons, dive into the water one after the other. They have made bets on who can dive the deepest.

Wiwi'aaḳ (Twins), a boy and a girl, are Castor and Pollux in GEMINI. When the Klamath Twins appear in the eastern sky at night in December, they look over the lake and freeze it with their glance. When they rise higher, then spring is on its way. They hold a bow, several stars in GEMINI and another one in CHARIOTEER.

Kai (Rabbit) is a star that appears in the southeastern sky before dawn in May and June. Kai can also freeze the lake by looking at it. Fomalhaut, which rises in the southeast before dawn in May and June, can be seen throughout most of the United States except Alaska. This star is the brightest in this part of the sky and seems a likely candidate for Rabbit.

Endandŏksni, PLEIADES, are children who proclaim the dawn's arrival when they appear on the eastern horizon.

The constellation *Slŭks* (untranslated) is composed of six stars, the three in ORION's belt and the three in the sword.

The Milky Way is a creek or river, or a ghosts' road.

CONSTELLATIONS OF THE PLATEAU

Culture Group	Star or Constellation	Classical Equivalent
Kootenai	Big Star	Evening Star, Morning Star
Kootenai	Trail of the Dog	Milky Way
Kootenai	Grizzly Bear	North Star
Kootenai	*Aqkitl kanka*	PLEIADES
Thompson	Dog Following Bunch's Trail	Aldebaran?
Thompson	Grizzly Bear	BIG DIPPER's bowl
Thompson	Three Hunters	BIG DIPPER handle
Thompson	Women Engaged in Roasting Roots	CHARIOTEER?
Thompson	Trail of Stars; Tracks of the Dead; Gray Trail	Milky Way
Thompson	Bright Face; Bringing in the Daybreak	Morning Star
Thompson	Bunch; Cluster	PLEIADES
Thompson	Bark Canoe	ORION
Thompson	Canoe	Unidentified
Thompson	Swan; Lake with a Swan on It	Unidentified
Thompson	Fishermen Fishing with Hook and Line	Unidentified
Thompson	Weasel's Tracks	Unidentified
Thompson	Arrows Slung on the Body	Unidentified
Thompson	Following Each Other	(Three stars in a line)
Thompson	The Racers	(Two stars)
Coeur d'Alene	Badger	Aldebaran?
Coeur d'Alene	Three Persons Running a Race	Altair and two stars?
Coeur d'Alene	Grizzly Bear	BIG DIPPER's bowl
Coeur d'Alene	Three Hunters	BIG DIPPER's handle
Coeur d'Alene	Women Cooking Camas Roots	CHARIOTEER?
Coeur d'Alene	Dusty Road	Milky Way

CONSTELLATIONS OF THE PLATEAU (*cont.*)

Culture Group	Star or Constellation	Classical Equivalent
Coeur d'Alene	Bringing the Day	Morning Star
Coeur d'Alene	The Canoe	ORION?
Coeur d'Alene	Cluster	PLEIADES
Coeur d'Alene	Lake with a Bird on It	Unidentified
Flathead	Five Canoe Builders	ORION?
Nez Perce	Sisters	PLEIADES
Spokane	Coyote's Eyeball	Aldebaran
Wasco	Wolf Brothers and Grizzlies	BIG DIPPER
Wasco	Big White Road	Milky Way
Wasco	Two Canoes	ORION's belt and sword
Wasco	Bunch of Stars (put up by Coyote)	PLEIADES
Wasco	Knife Stars	Unidentified
Klamath	The Divers (or Loons)	BIG DIPPER
Klamath	Twins (boy and girl)	Castor and Pollux
Klamath	Rabbit	Fomalhaut?
Klamath	River; Ghosts' Road	Milky Way
Klamath	Children Who Announce Dawn	PLEIADES
Klamath	Bow of Twins	GEMINI and CHARIOTEER

Chapter eight

Chasing Rabbits into a Heavenly Net

Stars of the Great Basin

The Great Basin is a vast bowl-shaped desert centered in Nevada and Utah. The Great Basin culture area reaches out from this center to include sections of Oregon, Idaho, Wyoming, Colorado, Arizona and California, plus small parts of New Mexico and Montana.

Ten thousand years ago, at the end of the last Ice Age, the Great Basin was filled with lakes, streams, and marshes. But with the retreat of the glacier and other changes in climate, the area dried up and became a desert. The scarcity of water imposed limits on food and dictated the lifestyle of the people who lived there. There was little agriculture. Instead, small groups moved from place to place in search of food, securing small game, fruit, seeds, roots, lizards, and—in some areas—insects. They occasionally supplemented this daily fare with antelope, mountain sheep, and other large game. The people of this area used digging sticks to gather food and were often called, pejoratively, "Digger Indians."

The Native Americans of the Great Basin area lived in bands affiliated by language and geography. Most Native peoples in this area spoke languages in the Western subgroup (Northern Paiute and Mono), the Central subgroup (Shoshone and Panamint), and the Southern subgroup (Ute, Southern Paiute, Kawaiisu) of the Numic branch of the Uto-Aztecan language family.

Storytelling was confined to the nights of late fall and winter, after the piñon harvest. Native peoples in the northern Great Basin believed that a storyteller who disregarded the taboo would cause a severe storm, while people in the southern area believed that such a storyteller would get bitten by a rattlesnake. Among the Southern Paiute only men told stories, and these tales served both as entertainment and as a means of instilling values.

Two mythical characters, Wolf and Coyote, figure prominently in the legends of the Great Basin. Wolf creates heaven and earth, so he is the most important. Coyote, his brother, is a trickster who frees game from a cave, gives fire and pine nuts to people, and causes death to come among them. Coyote also lusts after his daughters, causing his family to flee to the sky.

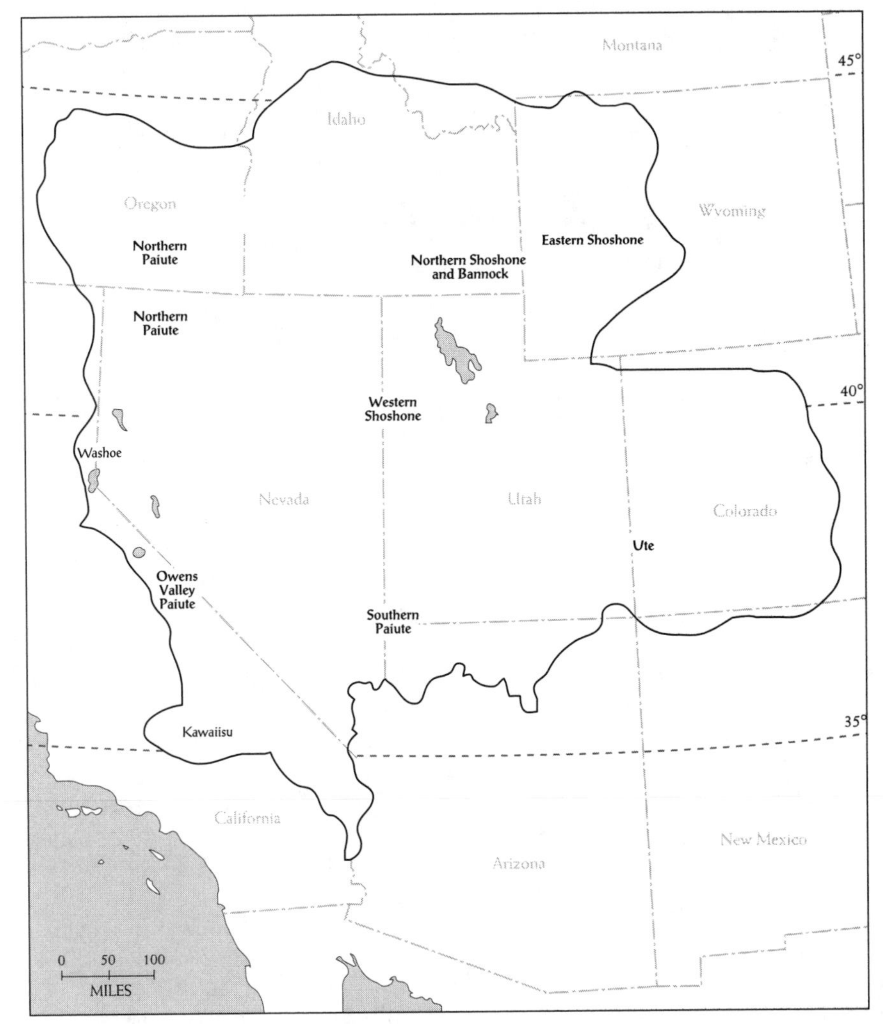

Fig. 8.1. Tribes of the Great Basin culture area and their approximate territories in the 1800s; some groups are shown slightly later. The tribes indicated in **boldface type** are included in this chapter. Not all tribes who inhabited this area are shown. (After D'Azevedo, *Handbook of North American Indians* Vol. 11, ix.)

Tribes of this area note the movement of four planets and tell stories about the BIG DIPPER, LITTLE DIPPER, Milky Way, PLEIADES, and ORION. Whereas in many other areas of the country the BIG DIPPER is described as a bear, in the Great Basin it is a mountain sheep or men chasing rabbits into a net, rabbits and mountain sheep both being important sources of food in the region.

Northern Paiute

The territory of the Northern Paiute stretched from western Nevada across a slice of California and into Oregon and Idaho; today the Northern Paiute live in Nevada,

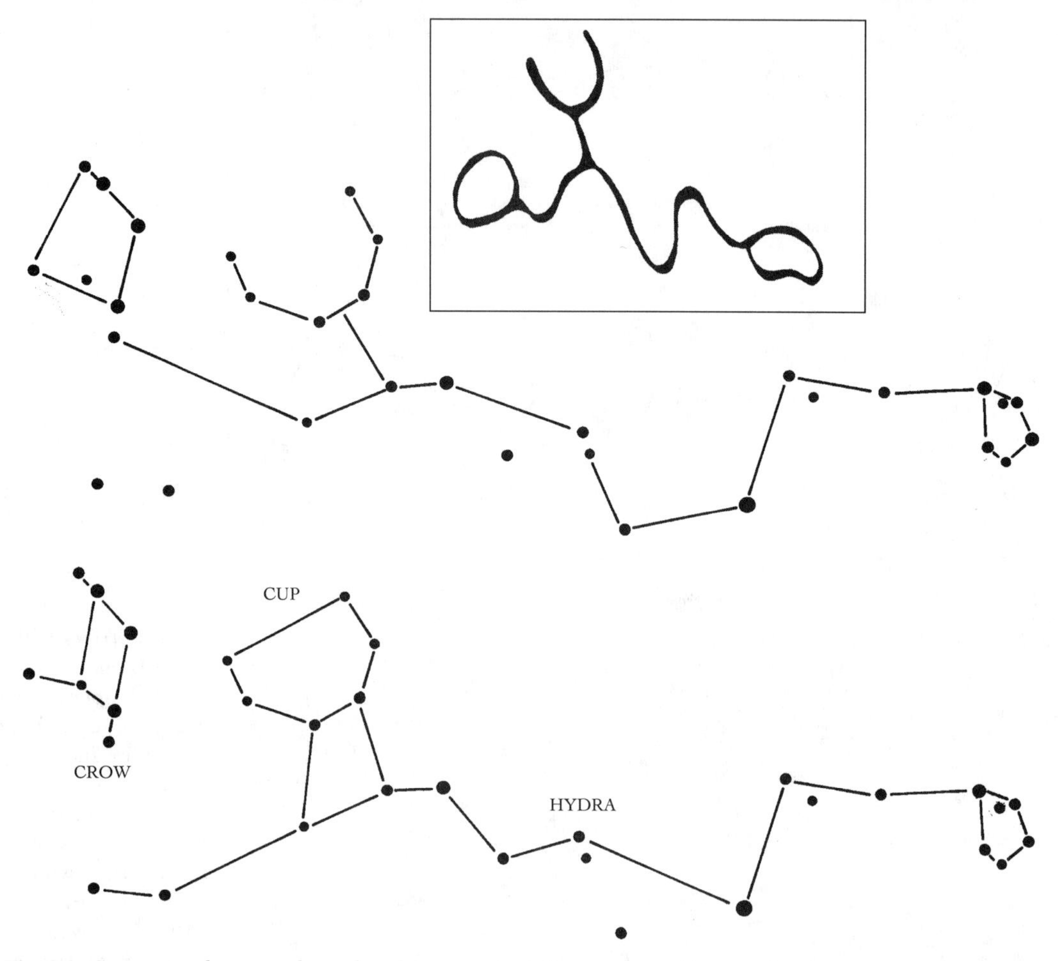

Fig. 8.2. One researcher speculates that this petroglyph from western Nevada (*inset*) represents several reconfigured constellations (*top*). The classical constellations HYDRA, CUP, and CROW (*bottom*) are the base stars in this drawing. Not all of HYDRA is shown; the constellation extends left, beyond CROW. (Illustration after Mayer, Star-Patterns in Great Basin Petroglyphs, 129.)

Oregon, and California. Although some areas along rivers and lakes have ample water, most of the territory is high desert. The Northern Paiute lived in conical or dome-shaped structures covered with mats of grass in the winter and used shaded areas in the summer. Although there were temporary leaders who might direct game drives or fishing Pefforts, there was no well-developed social structure.

The Northern Paiute believed they could find *puha* (power) in natural objects such as plants, animals, the sun, the moon, and the stars, as well as in natural phenomena such as wind and thunder. Although individuals could search for power, only shamans (men or women) could secure enough of it to perform good or evil. Shamans were healers; sickness occurred when the soul left a person's body, or when some outside element entered a person's body.

Milky Way: *Kus·ipo·*, Dusty Trail

The Milky Way, variously called *Kus·ipo·* (Dusty Trail) or *Nümü-po* (People's Trail), is the path along which the souls of the dead travel to another world in the south. These souls proceed past Coyote, who tries to delay them, and then they go past two clouds that are continually opening and closing. Beyond lies a land of abundant game and a life with ample time for dancing and gambling.

BIG DIPPER: Ta·noa´di, Men Chasing Rabbits into a Net

In the fall, the Native people of the Great Basin hunted jackrabbits and cottontails in communal drives. Rabbits were very important for sustenance and clothing; rabbit-skin robes kept people warm in bitterly cold winter weather. Hunters set out a net three hundred or more feet long in a semicircle or V formation, and then men, women, and children drove the quarry into it.

This hunting technique has been preserved in the sky. According to the Northern Paiute in Surprise Valley, in northern California, *Ta·noa´di* (Men Chasing Rabbits into a Net) is the BIG DIPPER. The five stars that make up the handle and adjoining bowl are men who are setting the net correctly. The bowl's pointer stars are two men who don't know how to hold the net or don't know where they are going. These two stars are "the Failures" because they have not set their end of the net properly.

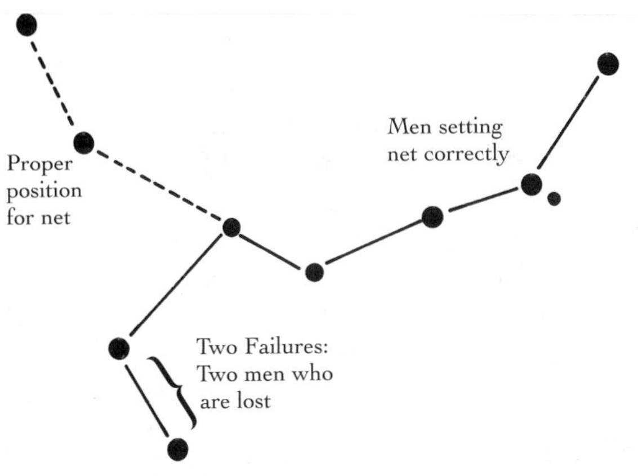

Proper position for net
Men setting net correctly
Two Failures: Two men who are lost

Fig. 8.3. The Northern Paiute constellation Men Chasing Rabbits into a Net (the BIG DIPPER). If the two men who are lost (the Pointer Stars of the BIG DIPPER) were to move into a semicircle with the other men, then the net would be set properly to trap rabbits. These lost men are also called the Failures, because they have failed to bring their end of the net into position.

ORION's belt: Mountain Sheep Husbands
Sirius: The Chaser

The Northern Paiute hunted deer, antelope, and desert bighorn sheep with bow and arrow as well as traps or corrals. Women collected wild foods, including seeds, which they processed and ground into flour for mush or dried in the sun and stored. Harvesting the seeds of piñon and

Fig. 8.4. In Great Basin myth, mountain sheep appear in the sky as ORION (Northern Paiute and Western Shoshone), and as the BIG DIPPER, the LITTLE DIPPER, the North Star, and ORION's belt (Southern Paiute). Petroglyphs of large game like mountain sheep are found along Great Basin game trails or in game wintering areas. It is thought that these rock drawings were not simply artistic expressions but were related to the ritual or supernatural process of securing game. This Great Basin petroglyph from eastern California shows mountain sheep and hunters with bow and arrow and an *atlatl* (spear-thrower). (Illustration from Heizer and Clewlow, *Prehistoric Rock Art of California,* fig. 24a.) Reprinted with permission.

other pine was a regular and vital part of the annual food cycle, and these seeds were an important part of the diet.

This search for food provides the context for a tale about people becoming stars. According to the Surprise Valley Paiute, the three stars of ORION's belt are *Koipü* (Mountain Sheep); the Northern Paiute call them *Kóipa kumámɨ* (Mountain Sheep Husbands) and the bright star Sirius a woman, *Tɨnagidɨ* (The Chaser). In the early 1900s, Blind Tom, a Paiute who lived along the Walker River in what is now Nevada, narrated a story in which a woman's husband, brother-in-law, and son become the belt of ORION.

A man and his brother, the man's wife, and their son lived in a camp. The two men went out hunting while the wife gathered seed. Each day the wife asked for a piece of sinew to give the baby to chew when it cried.

The men tracked a mountain sheep around and around the mountain but they could not catch it. "I will follow the sheep and you sit here and watch. It is strange that the sheep does not get tired," said the older brother. The younger brother watched and saw that the mountain sheep was actually his sister-in-law, who carried her baby on her head between sheep's horns. The husband was angry that his wife had tricked them and wasted their time.

The next day, the two brothers said that they would not go

out hunting, but that the woman should gather seed as usual. She left the child with them. The men made up their blankets as though they were sleeping and then went to a spring and disguised themselves as mountain sheep. When the woman returned to the camp and found no water, she kicked the water-basket and marched off to the spring. There she saw three sheep. She hurried back to camp to awaken the lazy men. "Get up!" she cried. "I have seen sheep!" She pulled back the blankets and found only sticks underneath. She realized that this time they had tricked her.

She pursued the men and her son, but they ran on before her. "Don't look back!" warned the husband. But the three ran into Coyote, who killed them. The wife caught up and made them alive again. She warned them not to turn into mountain sheep and told them to keep going. "I will follow you," she said. The man and his brother and the son became the three stars of Orion.

(Adapted from Curtis, *The North American Indian*, 147–148.)

PLEIADES: Coyote's Daughters; Women Fighting

The PLEIADES are called *Izá'a padímɨ* (Coyote's Daughters), probably referring to a widespread North American tale in which Coyote commits incest with his daughters and the daughters flee to the sky, where they become stars.

The Northern Paiute of Surprise Valley describe a summer constellation of about ten stars seen in the east around 3 A.M. as a group of women fighting. The ethnologist who recorded this information suggests that the constellation is PLEIADES, which does appear in the eastern sky at this time. However, the number of stars (ten) may indicate another group.

In 1914, Tom Austin of Fallon, Nevada, told another story about women fighting, one that explains how people learned about jealousy.

The *Kusi´tawā´qari* (PLEIADES) are sisters who are fighting one another over a man. The man's mother wants them to stop fighting because her son's knife is loose between them and she fears they might break it. There are many people (stars) standing around, watching.

(Adapted from Lowie, Shoshonean Tales, 234.)

Other Stars and Constellations

The North Star is *Kái yɨcɨŋ·adɨ pá·tusuba* (Not Moving Star).

A "large single star which shines in the east about 3 o'clock summer mornings" is *Paba´i-yü-mogo´tni* (Big Woman). Fomalhaut, which is a large, bright star that lies in the east in the pre-dawn sky in June and July, fits this description.

The unidentified constellation *Müza´a* is a group of Paiute women that became a band of mountain sheep; this figure is separate from the mountain sheep of ORION. Other unidentified constellations include two women fighting (separate from the PLEIADES), a man holding a match to watch the fight, houseflies, an eagle, and a weasel.

Southern Paiute

The Southern Paiute occupied what is now southern Nevada and adjacent regions in California, Arizona, and southern Utah; they currently live in portions of these areas. They used conical or semiconical structures covered with bark, grass, or other vegetation in the winter and slept under trees or shade structures in the summer.

In various tales Coyote steals fire for humans, shows them how to make pottery, and gives certain restrictions when hunting. He is also a seducer of his daughters and other women. The stars are considered to be Coyote's children. They hide behind clouds and sleep during the day, then wake up as the sky gets dark. The biggest stars are gods and the smaller ones are their families.

Stars and Constellations

The Milky Way is Ghost Road or Sky Path.

A winter month, the time of the greatest frost (December or January), is named after *Qaŋa* (Morning Star). Children were urged to go outside in the dawn to watch the star rise in the east.

The stars of ORION's belt are mountain sheep.

PLEIADES: Coyote's Family

The myth of the trickster who seduces a daughter, niece, or other female relative is widespread, told by tribes in the Plains, the Great Basin, the Plateau, the Southwest, California, and the central Subarctic. The major character—whether Coyote, Raven, Spider, or some other being—lusts after a female relative, pretends to die, returns in disguise, and marries the relative. When the family discovers what he has done, they are horrified and flee to the sky. Stories of Coyote as seducer in the Great Basin are found among the Ute, Northern Shoshone, and Southern Paiute. Coyote's family becomes a group of stars that both the Northern and Southern Paiute identify as the PLEIADES.

BIG DIPPER and LITTLE DIPPER: Mountain Sheep

A tale from the Coal Creek Band of Southern Paiute in Utah describes the North Star, the BIG DIPPER, and the LITTLE DIPPER as mountain sheep. Almost all Southern Paiute bands hunted mountain sheep, so it is not surprising that this animal appears in several constellations of these Native Americans.

The story mentions but does not identify other constellations, including an eagle, a bison, a deer, a horse, and several birds. The Southern Paiute did not hunt bison, which lived to the east of them, but these Native people did know about the animals, and bison appear in various Southern Paiute tales. Note how the mammals in the sky are traveling, following the grass, just as the Southern Paiute themselves moved from place to place in search of food.

William Palmer, who recorded this story, identifies Shinob as the younger of two brothers who are gods. Tobats is the elder brother.

The sky is a world like this one, with rivers, plants, animals, and seasons. The stars that we see are animals. The buffalo, the deer, and the horse are there, looking for good grass and fine weather. The eagle is there and so are birds that fly away when it gets cold and return when it gets warm. All of these animals move through the sky making paths, but there is one animal that does not move. That star is *Qui-am-i Wintook*, the North Star. Here is how the North Star came to rest in the sky.

There was once a mountain sheep, Na-gah, that lived in this world. This mountain sheep was the son of Shinob, who was so proud of this son that he gave him beautiful earrings [his horns] to enhance his stature. Na-gah loved to climb and scaled every mountain that he encountered. One day he came upon a massive mountain that stretched into the sky itself. He decided to climb it but could not find a trail.

What? A mountain that Na-gah could not climb? He searched harder and harder. Finally he found a crevice that led down into a cave. He climbed downward, then upward. It was so dark that he could not see at all. Then he heard a huge rumble that swelled from below him, as if the cave had closed in. And it had—he could not return the way he had come.

He continued, higher and higher, until he at last arrived at the pinnacle of the mountain. Below him stretched cliffs on every side, and far below him lay the earth. He was trapped, for he could not go up and he could not go down.

Shinob, his father, saw him there and was sad that his fine son would die on that tiny patch of ground so high on the mountain, never be able to climb again. Shinob decided to save his son by turning him into a star that would serve as a guide to all those who looked up to the sky. The son is *Qui-am-i Wintook Poot-see*, the Star That Does Not Move.

Other mountain sheep wanted to join Na-gah, and Shinob turned them into stars, too. They are the Big Dipper and the Little Dipper. They travel around and around the North Star, searching for the trail that leads to the top of the mountain.

(Adapted from Palmer, *Why the North Star Stands Still and Other Indian Legends*, 71–75.)

Shoshone

The Shoshone occupied a large swath stretching across the Great Basin into the Wyoming Basin. The descendants of the Shoshoni live with Paiutes on two reservations in Nevada, and the Wind River Shoshoni live on or near the Wind River Reservation in Wyoming. The Western Shoshone, like the Paiute, relied heavily on wild plant foods, whereas the Eastern Shoshone acquired horses and developed some elements of a Plains culture. The Bannock, which are classified with the Northern Shoshone, were Paiute-speaking people who moved east into Shoshone territory. Sacajawea, who guided Lewis and Clark, is the most famous Shoshone.

Stars and Constellations

The stars are Coyote's children. A grizzly bear (the stars of SWAN and FOX) races up a snow-covered mountain, kicking off ice crystals that become a trail (the Milky Way). The Shoshone also call the Milky Way Backbone of the Sky, Dusty Road, and Smoke from a Fire.

The Panamint, a subgroup of the Western Shoshone, identify the constellations Mountain Sheep and Arrow (ORION), Coyote's Family (PLEIADES), Rabbit Net (BIG DIPPER), Smoke from Rabbit-Driving Fire (the Milky Way), and Backbone of the Sky (also the Milky Way). Duck and Y-Frame Cradle are star groups whose classical identities were not recorded.

Ute

The Ute called themselves *Nu cI* (Ute, Person). Euro-American settlers called them *Utahs,* which was shortened to Ute. The Ute occupied the northern half of the Colorado Plateau in what is now Utah and Colorado, and the area east into the Rocky Mountains. As a whole, this part of the Great Basin culture area had more food and of a wider variety than did the desert of the Great Basin. Bison, for example, ranged through most of Colorado, providing a source of food and hides.

Stars and Constellations

There are few recorded Ute constellations. One group of stars represents women with children (PLEIADES as Coyote's family?), and another group is called either Jackrabbit or Mountain Sheep. One Ute band calls the Milky Way the Ghost Road.

CONSTELLATIONS OF THE GREAT BASIN

Culture Group	Star or Constellation	Classical Equivalent
Northern Paiute	Men Chasing Rabbits into a Net	BIG DIPPER
Northern Paiute	Big Woman	Fomalhaut?
Northern Paiute	Dusty or People's Trail; Ghost Road	Milky Way
Northern Paiute	Not Moving Star	North Star
Northern Paiute	Mountain Sheep; Mountain Sheep Husbands	ORION's belt
Northern Paiute	Women Fighting	PLEIADES?
Northern Paiute	Coyote's Daughters; Coyote's Family	PLEIADES
Northern Paiute	The Chaser	Sirius
Northern Paiute	Two Women Fighting	Unidentified
Northern Paiute	Man Holding Match	Unidentified
Northern Paiute	Houseflies	Unidentified
Northern Paiute	Eagle	Unidentified
Northern Paiute	Weasel	Unidentified
Northern Paiute	Band of Mountain Sheep	Unidentified
Southern Paiute	Mountain Sheep	BIG DIPPER
Southern Paiute	Mountain Sheep	LITTLE DIPPER
Southern Paiute	Ghost Road; Sky Path	Milky Way
Southern Paiute	*Qaŋa*	Morning Star
Southern Paiute	Star That Does Not Move (a mountain sheep)	North Star
Southern Paiute	Coyote's Family	PLEIADES
Southern Paiute	Mountain Sheep	ORION's belt
Southern Paiute	Eagle	Unidentified
Southern Paiute	Bison	Unidentified
Southern Paiute	Deer	Unidentified
Southern Paiute	Horse	Unidentified
Shoshone	Ice Crystal Trail; Dusty Road; Smoke; Sky Backbone	Milky Way
Shoshone	Grizzly Bear	SWAN, FOX
Western Shoshone	Rabbit Net	BIG DIPPER
Western Shoshone	Smoke from Rabbit-Driving Fire; Sky Backbone	Milky Way
Western Shoshone	Mountain Sheep and Arrow	ORION
Western Shoshone	Coyote's Family	PLEIADES
Western Shoshone	Duck	Unidentified
Western Shoshone	Y-Frame Cradle	Unidentified
Ute	Ghost Road	Milky Way
Ute	Women with Children	PLEIADES?
Ute	Jackrabbit or Mountain Sheep	Unidentified

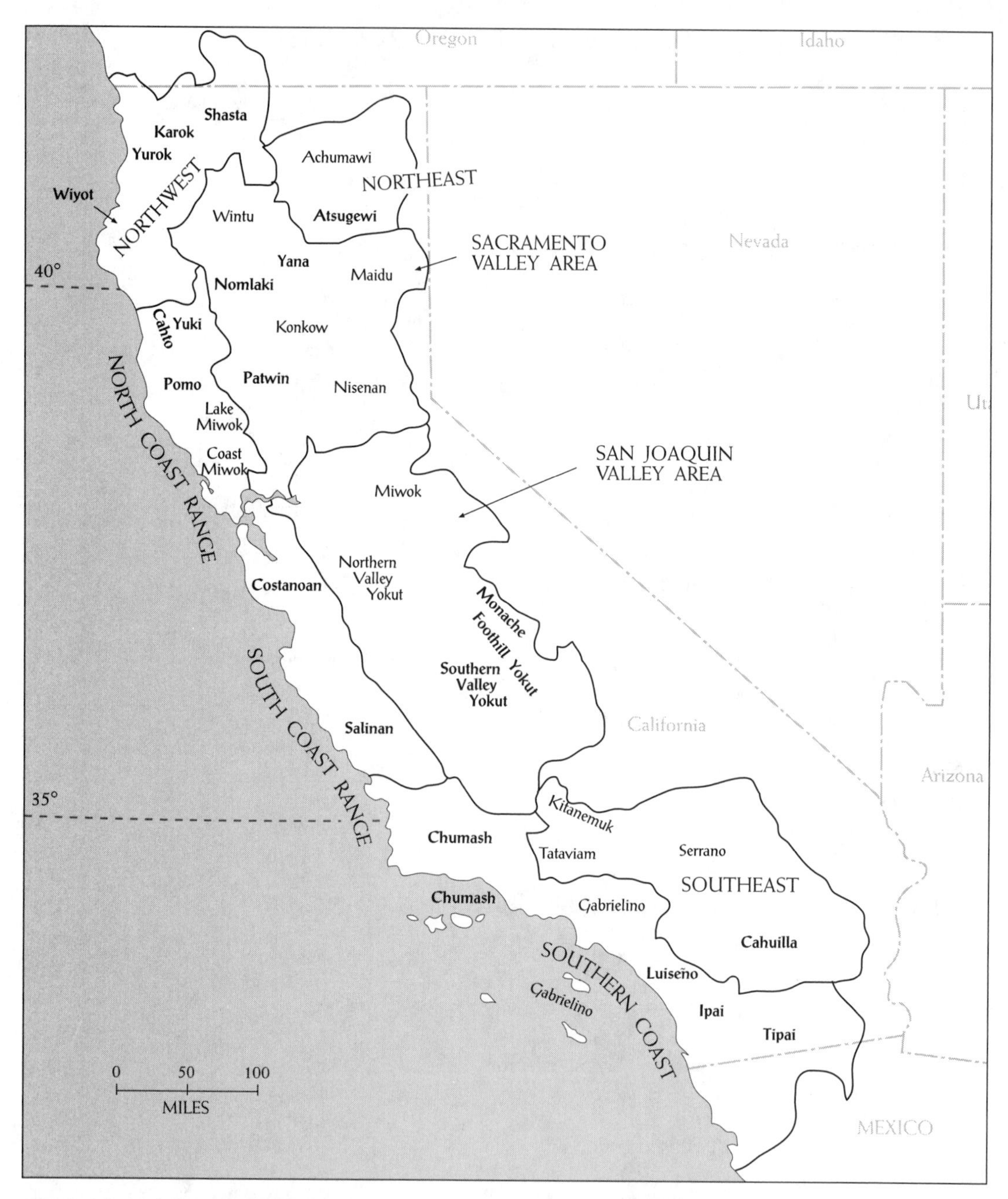

Fig. 9.1. Tribes of the California culture area and their approximate territories at the earliest time for which information is available. The tribes indicated in **boldface type** are included in this chapter. Not all tribes that inhabited this area are shown. (After Heizer, *Handbook of North American Indians*, Vol. 8, ix.)

Chapter nine

Knocking off Acorns

Stars of California

Native Americans in California generally lived their entire lives within the confines of their villages' territory. "Home" was an area of known dimensions, a relatively safe place to live. The physical world merged with the mythical world and both were considered real. Yurok and Ipai maps of the universe show the local area, marked by mountains, rivers, or other physical features, as surrounded by circles representing the edge of the sky, the horizon, or the boundary of the universe. Ceremonies and cultural taboos clarified aspects of life that touched on the supernatural.

Although the details of life varied greatly, one aspect remained consistent: Almost all Native peoples in California considered acorns as an essential part of their diet. They preferred acorns from tan oak, black oak, blue oak, and valley oak, but other species sufficed when necessary. Acorns were important even in areas where oaks were not abundant, such as in the redwoods and the Modoc Plateau in the northern tier. People traveled for some distance if there were no oaks nearby or if the trees were of an undesirable species. Women ground acorns and then boiled or rinsed the meal in several changes of water to leach out tannin, a substance that makes the meal bitter. They then added the meal, which is rich in protein and fat, to soup or bread, or made it into mush. In light of the importance of acorns, it is not surprising that this food appears in celestial stories. The Patwin, in the Sacramento Valley region, call the BIG DIPPER *Lacto´*, a stick for knocking off acorns. These Native Americans also say that the northern lights are older people in camp cracking the acorn nuts.

The other two mainstays in the diet were fish, including salmon and other anadromous fish that spawned in area streams, and large game such as deer, elk, and antelope. Coastal groups had access to the resources of the ocean, including marine mammals, fish, mollusks, and seaweed.

In California, the word "tribe" is used to designate a group of linguistically related *tribelets*. A tribelet is a self-governing social and political group. A tribelet may include one or more (sometimes as many as a dozen) villages with one chief. There were, for example thirty-four Pomo tribelets, each of which acted independently. The cultural picture, therefore, is very complicated. In addition, Native peoples in California spoke many languages, perhaps as few as sixty-four or as many as eighty, in six major language groups.

Because of this decentralization and diversity, the Native people in California were particularly vulnerable to Euro-American influence. Settlers and gold-rush miners overwhelmed some Native groups, which became so dispersed that they no longer existed as a tribe. Although members of some tribes now live on reservations (including Hoopa Valley, Round Valley, Tule River, Santa Ynez, Colorado River, Fort Yuma, Pit River, Jamul, Diegueño, and numerous small rancherias) the descendants of many others are scattered throughout California.

Most Native people in the California area practiced one of three major religious systems: the World Renewal cult, the Kuksu cult, or the toloache cult. Inhabitants of northwestern California, including the Karok, Yurok, and perhaps the Wiyot (as well as groups in the Pacific Northwest) engaged in annual World Renewal ceremonies to rejuvenate the earth and ensure that the coming year would provide good fortune. Native Americans in the North Coast Range and part of the Sacramento Valley area participated in the Kuksu cult, in which trained spiritual leaders led elaborate ceremonies whose goal was to bring people back into the sacred state that was present at the time of the creation. Of the groups addressed in this book, the Pomo, Yuki, Patwin, and Maidu participated in the Kuksu cult.

Native Americans in central and southern California (including the Yokut, Cahuilla, Chumash, Gabrielino, Luiseño, Ipai, and Tipai) had long used the hallucinogenc plant datura, also called *toloache* or jimsonweed, in ritual and medicine. Two toloache subsystems that developed along the coast, the *ʔantap* religion among the Chumash, and the Chingichngish religion among the Luiseño, Ipai, and Tipai, refined the use of this plant. Among the Chumash, shamans took toloache when interceding with the supernatural, and there is one Chumash word for both toloache and an important Chumash deity. Chingichngish is a mythical hero who gives the people laws and rituals, including the observances for drinking datura. The Chingichngish cult, which incorporated previous religious beliefs of the Gabrielino, Luiseño, and other groups, is thought to have developed after Spanish contact.

The great diversity in tribes, languages, and geographic areas contribute to a great diversity in constellations. One commonality is that tribes of central and southern California (including the Yokut, Cahuilla, Chumash, Luiseño, and Ipai) consider Coyote to be the star Aldebaran. Several of these tribes also consider Buzzard to be the star Altair. The Luiseño, moreover, believe Buzzard to be one of five chiefs who reside in the sky. Sky beings are also important to the Chumash, who believe that Eagle and the Sun play a gambling game against Sky Coyote and the Morning Star to determine whether food will be plentiful on Earth in the coming year.

The California culture area covers a vast range of habitats, from the rain forests of the Pacific Northwest to the deserts of Baja. The principle natural features include the Klamath Mountains, Cascade Range, and Modoc Plateau in the north; the Coast Ranges, Great Central Valley (including the Sacramento Valley and the San Joaquin Valley), and Sierra Nevada in the central part of the state; the Transverse and Peninsular Ranges and Mojave Desert in the south; and a coastline that stretches almost one thousand miles.

Because of the complexity of the California culture area, this chapter addresses tribes as they occur within geographic regions: the Northwest, the Northeast, the Sacramento Valley, the North Coast Range, the San Joaquin Valley, the South Coast Range, the Southern Coast, and the Southeast.

NORTHWEST CALIFORNIA

Northwest California is similar in many ways to the Pacific Northwest culture area. Like their neighbors to the north, many tribes built plank houses in permanent settlements, used various dentalium (shells) as currency, and lived in a society in which a wealthy elite ruled.

Yurok, Karok, and Wiyot

The Yurok lived in villages along the Pacific Ocean and inland along the lower Klamath River. They used redwood to make dugout canoes, paddles, houses, and some household items and had extensive social laws to regulate conduct between people. To the Yurok, Coyote is a trickster but not a culture hero, for he does not create people, game, or rituals. The Karok lived upstream from the Yurok, and though the two tribes spoke different languages, their cultures were almost indistinguishable.

The Wiyot, who lived south of the Yurok along the Pacific, engaged in different customs than did their neighbors. The Yurok refrained from sexual relations before fishing, for example, and performed a ritual to mark the first salmon caught each season; the Wiyot did neither.

PLEIADES: *Teinem,* Many
PLEIADES: *Atairam Tunueich,* Little Stars

Dancing was an important part of Yurok culture. The Deerskin Dance and the Jumping Dance, the two most important ceremonies, were ways of showing wealth and were related to World Renewal, the most elaborate of the Northwest California rituals. World Renewal ceremonies ensured nature's renewal, and the rituals in these ceremonies were intended to "firm up the earth" and ensure good fortune. The annual ceremonies played an important part in the spiritual lives of tribal members.

During the Deerskin Dance, deer skins were put on display, and during the Jumping Dance, dancers wore headdresses made of red-headed woodpecker scalps, which were items of wealth. The Karok also held dancing ceremonies, including the Adolescence Dance that occurred at puberty. The girl who was gaining womanhood wore a dress made of maple bark and a band of blue jay feathers on her head. She held a deer-hoof rattle and danced in front of a line of other dancers.

The Yurok called PLEIADES *Teinem* (Many), and the Karok called the cluster *Atairam tunueich* (Little Stars). Both tribes tell stories about Coyote joining women in the puberty dance. In the Yurok tale, Coyote is not able to dance through the night and dies. He falls from the sky to the Earth, where he comes to life again. In the Karok version:

> PLEIADES, seven little stars, are holding hands and dancing across the sky during an Adolescence Dance, when Coyote wants to join them. They warn him not to dance too far or too fast because he will tire before the dance is done, but he assures them that he is able to do it all. But then, he wants to stop to urinate, and then to defecate. "No, do not stop!" cry the girls. "We must all continue dancing." Gradually, bits of Coyote drop off of his body until there is only a small piece of him remaining in the sky.
>
> (Adapted from Kroeber and Gifford, *Karok Myths*, 16.)

Other Stars and Constellations

The Wiyot describe PLEIADES as six people in a boat and associate the Milky Way with the ocean, probably the foam. Although the Wiyot were not oceangoing people, they hunted sea lions, fished for saltwater species, and secured meat from stranded whales. It is not surprising that the ocean played a part in their celestial beliefs.

The Karok believe that when the Milky Way appears in the sky, someone has died and the road to the land of the dead is being prepared for the spirit. They also say that the Evening Star sings a song to bring back her lover; other women can sing the song, too, to bring back men who have left them.

Shasta

The Shasta lived inland, occupying the Mount Shasta valley north to the Klamath River. The people of this tribe used permanent planked houses in the winter and various other structures as they moved from place to place the rest of the year to collect food. Deer and acorns were important food items supplemented by salmon and other fish, mussels, bear, small mammals, fowl, grasshoppers, nuts, seeds, and fruit.

In Shasta villages, myths were told in the winter, lest Rattlesnake be offended. The stories were directed primarily at children, who repeated each sentence after the storyteller. Sometimes characters have conversations with what the Western worldview would consider inanimate objects. In the following story about Raccoon's children, for example, both the children and Coyote talk with objects in Coyote's house, and Ashes (the ashes from a fire) tells Coyote what he wants to know.

PLEIADES: Raccoon's Children and Coyote's Youngest Son

The PLEIADES were well known to the Shasta, for they used this constellation's reappearance at dawn to mark the beginning of autumn. Coyote is an important figure, both a benefactor and a trickster in turn, though in the following tale he reaps a bitter reward for his jealousy. There is no small red star "behind" the PLEIADES, but a large red star—Aldebran—rises after the star group. This tale was recorded on the Siletz Reservation in northwestern Oregon in 1900.

Coyote and Tcinake, the Raccoon, were living together. Each had five children. One day Coyote said, "A feast is taking place not far from here. Let us go there!" to which Coon replied, "All right!" They went to the fair and had a good time. Coyote fell in love with two girls; but they preferred Coon and paid little attention to Coyote. Towards evening Coyote said to Coon, "I am going away for a little while. I'll be back soon. Do watch those two girls!"

While Coyote was gone, the two girls invited Coon to go with them, telling him that they did not care for Coyote. Coyote returned and looked for his friend. In vain he called Coon's name repeatedly, but he could not find him. At last Coon appeared, and Coyote asked him, "Where have you been? Where are the girls?" Coon told him that the girls were in the woods, whereupon Coyote accused him of having taken them. Coyote was very angry.

After a while they started home. On their way they saw a squirrel running into a tree-hole. Coyote asked Coon to put his hand into one end of the hole, so as to scare the squirrel and drive it to the other side of the opening, where he (Coyote) was waiting for it. Coon reached into the hole with his hand and Coyote seized and began to pull it. Coon shouted, "Hold on! This is my arm."

"No," said Coyote, "this is the squirrel." And he kept on pulling until the arm came off and Coon died.

Then Coyote went home, carrying Coon's body. Upon his arrival home, he distributed the meat among his children. But the youngest boy, angry because he was not given an equal share, ran over to Coon's children and said, "My father has killed your father. He did not bring home all the meat. Tomorrow he is going for more."

Coon's children said, "All right! Tomorrow we shall kill your brothers, but we will spare you. We shall take you with us." The next day, while Coyote was away, they killed his four children and left them on the floor. Then they ran away, enjoining everything in the house not to tell Coyote where they had gone. They forgot, however, to caution Ashes.

Coyote came home and tried to wake his children, but they were dead. He asked everything in the house to tell him where the murderers of his boys had gone. No one knew. Finally he asked Ashes. The Ashes flew skyward and Coyote followed their flight with his eyes. Before they were halfway up the sky, Coyote saw Coon's children and his own boy trailing behind them. He wept, and called to them to come back, but they would not listen to him. Then he tried to catch them. He could not overtake them.

The children remain in the sky as stars. They are the Pleiades. The five big stars are Coon's children. The smaller star behind them, the red star, is Coyote's boy.

(Reprinted from Farrand, Shasta and Athapascan Myths from Oregon, 220–221.)

Milky Way: Path of the Soul

Like many other people across North America, the Shasta conceived of the Milky Way as a path across which the soul journeys after death. Shastan souls travel from east to west, arriving at the home of Mockingbird, who flies up the Klamath River in the spring and down it in the fall.

NORTHEAST CALIFORNIA

The culture of the Plateau reaches into the northeast corner of California in the Cascade Mountains and Modoc Plateau, home of the Atsugewi and Achumawi. Native peoples here ate camas roots and built lodges, for example, just as Native Americans did in the Plateau area.

Atsugewi

Each day at dawn the Atsugewi village chief woke up his people by exhorting them to get up and be productive. Hard work was believed to be the means of achieving success. The Atsugewi lived in earth lodges or bark houses and ate fish and acorns as staples, along with deer, bear, rabbit, duck, and other game.

The Atsugewi had few ceremonies, but they did celebrate the time when each boy and girl reached puberty. Three dances were performed during the girl's ceremony: a women's circle dance, a dance for men and women, and the girl's dance. In the third dance, the girl carried a cane and danced near a fire, facing east. She was expected to be energetic; if she was tired and could not perform well, it was a sign that she would be lazy in the future. Sexual restrictions were loosened somewhat during the ceremony, and at that time couples drifted off into the bush without censure.

Stars and Constellations

Only a few Atsugewi constellations have been recorded, but one of them appears to be associated with the puberty dance. It is said that *Wĭrɛtisu* (Seven Sisters)—probably PLEIADES—attended a puberty dance where a little rabbit boy seduced them. They were embarrassed and fled to the sky to become stars.

The Milky Way is *Máskawi makuri* (Devil's Trail), or a dead person's trail. The BIG DIPPER is Coyote's Cane. Tribes in the Sacramento Valley and the North Coast Range, to the south and west, also considered the BIG DIPPER a stick, or a long pole with a hook.

SACRAMENTO VALLEY

The Native Americans of the Sacramento Valley had access to streams in the grasslands of the Central Valley, oak woodlands, and pine-fir forests of the Sierra Nevada. They were thus able to secure fish, acorns, and deer, the three foods considered staples of Native Americans of California.

The northern California Sierra, which lie east of the Sacramento Valley, are the source of three Native American constellations. The DOLPHIN is a paw print, faint stars in AQUARIUS and PEGASUS are Cottontail Man, and the PLEIADES may be Grizzly Sisters. The Grizzly Sisters play with Deer Sisters in a cave until Grizzly Mother kills and eats Deer Mother. Deer Sisters lock their former friends in the cave, killing them.

Patwin

The Patwin ("People") lived in the area west of the Sacramento River and north of San Pablo and Suisun Bays. Their territory approached but did not include Clear Lake. They made their homes in semisubterranean structures covered with earth and used similar structures for ceremonial dance houses and sweat houses.

Stars and Constellations

People hunted or fished individually or in small groups, using spears or nets to catch fish. The Evening Star is West Star Now Comes, as well as *Si´wa-men* (Seining in the Slough). It is not certain whether the appearance of the star relates to the timing of seining season or has some other meaning.

Fig. 9.2. The paw print constellation (*left*) from the Northern Sierra, compared to a cottontail rabbit paw print (*right*).

The BIG DIPPER is *Lacto´* (Stick for Knocking off Acorns). Acorns also figure in the interpretation of the northern lights, which are considered to be old people in camp cracking these nuts.

The Milky Way is called both *Kano yeme* (Antelope's Road) and *Put* (Ashes).

ORION is *Se´deu notiLi* (Coyote Carries on Head), though it is not clear whether the stars represent Coyote or the object carried on his head. (The Yana, another tribe in the Sacramento Valley, call ORION's belt Coyote's Arrow.)

The PLEIADES are *Poto´ni* (not translated).

Maidu

The Maidu (also called the Northeastern Maidu) lived in an area of high mountain meadows south and east of Lassen Peak that includes present-day Susanville. The meandering creeks provided fish and waterfowl, the oaks provided acorns, and the mountains provided habitat for elk, deer, quail, rabbit, and squirrel. The Maidu lived in communities composed of several villages. Each community acted autonomously and had its own leader, who served as an advisor and intermediary with other communities.

The Maidu believe that a multitude of spirits inhabit mountains, lakes, waterfalls, and other natural features, as well as the sky and areas under the Earth. They also speak of a huge bird (perhaps the California condor) that can swoop down and kill a person with a whack from its wing.

Stars and Constellations

According to the Maidu, the stars are made of a soft material something like buckskin. The BIG DIPPER is *O´koikö* (Looking Round), perhaps meaning that the creator is looking around from a vantage point in the sky. (The Pomo in the North Coast Range call the North Star the Eye of the Creator.)

La´idam-lülü bō (Morning Star's Trail) is the Milky Way, *Do´todoto* (untranslated) is the PLEIADES, and *He´muimū* (To Roast) is DOLPHIN. The ethnographer who recorded this information did not say what "To Roast" means. The Maidu cooked food by baking it in a pit lined with rocks or roasting it on top of hot coals.

NORTH COAST RANGE

The North Coast Range is a densely populated area that receives abundant rainfall and has a rich, varied vegetation. The Pomo, Yuki, Cahto, Coast and Lake Miwok, and Wappo lived in this region. Most Native Americans in the North Coast Range area, as well as some groups in the adjoining Sacramento Valley area, observed practices of the Kuksu cult, in which secret societies of men and sometimes women were carefully trained to lead singing, dancing and curing ceremonies. These ceremonies were designed to reestablish the sacred time of the creation. Each society focused on its own goals, such as healing or holding rites commemorating the first fruits of a harvest. In some ceremonies members wore intricate costumes and represented spirits.

Coyote appears as one of two creators in many myths of the North Coast Range. The Cahto, a small group in the northern section of this area, believe that Coyote stole the sun from the house of an old woman and then cut it up to make the moon, the stars, and the sun.

Pomo

The Pomo included several dozen tribelets located about fifty miles north of San Francisco along the Pacific Coast north of the Russian River and around Clear Lake (thus, neighboring the Patwin). Language and culture varied considerably among these tribelets. The Pomo lived in the redwood forest along the coast and in the valleys and hills inland. Coastal Pomo lived in single-family conical structures of redwood bark, and inland groups lived in multifamily structures. There were also semisubterranean buildings for ceremonies and dancing. The head of the *yomta,* a Pomo secret society, said prayers to the sacred stars and watched the night sky carefully to help determine the calendar.

The Pomo relied on fish and acorn mush as their daily fare, adding to it other nuts, berries, seeds, roots, bulbs, large game (deer, elk, antelope, and bear, when available), small animals, birds, insects, and—along the coast—sea lion, seal, and seaweed.

William Benson, Jim Pumpkin, Charley Bowen, Boston, and Drew—all of whom were Pomo—provided information about the stars to ethnologist Edwin Loeb in 1924 and 1925.

Stars and Constellations

According to the Eastern Pomo, the Milky Way is the track of a bear. When the bear meets the sun in the sky, the sun refuses to stand aside, and the two fight; this fight produces an eclipse of the sun. When the bear encounters the moon, the sun's sister, they also fight, producing an eclipse of the moon.

The Eastern Pomo call PLEIADES *Latsó* (Buckeyes Bunched Up). Members of the *yomta,* the secret society, knew when the PLEIADES would appear in the sky but never prayed to these stars. In some parts of California buckeyes were considered starvation food and eaten only in times of emergency because the nuts contain prussic acid, which is difficult to leach out of the nut meat, and the acid makes the nut meat poisonous. The Pomo, however, collected and stored buckeyes for year-round use.

The Pomo called three stars near PLEIADES *Camul ibũi* (Strong Sucker). Suckers were the first fish to ascend creeks around Clear Lake each year, going upstream when the current was still strong in February or March, and presumably the fish had to be strong to succeed. The Pomo caught them in basket traps and stored the dried meat for later use. Strong Sucker, described as "three stars under the PLEIADES," is likely ORION's belt.

The BIG DIPPER is *Baghal,* a stick with a hook that was used to pull shriveled limbs from trees. The Eastern Pomo could tell by the position of the BIG DIPPER what time of night it was, and therefore what type of fish would be biting.

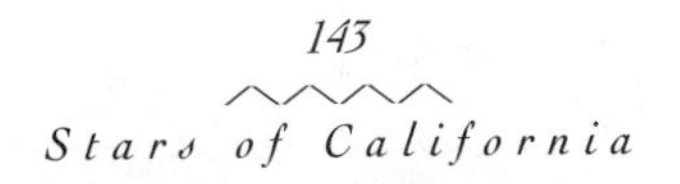

The Eastern Pomo used the North Star as a direction marker. *Gubula uiaxo* (the North Star) is the eye of Marumda, the creator, who has retired to the sky. When the Earth was younger, Marumda's younger brother, Coyote, changed monsters into animals and stole the sun to give people daylight.

Morning Star: Day Woman, Morning Eye Fire, Big Star
Evening Star: Night Woman

The Eastern Pomo call the Morning Star *Xa'a da* (Day Woman) or *Xa'a uia xo* (Morning Eye Fire), and they call the Evening Star, *Duwe da* (Night Woman), Day Woman's younger sister. The Coast Pomo call Morning Star *Ka amul bate* (Big Star).

In Pomo mythology, Day Woman becomes disgusted when people commit incest and other wicked acts. When these people die in a flood, Day Woman saves herself by fleeing to the sky, where she appears every morning. The Pomo chanted prayers to her in the morning, asking her to assist them in hunting and other activities.

In another Pomo story, Morning Star and her husband, a hawk, live happily on Clear Lake. The people of their village, however, fear Morning Star because she comes from Star Land, where an evil eagle rules. This eagle sweeps down and kills people, taking them back to Star Land to eat. To allay the fears of the villagers, Morning Star goes to Star Land and kills the eagle, then returns to Clear Lake, where she and her husband live in honor and affection.

The Pomo also describe the Morning Star as a companion to the departing spirit as it travels to the "happy isles" in the west after death.

The Yuki, a group that lived north of the Pomo, likewise consider Morning Star to be a young woman. Coyote has placed her in the morning sky near the horizon to announce the sunrise.

SAN JOAQUIN VALLEY

The San Joaquin Valley is similar to the Sacramento Valley except that it is much dryer, receiving less than ten inches of rain a year. Still, the marshlands along the rivers provide habitat for a great variety of plants and animals. Tribes in this region include the Miwok, the Yokut, the Monache, and the Tubatulabal.

Residents of the San Joaquin Valley ate fish, waterfowl, and shellfish, as well as roots, seeds, small mammals, and birds. Groups living in the northern part of the valley had greater access to acorns than did those farther south, but large game was probably not a major part of the diet. People living in the western foothills of the Sierra Nevada relied more on deer, quail, acorns, and other plant foods than on fish.

Yokut and Monache

The Yokut, who lived in the San Joaquin Valley and its foothills, encompassed as many as fifty individual tribelets whose affiliation was through language rather than political

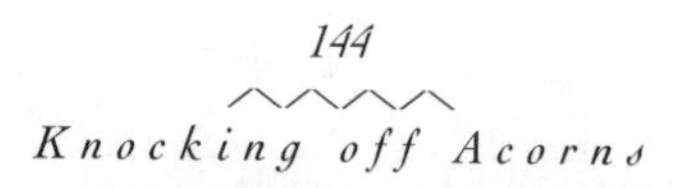

organization. Most Valley Yokut lived in single-family structures covered with tule (grass) mats, though there were some larger communal dwellings. Foothill Yokut used conical dwellings as well as structures to provide shade. Both Valley and Foothill Yokut made grass baskets of great variety.

The Monache, to the east of the Yokut, included six or more independent tribal groups. The Monache inhabited the upper western slope of the Sierra Nevada, from about three thousand to seven thousand feet in elevation; the Yokut inhabited the area up to about three thousand feet. Though the Monache spoke a language similar to the Owens Valley Paiute (who lived on the other side of the mountains in the Great Basin), the culture of the Monache was in many ways closer to that of the Yokut. Many constellation stories are attributed to both the Monache and Yokut.

Yokut stories are populated by animals such as Falcon, Duck, Roadrunner, Crow, Wolf, and Crane. Coyote appears as an assistant to Eagle, who is the chief and the creator of the world. The large number of independent Yokut bands means that there are often different versions of stories about well-known constellations like the PLEIADES and ORION.

For the Yokut, the stars are part of the yearly cycle of natural events, and activities in everyday life correspond to the appearance of constellations. When the PLEIADES rose above the western horizon at dusk in May, for example, it was time to prepare for the salmon run.

PLEIADES: Six Young Women

The Yokut and Monache tell several variations on the theme of six women ascending to the sky to become the PLEIADES. These women leave the earth because they become cold, they grow bored with their lovers, they are lectured by their mothers, they dislike their husbands, or they offend their husbands by eating raw onions. In some of the stories, including the first of those that follow, the husbands pursue them and become a constellation called the Young Men, identified with ORION (perhaps the belt) or the HYADES, the distinct V-shaped cluster of stars in TAURUS.

The stories embody various aspects of Yokut culture. The mush mentioned in the first tale that follows, for example, was likely a mush that the Yokut prepared from a starchy flour processed from the tule root. The Yokut considered eagles to be sacred birds; only a chief could sanction the killing of an eagle. Valley Yokut twisted downy eagle feathers with milkweed fibers to make skirts used in group ceremonies. These skirts, and ropes of similar materials, had supernatural powers.

American Joe, a Monache who recounted myths about the stars in 1927, said that if a man wanted to be lucky at gambling, he should hold up his eagle-down rope and look through it at the Young Men when they rise in the sky. The Young Men can seduce women on Earth, and women who have menstrual cramps have been so seduced. The girls forming the PLEIADES, he said, are dangerous and can charm a man until he becomes crazy.

Sam Osborn, a Southern Valley Yokut, narrated the following story in 1928. Like other Yokut, he localized the plot, making reference to familiar areas such as the picnic site. There was probably also a stone spring and a large flat rock in his vicinity.

Seven young wives lived with their husbands in a village. One of these women was smaller than the others; another one had a child. All seven of these women were homosexual and daily went off together to picnic on Koiwuniu while their husbands were out hunting. It was spring, and clover and wild onion were beginning to grow.

The young women did not know how to avoid their husbands, who were not attractive to them. When it came time to go home they rubbed wild onion on their mouths. That night when they went to bed, all the husbands were repulsed by the odor and told their wives to turn over the other way or go to sleep elsewhere. The girls pretended indignation and the next day went to their parents' homes. But they were at once sent back to their men, being told to go and eat with them.

They went home and fed their men meat and mush, but they did not eat with them. They kept this up for six days and ate no food; they were planning to leave their men permanently.

On the last day they all went up the hill again. Each of them had eagle-down ropes, which were their talismans. The next day they went up the hill again, but on the way up the youngest girl's nose began to bleed. She asked the others to wait for her. She sat down to rest and her friends sang to her. She turned to stone. She can still be seen sitting there. A little stream of water flows out of her stone nose; those who have tasted it say that it is bitter.

Then the rest went on up to the large rock. They laid out all their talismans in a single line and stood on them. They stood in a row, all facing the same way. The mother had brought her little girl with her. Then the leader of the group threw her eagle-down rope up in the sky. It caught in the center so that both ends hung down to the ground. These ends the woman tied around the ropes on the ground.

Then the girls all clasped hands, called upon their talismans to help them, and began to sing. Slowly the ropes on which they were standing began to rise and swing slowly around and around like a buzzard. It swung in bigger and bigger circles. They sailed over their village and their parents saw them. The people below rushed out with beads and belts to try and bribe the girls to come down, but to no avail. The husbands saw their wives up there and scolded their parents-in-law for letting the girls get away.

Now the men had eagle-rope talismans just like their wives, and they decided to follow. They ran up the hill and saw the stone woman sitting there, and they knew what had happened to her. They put down their ropes as the girls had done and were soon sailing out in the sky and over the village. Again the old people came out and begged their sons to come back. By this time the women had reached the spot in the sky where we see them now. They looked back and saw their men coming after them. They talked about whether or not they should let them come on. But they stood by their first plan: They wanted to be alone in the sky.

As soon as the men got close enough, the girls shouted to them to stop right where they were. There they stayed, and you can see them now a little way back of their wives. They did not want to go home again.

(Reprinted from Gayton and Newman, Yokuts and Western Mono Myths, 34–35.)

The next intriguing Yokut version of the six women story has the women drinking datura and then flying through the air like condors. Here the women set out to gather seed to make meal, part of a daily and ongoing effort to provide themselves and their families with food.

When Eagle made the original animals and people, he did not make fire for them, so they were cold all of the time. When six young women went out to gather grass seeds on the bank of the San Joaquin River, they began shivering and agreed that they must find someplace warm. They found a shallow cave whose rocks had absorbed the heat of the sun, and they sat down there to grind their seeds. "The sun makes these rocks warm," they said, and began thinking of a way they could go to the sun.

A plant called *tah´-ni* (jimsonweed) grew nearby, and the women made some tea from its roots. That night they slept in the cave. The next morning they observed downy feathers growing on their arms. They drank more tea, slept in the cave, and watched the feathers grow some more. "We are like the Condor," they said. By the sixth morning they could fly around the cave, and by the seventh they flew into the air, over the river, and over their village.

"But we will not get to the sun this way," said one woman. "We need help." The women chanted a song to the Whirlwind:

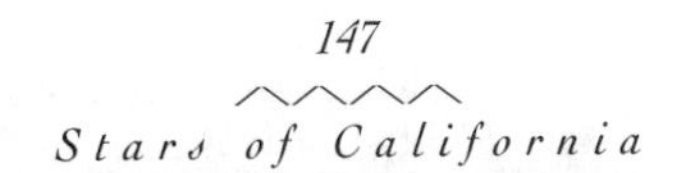

"Whirlwind, start me. Whirlwind, take me up." The Whirlwind picked them up and took them to the sun.

Meanwhile, the six husbands had begun searching for their wives. They found the cave and the tea, and began drinking the tea themselves. They, too, grew feathers and learned to fly. They, too, asked the Whirlwind for help in order to join their wives near the sun. The Whirlwind complied, and both the wives and the husbands are in the sky as the stars we call *Kotch-pih´-lah* (Pleiades).

Sometimes, in the spring, the wives and husbands want to return to the Earth. The flapping of their wings makes the thunder and the disturbance of the clouds makes the spring rains.

(Adapted from Latta, *California Indian Folklore*, 140–143.)

Native American stories often include humor, as does the next one. A Yokut man identified only as Tom who lived on Tulare Lake in southern San Joaquin Valley told this tale about five women who marry a flea in the early 1900s. He also told the two stories that follow it.

The Pleiades were five girls and a flea, *baakil*. The girls sang and played all night in the sky. The flea [apparently in the form of a man] constantly went with them. They did not like other men that came to them; they liked only him. When other men came they ran away, but the flea went with them. And they let him marry them. He married all five. Now he turned into a flea, and in summer became sick with the itch. The girls did not like him any longer. They said, "Let us run away. Where shall we go?" Then they agreed to go east together. "When shall we go?" they said. "As soon as he sleeps."

Now the flea slept and the five got up and went off. After they were far away the flea woke up and thought, "Where are my wives?" He found that they had gone away. He thought, "Where shall I go?" He went east. At last he came in sight of them, just before he reached the ocean [Tulare Lake].

"I will catch you," he said.

"He is coming. Let us go on," they said, and ran on. Then one asked, "Do you see him again?" Another said, "Yes, he is near." They said, "Let us go up into the air. Then he cannot come with us," and they went up.

But the flea rose, too. That is why there are five stars close together now in the Pleiades and one at the side. That one is he, the flea.

(Reprinted from Kroeber, Indian Myths of South Central California, 213–214.)

ORION's Belt: Crane and Her Sons

Various Yokut bands call the stars of ORION's belt Men Chasing Pleiades, Coyote Chasing Pleiades, Bear Chasing an Antelope, Three Babies, and Crane and Her Sons. The Crane story describes the violent demise of Wolf, Crane's husband.

Wolf constantly hunted but never gave his wife and two boys any meat. Once in the morning he went hunting. Then his wife, Crane, ran off. He returned and found her gone. He followed her. He was angry and wanted to kill her. He saw her and tried to shoot her, but she was high up in the air. Slowly she settled and at last lit far off. Then he shot and hit her. He went to her. With her bill she tried to stab him. He used an arrow to ward off her blows and tried to stab her. Then she pierced his breast and knocked him down. She stabbed him again and again, until she killed him. Then she went off with her boys. They turned into stars in the sky. She is in advance; her two boys are following her. They are called *Yibish*, the three stars of Orion.

(Reprinted from Kroeber, Indian Myths of South Central California, 214.)

Milky Way: A Racetrack

Why do deer in the San Joaquin Valley favor brushy areas in the hills while antelope frequent grassy plains? The following story answers this question and also explains why the Milky Way splits into two sections. This story takes place when the Milky Way is aligned roughly north-south in the evenings in summer and early fall. According to the Yokut, the Milky Way is a racetrack and the entrants are, in various stories, an antelope, a deer, an elk, a mountain sheep, a bear, a frog, a canvasback duck, and the Yokut hero Mih-kit´-tee (Falcon). Although the story specifies that Antelope ran on the wider western path, it appears that the eastern fork of the Milky Way is actually wider.

The antelope and the deer were together. The antelope said, "I can beat you running."

"I think not," said the deer.

"Well, let us try," replied the antelope.

"We shall run for six days. Let us go south and run northward," said the deer, and the antelope agreed.

Then they went far to the south across the ocean [Tulare

North

East

Deer's Path

Antelope's Path

West

South

Fig. 9.3. In Yokut mythology, the Milky Way is the racetrack of a deer and an antelope. This illustration shows the position of the Milky Way in July in the evening. Note the north-south alignment and split "track."

Lake] in order to run northward to the end of the world. The antelope said, "If I win, all this will be my country and you will have to hide in the brush."

"Very well, and if I win it will be the same for me," the deer answered.

The antelope said, "This will be my path on the west here. You take the path on the east." The deer agreed and they started. Their path was the Milky Way. On the side where the antelope ran there is a wide path; on the other side there are patches. That is where the deer jumped.

They ran and the antelope won. So now he has the plains to live in, but the deer hides in the brush.

(Reprinted from Kroeber, Indian Myths of South Central California, 213.)

Other Stars and Constellations

The Yokut associate Altair with Buzzard and Aldebaran with Coyote. The North Star may be connected with Sky Coyote (a different being than Coyote), who plays a gambling game with the Lord of the Land-of-the-Dead. (See the Chumash story about this gambling game on page 154.)

The Monache call the BIG DIPPER Seven Boys.

Yokut and Monache star stories mention but do not identify Blackbird, Raven, Roadrunner, Dog, Duck, Hummingbird, and Y-Frame Cradle as stars or constellations. (One style of Yokut cradle has a wooden Y-frame with cross pieces covered with twined tules.) The Yokut say that Eagle is in the sky.

SOUTH COAST RANGE

The Costanoan, Esselan, and Salinan Indians lived along the South Coast Range in central California. Very little star lore from these groups has been recorded. The Costanoan recognize PLEIADES as a "bunch of little ones shaking," perhaps meaning children. The Salinan call PLEIADES a group of maidens, and the North Star the Star That Does Not Move. They describe ORION's belt as "three stars in a row."

SOUTHEAST CALIFORNIA

Southeast California, the home of the Kitanemuk, Serrano, and Cahuilla, is largely desert scrub with areas of chaparral, pine-fir forest, and woodland along the southwestern edge bordering the Southern Coast.

Cahuilla

The Cahuilla occupied the present-day area of Palm Springs. The tribal territory extended from the San Bernadino Mountains south to Borrengo Springs, from Riverside east past the Salton Sea, but most people made their homes in the desert rather than the mountains. The Cahuilla lived in circular or rectangular structures covered with brush, and they also built larger ceremonial houses.

The Cahuilla were primarily hunters and gatherers, though they did grow some corn, squash, beans, and melons. They hunted what they found in the desert: rabbits, mice, rats, insects, fish, birds, and large game such as deer, mountain sheep, and antelope. They gathered food from agave, cacti, yucca, Mariposa and desert lilies, mesquite, catsclaw, and screwbean from the low desert, and piñon nuts, acorns, juniper berries, wild onion, and other foods from the high desert.

The Cahuilla tell a story in which twins, Múkat and Tamaioit, emerge from two eggs that appear in a dark void. These twins make the Earth, sky, and water from their hearts, and then create the stars to keep the sky from shaking. Afterward, they make ants and other insects, creatures that live in water, people, the sun, and the moon.

Constellations and Stars

Each Cahuilla belonged to one of two nonterritorial tribal divisions (moieties) that governed marriage partners. A person could marry only into the opposite division. The two divisions were the Wildcats and the Coyotes. Wildcat and Coyote play a part in star myths as well. The Milky Way, for example, is the dust that Isil (Coyote) and Tukut (Wildcat) kick up as they compete in a race. The Cahuilla, like other tribes in southern California, associate Coyote with Aldebaran and Wildcat with stars—most likely the BIG DIPPER and LITTLE DIPPER—that whirl around the North Star.

ORION's belt is a group of three mountain sheep, the sword is an arrow, and the bright star Rigel is the hunter who has shot the arrow.

PLEIADES: The *Chehaum,* Three Little Girls

In the early 1900s, Ramon Garcia, a Cahuilla, told the following story in which three girls become *Chehaum* (PLEIADES). Although Garcia leaves us wondering how three girls could become seven stars, another storyteller, Chief Francisco Patencio, fills in the detail: The four other stars are jewels that the girls wear on their arms, necks, and heads. A pretty woman with a bright pin and two suitors become stars nearby. The pretty woman

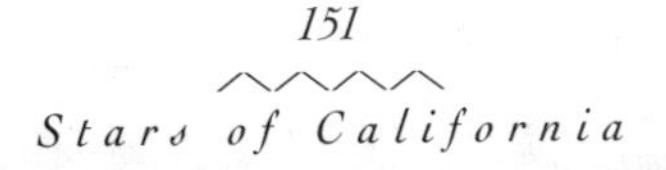

may be the North Star or perhaps stars nearer to PLEIADES and the suitors, Aldebaran and another nearby star.

When he told the story, Garcia added that the PLEIADES are used to predict the weather. When one can see the stars, good weather will follow, but if *Chehaum* is dim, then rain will follow. This prediction makes sense, for high, light cirrus clouds that would obscure PLEIADES often precede rain.

It is interesting to note that *"Chehaum"* closely resembles *"Chehaiyam,"* the Luiseño name for PLEIADES. The Luiseño bordered the Cahuilla on the southwest.

There once were three little girls, Moki, Kipi, and Tewe. They were small and not at all pretty, but were constantly laughing. Tukwishhemish was a large woman and very pretty. When she laughed, she never opened her mouth. This made the little girls very curious. One day they made her laugh very hard, and she opened her mouth. They then saw that instead of having one row of upper teeth she had two. The little girls thought this very funny and they laughed at her.

Tukwishhemish was ashamed, so she ascended to the sky and became a star. Soon the little girls became so lonely that they too went to the sky. Tukwishhemish can be seen to this day. She has her arms outstretched and wears a beautiful pin at her neck that shines very brightly in the sky. The three little girls are known as the *Chehaum*. They keep trying to look at Tukwishhemish, but she keeps turning away from them.

Up to the time these four women went to the sky there had never been a marriage, but the people were beginning to desire something of the kind. Two men, Isilihnup and Holinach by name, heard of these girls and decided to go to them and make them their wives. When they arrived at the home of the girls and found that they had left, they felt very badly. They looked all over the world for them but of course could not find them.

One night they slept in the big house. When Isilihnup woke up in the night, he could see through the smoke-hole. There he saw the *Chehaum* and knew that they were the little girls. He wondered how he could get near them. At last he spied a greasewood stick. He put it in the fire until it began to burn, then threw it through the smoke-hole, and it went to the sky. He followed it. As the stick went through the hole, some ashes fell off. Since then, whenever ashes fall, it is a sign that it is going to snow. Isilihnup became a star and still may be seen at one side of the *Chehaum*.

When Holinach woke up, he missed his partner and wondered where he could have gone. That night he slept in his brother's place. When he woke in the middle of the night, he too saw the *Chehaum*. He then knew where his brother had gone. He took a *naswit* branch, lit it, and threw it through the smoke-hole as his brother had done. Ashes fell just as they had done before. Holinach followed and went up to the other side of the *Chehaum*. There Isilihnup and Holinach can still be seen guarding the *Chehaum*.

(Reprinted from Hooper, The Cahuilla Indians, 365–366.)

Another tale describes the adventures of Isilihnup and Holinach's stepbrother and father, both of whom are transformed into stars. The stepbrother, Kunvachmal, becomes a bright star that "comes up at night just over the horizon for a little while and then goes right back again." One star that fits this description is Canopus, the second brightest star in the sky.

Unidentified Stars: The Funny Star

Múkat, one of the twin creators, put a star—*Papinut*—on the horizon. "This star twinkles more than the others and they call this jumping. It jumps all night. They say Múkat put it there to be funny, so it is spoken of as the funny star," wrote Lucile Hooper, who recorded Cahuilla customs and beliefs in 1918.

Hooper also recorded a brief account of Múkat's death, mentioning three sisters who became unidentified stars that are visible in winter.

Eventually Múkat grew sick and prepared to die. He told three sisters that when he died, frost would appear around his house. The sisters went to the sky and became stars. One morning they saw frost and knew that Múkat had died. They cried and could be heard from very far away. Now, whenever there is frost, those three stars are in the sky.

(Adapted from Hooper, The Cahuilla Indians, 363.)

SOUTHERN COAST

The Southern Coast is largely chaparral, with regions of grassland and pine-fir forest in the north and oak woodland in the central region. Acorns are less plentiful because of the low rainfall, but they were still an important part of the Native diet here. Bands along the coast took advantage of the rich resources of the ocean.

Although datura was used in other parts of California, along the Southern Coast it became part of formalized religious cults, *ʔantap* among the Chumash and the Chingichngish religion among the Gabrielino, Luiseño, Ipai, and Tipai.

Chumash

The Chumash lived in the area that is now Santa Barbara and the Channel Islands. The mild climate and ample food supply, especially along the coast, allowed for an easy life with leisure time for dancing, singing, and games.

The Chumash gathered acorns, pine nuts, berries, mushrooms, bulbs, grass seeds, cattail, cress, and other wild foods. They also hunted mule deer, coyote, and fox, as well as smaller animals, and they fished in the rivers. Coastal resources included seal, sea otter, porpoise, mollusks, and saltwater fish.

The Chumash made beautiful baskets and stone bowls. They and the Gabrielino to the south were the only groups in North America to use plank boats rather than skin-covered, bark-covered, or dugout canoes. The Chumash used their boats to travel to and from the Channel Islands and to hunt at sea for fish and marine mammals.

The Chumash are known for the imaginative, sometimes abstract rock art that has been found on cave walls and ceilings in their area. Tribal members told anthropologists that shamans made these drawings or supervised their creation during religious ceremonies. Some rock paintings show men or bird-men with fringed wings; these pictographs may represent ceremonies honoring the California condor, a bird held sacred by several tribes in southern California.

The Chumash say that the world is divided into three spheres: the middle world, where humans live; the first world below, where dangerous creatures live; and the first world above, where powerful supernatural beings—including celestial beings—live. The universe is orderly but unpredictable and supernatural beings can influence the fate of humans.

The Chumash tell of a gambling game played in the first world above that determines whether there will be abundant food and ample rain in the middle world, Earth. The game, which was also a favorite among the Chumash, involves guessing which hand holds a white bone marker, called a *peon;* whoever guesses correctly wins the *peon*. In the first world above, Sun and *Slo'w* (Eagle) make up one team and Sky Coyote and Morning Star make up the other, with Moon serving as referee. These teams play each night, and at the end of the year, at the winter solstice, they tally the score. If Sky Coyote's team wins, then there will be an immediate tumbling of food from the sky and the following year will be rainy and therefore fruitful. If the Sun's team wins, then human lives must be forfeit.

This uncertain and possibly dangerous universe required the skills of people who could maintain the necessary balance in the cosmos, and members of the *'antap* religious cult fulfilled this role. These people, who included the *wot* (political chief), shamans, and other individuals with particular knowledge, conducted ceremonies, administered datura, healed the sick, offered advice, and looked after the welfare of the people. One official, who, it was believed, obtained power from celestial beings and could foretell the future, served as the astronomer/astrologer. This person named children according to signs in the stars, foretold the weather, and fulfilled other important duties.

Fernando Librado *Kitsepawit*, a Chumash knowledgeable about the Ventureño and

Isleño bands, narrated many of the stories presented here. María Solares, of Chumash and Yokut heritage, also provided stories and information about Chumash star beliefs.

Those who would like a closer look at Chumash astronomy should consult *Crystals in the Sky: An Intellectual Odyssey Involving Chumash Astronomy, Cosmology and Rock Art,* by Travis Hudson and Ernest Underhay.

LITTLE DIPPER: *'Ilɨhɨy,* Wise Woman or Fox

Fernando Librado identifies a constellation called *'Ilɨhɨy,* which he calls "guardian stars" because they guard the North Star. Hudson and Underhay speculate that this figure is likely the bowl of the LITTLE DIPPER. The Chumash word *'ilɨhɨy* has several different meanings: a political office held by a woman; a perceptive woman; and a female fox. The authors suggest that the four stars of the bowl represent either a fox or a wise female, perhaps one who holds a high office among the Sky People.

BIG DIPPER ?: *Manoxonox Awawaw,* Seven Geese

Hudson and Underhay suggest that the unidentified Chumash constellation *Manoxonox awawaw* is the BIG DIPPER, and that these stars were once seven boys who turned into geese. María Solares says that the event occurred "long ago, when animals were people."

There was a woman who remarried because her husband had left her. She had a little boy, but she and her new husband did not give him anything to eat. He went out and dug some roots because he was hungry. He met another boy who was hungry for the same reason, and another, and another. Finally, there were seven boys and Raccoon, who helped them find food.

The seven boys decided that since their families did not want them, they would go north. The oldest boy placed goose-feather down on the head and shoulders of each boy and of Raccoon, who agreed to go with them. The oldest boy led the others in a song, and gradually their feet lifted from the ground and they were flying, all except Raccoon. No matter how much down they put on Raccoon, he could not fly.

The boys flew higher in the air and changed to geese. Their mothers came out and cried to them to return, but they kept circling and climbing higher.

(Adapted from Blackburn, *December's Child: A Book of Chumash Oral Narratives*, 245–248.)

North Star: The Star That Never Moves

The Chumash call the North Star the Star That Never Moves. One of the Chumash names for it, *Miwalaqsh* (To Separate in the Middle), implies that the North Star divided the sun's east-west transit in half.

The North Star is not definitely identified with any sky person, though Hudson and Underhay suggest that it might have been linked with Sky Coyote, who appears in several myths and plays the game of *peon* with the Sun.

PLEIADES: Eight Wise Men

A brief story that Fernando Librado tells about PLEIADES suggests that the Chumash watched the heavens carefully and noted the dimming phase of the variable star Pleione in their oral tradition. Although most Native American groups identify six stars in PLEIADES, or seven stars with one fading or hiding, some groups identify other numbers. In one story, the Delaware say the PLEIADES is seven prophets, and in another ten young men. The Arikara say that the constellation is a girl and four brothers. Clearly, different tribes "saw" the PLEIADES in different ways.

> Eight wise men decided to live together in order to devote their knowledge and skills to the collective good. This arrangement worked well, but one man vanished unaccountably. No explanation was ever found for his disappearance, and only seven wise men can be seen in the sky now. The Pleiades are these wise men, "distant ones" who went away "where not everyone will go."
>
> (Adapted from Hudson et al., *The Eye of the Flute*, 35.)

Sirius and HARE?: *Mech*

Fernando Librado speaks of a constellation made up of seven stars near ORION that resembles the LITTLE DIPPER. The constellation does not appear in any known Chumash myths or rituals. Hudson and Underhay suggest that this constellation, *Mech*, is composed of the bright star Sirius in BIG DOG and several stars in the HARE. Sirius is the brightest star in the sky and its prominent position in the southern sky would certainly have been noted by Chumash astronomers.

Morning Star, Evening Star, and Mars

Fernando Librado identifies Morning Star, *'Alna*hy*ɨt 'i'aqiwɨ* and Evening Star, *Sma'ayɨ i'aquiwɨ*, as being married to the sun. María Solares, however, relates a myth in which the sun is a widower who lives alone with his two daughters and pets. She says that Morning Star is a male, a teammate of Sky Coyote in the game of *peon*. The Morning Star's job is to light the dawn, after which the Sun lights the day.

Venus may also represent the Two Thunderers, brothers who are supernatural beings in the upper world. The Thunderers shoot lightning from their eyes and create thunder when playing the hoop and pole game, a traditional Native American game involving throwing a pole into a rolling hoop (see page 197). They rise in the morning, travel around the world each day, and return to the sky before dark.

Hudson and Underhay suggest that Venus as the Evening Star may be the supernatural being *Slo'w* (Eagle). Eagle plays the game of *peon* on the side of the Sun and

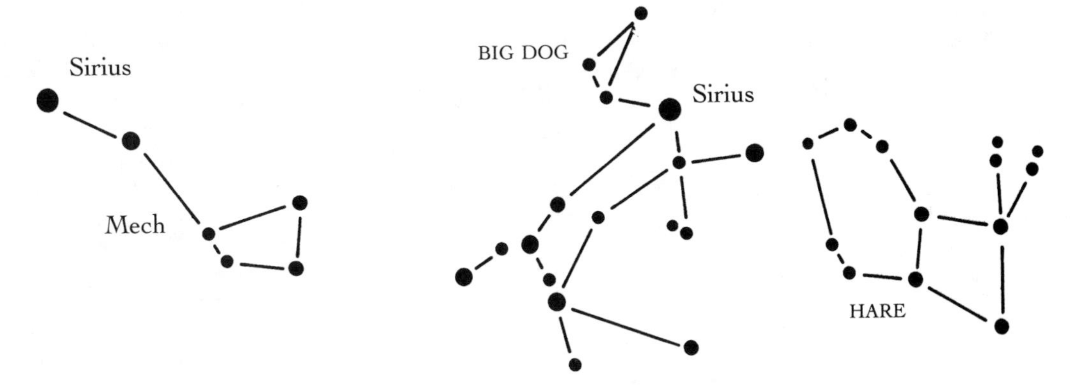

Fig. 9.4. The Chumash constellation *Mech* (left), thought to be composed of stars in BIG DOG (center) and HARE (right), resembles the LITTLE DIPPER.

may be the chief of the Land-of-the-Dead. The chief of the dead lives in a crystal house in the west.

Hudson and Underhay propose that Mars is the supernatural being Xolxol—possibly a condor—who also lives in the sky. María Solares describes Xolxol as having notable powers. He wears clothing and carries two sticks that enable him to jump great distances in a short time. Once, when Coyote (not Sky Coyote) wants to find Eagle, Coyote takes Xolxol's clothes and sticks, quickly locates Eagle, and then returns Xolxol's belongings.

Mars is a logical choice for Xolxol. As a planet, it travels through the constellations and so may have been attributed supernatural powers. Its red-orange color may also have been linked with the condor's orange-colored head.

Other Stars and Constellations

One group of Chumash, like other tribes in southern California, identify the bright star Aldebaran as Coyote. Coyote visits the Sky World on several occasions, including the time he sets out after Eagle, and he also holds the important job of administering datura to all eight-year-old boys in the village. The datura rite occurs in winter, when Aldebaran is high in the sky.

The Sun has two female cousins who leave the Earth to visit him in the sky. The sisters live in a house near him. María Solares, who presents this information, says that they are still in the sky as stars near the moon; the larger star looks out for the smaller star, her younger sister. Hudson and Underhay suggest that these two stars are Castor and Pollux, the bright pair in the constellation GEMINI.

The belt of ORION may have been associated with a bear, one of Sun's pets.

Milky Way: Journey of the Piñon Gatherers

Fernando Librado calls the Milky Way *Suyapo'osh* (Journey of the Piñon Gathers), explaining that piñon nuts are white inside, just like the Milky Way. Another Chumash group calls the Milky Way Night's Backbone, and there is some indication

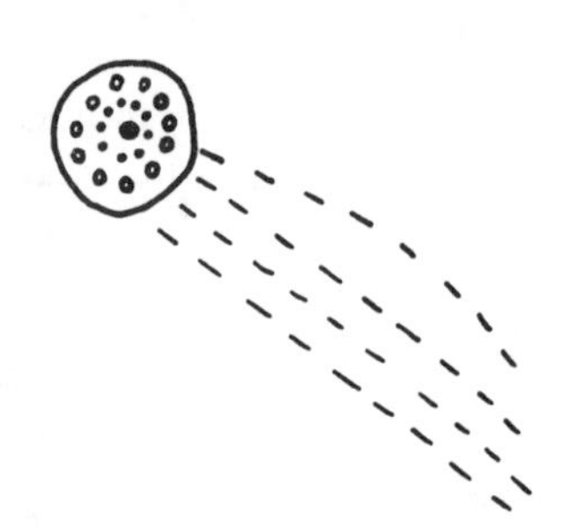

Fig. 9.5. Possible star motifs and a comet (*right*) in Chumash rock art. (Illustrations from Grant, *The Rock Painting of the Chumash,* figs. 75, 76, and plate 25.)

in myth that the Milky Way is—at least in winter—considered to be a road to the Land-of-the-Dead.

Altair and Two Stars: Three Steady Persons in their Place

Fernando Librado says that there are three stars called the Three Kings. These stars represent "three steady persons in their place." There is a tree to the north of these people and one to the south; the trees, called *contrición*, serve as flagpoles during ceremonies.

These three stately persons are probably Altair and the other two stars in the head of the EAGLE. Altair may have marked the time of the winter solstice; it appears on the western horizon after sunset in late December. Altair may also be associated with death, since it is in the west (the location of the Chumash Land-of-the-Dead) when the ceremonies honoring the dead are performed in the winter. The Chumash used poles decorated with bird feathers during such ceremonies.

Other Constellations Near the Land-of-the-Dead

At the time of the winter solstice, the Milky Way lies in an east-west position. In Chumash myth, *Šimilaqša* (the Land-of-the-Dead) lies in the west, and souls must pass several landmarks in the Upper World on the way there. These landmarks may be represented by constellations along the Milky Way. María Solares narrates this myth about a man who tries to follow his wife to the Land-of-the-Dead; the myth provides details about the trip.

The Land-of-the-Widows was the soul's first stopping place as it journeyed to the Land-of-the-Dead. The soul was greeted by widows, who bathed in a spring that preserved their youth. The widows did not need to eat food, but simply inhaled it. The soul next traveled through a deep, rocky ravine, where two ravens pecked out the eyes of all who passed. Each soul picked poppies to replace their eyes.

When the soul moved beyond the ravine, it came upon Scorpion Woman, "She Who Thunders." Scorpion Woman is tall and white. Souls tried to pass by Scorpion Woman quickly lest

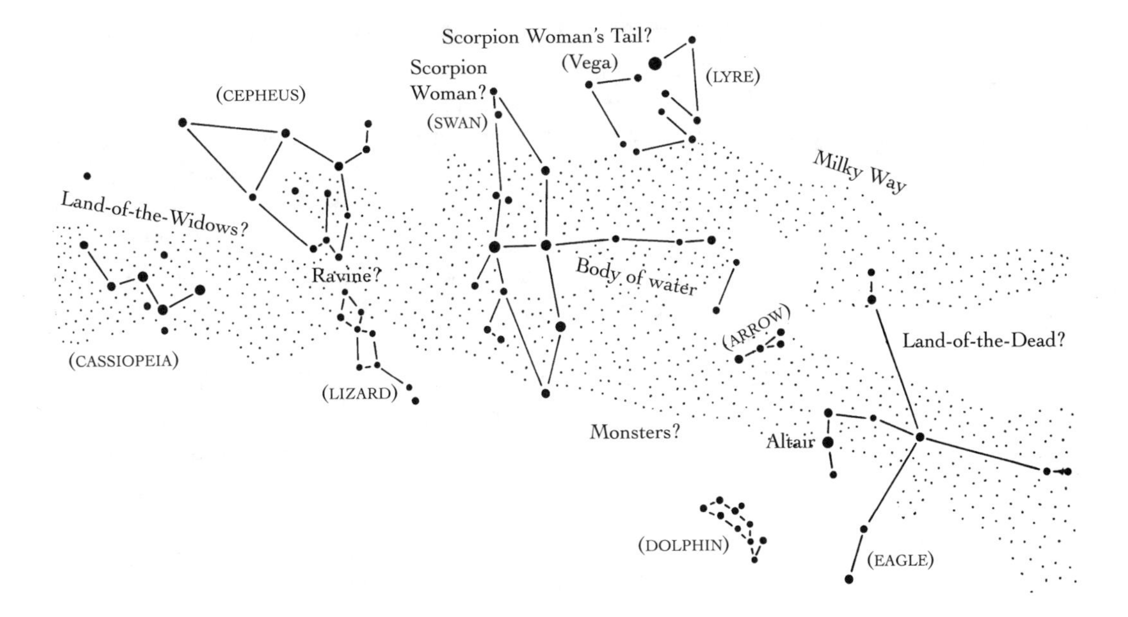

Fig. 9.6. Stars near the Milky Way may represent Chumash constellations encountered en route to the Land-of-the-Dead: Land-of-the-Widows (CASSIOPEIA?); Ravine (CEPHEUS and LIZARD?); Scorpion Woman (LYRE and SWAN?); Stinger in Scorpion Woman's tail (Vega?); body of water and location of bridge (cleft in the Milky Way?); and Monsters (SWAN, ARROW, or DOLPHIN?).

> she sting them with her tail. Beyond Scorpion Woman lay a body of water with a bridge guarded by two monsters, one on each end. Good souls were able to cross the bridge safely.
>
> (Adapted from Blackburn, *December's Child: A Book of Chumash Oral Narratives*, 249–251.)

Hudson and Underhay propose that constellations along the Milky Way are places through which the soul passes on the way to the Land-of-the-Dead. They suggest part or all of CASSIOPEIA plus some nearby stars as the Land-of-the-Widows; stars in CEPHEUS and LIZARD as the ravine; the star Vega in LYRE as the stinger in Scorpion Woman's tail; other stars in LYRE and the SWAN as the rest of Scorpion Woman; the dark area where the Milky Way splits near SWAN as the body of water and the beginning of the bridge; and stars in SWAN, ARROW, or DOLPHIN as the monsters. Because Altair and the other two stars in the head of EAGLE may be associated with the Land-of-the-Dead, and because EAGLE lies just past SWAN and ARROW on the western horizon at night in December, the myth and these constellations do seem to fit together.

Unidentified Stars and Constellations

According to the Chumash, if it weren't for Lizard, we would all have paws instead of hands. María Solares does not identify the Hand constellation in the following story, but perhaps it is the same figure as the Tipai Hand, thought to be the NORTHERN CROWN, stars in HERDSMAN, or the sickle in LEO.

Sun, Moon, Sky Coyote, Morning Star, and *Slo'w* (Eagle) were deciding how to make man when Sky Coyote announced that his hands, being the finest, should be the model for human hands. The next day they all gathered around a smooth rock table in the sky that would take the impression of whatever was laid upon it. But before Sky Coyote could place his hand upon it, Lizard reached out and left an exact handprint on the surface. Sky Coyote was very angry, but Sun and *Slo'w* sanctioned the change, and the rock still stands in the sky with the handprint on it. If Sky Coyote had acted first, we might have hands like a coyote.

(Adapted from Blackburn, *December's Child: A Book of Chumash Oral Narratives*, 95.)

Luiseño

Each Luiseño village was an independent tribelet with its own chief and shamans and its own hunting, fishing, and food-collection areas. The lands were controlled by individuals, by families, by the village chief, or by the village as a whole. Generally, women collected wild edible plants while men hunted, but there times when the tasks overlapped.

Like the Chumash, the Luiseño relied on acorns as an important source of food. Women collected seeds; pods and fruit from cactus; buds, blossoms, and pods from yucca; tubers, roots, and bulbs; mushrooms; and other wild foods. Men hunted deer, antelope, rabbit, jackrabbit, ground squirrel (but not tree squirrel), ducks, and other birds, but they avoided most predators. The Luiseño also fished for trout and other fish, and used balsa or dugout canoes to take advantage of the sea mammals, fish, crustaceans, and mollusks that lived along the coast.

The Luiseño lived in conical semisubterranean houses thatched with reeds, bark, or brush, and each village had a central ceremonial structure. There were important rituals that marked puberty, marriage, fertility, and death, as well as rituals for purification before hunting and for the drinking of datura. These rituals were carefully supervised by the religious elite.

In the Luiseño creation myth, two clouds appeared in a void. One cloud became a female and the other a male, and they produced the first people. A powerful god, *Wiyot,* ruled but then died and became the moon. After his death, the First People went to the sky to escape death themselves.

The Chingichngish cult, which is thought to have developed after Spanish contact, incorporated precontact Luiseño beliefs. Chingichngish was a mythical hero who gave the Luiseño laws and rituals, including the observances for drinking datura. Constance DuBois, who recorded ethnological information about the Luiseño in 1906, wrote that Chingichngish was a religion of fear: The sun and the North Star are in the sky to watch the people to see if they do anything wrong. She added that "in the old times" the Luiseño knew much more about the stars than they did when she was with them. Ritual songs mentioned stars that people could no longer identify for her.

Stars and Constellations

The chiefs among the First People took their relatives and compatriots with them when they went to the sky. The bright stars are chiefs and the surrounding stars are their relatives and friends, though they are not individually named.

Altair is one of the oldest and most important star chiefs, *Yungavish* (Buzzard). Vega is Buzzard's raised right hand, *Yungavish po-ma* (Yungavish His-Hand); a nearby star (perhaps Deneb or the stars in DOLPHIN?) is Buzzard's headdress, *Pecheya Yungavish* (Buzzard His Headdress). Antares is the second most important chief, *Nükülish*; Arcturus is his right hand, *Nükülish po-ma* (Nükülish His-Hand). Other sky chiefs include Spica, *Waonesh*; Fomalhaut, *Nawiwit Chawachwish*; and the North Star, *Tukmishwut* (Heart of the Night Wolves or Heart of the Coyotes).

Venus, the only planet named to DuBois, is called *Aylucha* (Leavings, or That Which Is Left Over from Evening Until Morning). Perhaps this name reflects Venus's role as both an evening and a morning star.

The Luiseño believe that the dead rise into the sky and become stars, thus accounting for the many stars that appear in the heavens.

North Star: *Tukmishwut*

DuBois notes that the North Star, as a chief, remains motionless while his people move in a circle around him. "This is the reason the [Luiseño's] dancing and marching are in a circle around the sacred enclosure, the fire," she writes.

She also records that the grandfather of Albañas, one of her storytellers, showed him a constellation of the northern sky. "Albañas' grandfather taught him the outlines of this constellation of the North Star in the evenings when the little boy sat by the hearth fire, tracing the figure in the sparks of little live coals upon the earth floor of the hut," she writes. The constellation involves some small stars that can be seen only when the sky is very clear, and it includes four fingers—the blunt one having been bitten off by a rattlesnake, as described in the following story. DuBois notes that being bitten by a rattlesnake is a sign that the person has offended Chingichngish, and it is necessary to perform a dance and ceremonies to ask for the god's forgiveness.

This tale describes the North Star before it ascended to the sky.

While they were living at Temecula, the rattlesnake was there, and because he had no arms or legs the others would make fun of him. The North Star, especially, who was then a person, was the leader in this abuse. He would fling dirt in Rattlesnake's

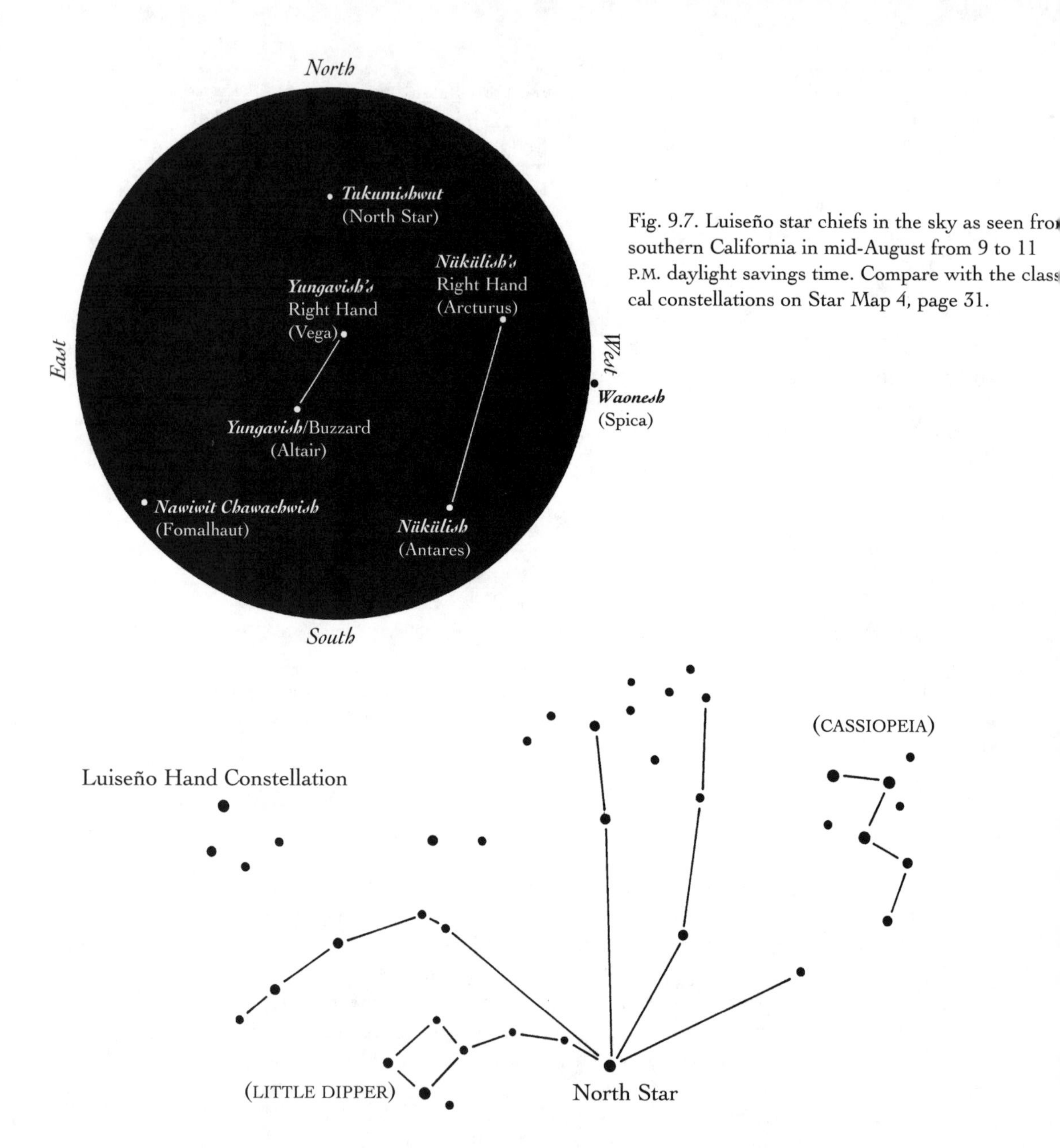

Fig. 9.7. Luiseño star chiefs in the sky as seen from southern California in mid-August from 9 to 11 P.M. daylight savings time. Compare with the classical constellations on Star Map 4, page 31.

Fig. 9.8. Constance Goddard DuBois, an ethnologist, described the Luiseño hand star constellation in 1906: "Starting from the North Star as a centre, there is a vortex of small stars, which in the clear air of the southwest are very plainly seen. They may easily appear as the five fingers of a hand; a line of three or four stars for the thumb, with several curving lines for the fingers, of which the last, a straight line shorter than the rest, and pointing towards Cassiopeia, is the one bitten off by the rattlesnake." The above illustration is one possible interpretation of the stars in this constellation. (Compare with the illustration of the LITTLE DIPPER and nearby constellations in Fig. 2.2, page 11.)

face, throw him down, and drag him about by the hair. So the rattlesnake went to the Earth-Mother and complained of this treatment, wishing to avenge himself on *Tukmishwut,* the North Star. So the Earth-Mother gave the rattlesnake two sharp-pointed sticks with which he might defend himself against any who disturbed him. So the next time when the North Star came and began to torment him, the rattlesnake used the sticks (his fangs) and bit off one of *Tukmishwut*'s fingers, as you may still see in the sky.

The Earth-Mother further contrived that, in order to make the bite of the rattlesnake effective, it should be followed by three intensely hot days; and at the present time, when three hot days come in succession, you may know that some man has been bitten by a rattlesnake.

(Reprinted from DuBois, Mythology of the Mission Indians, 54–55.)

PLEIADES: *Chehaiyam,* Seven Young Women; Hearts of the First People ORION's belt: *Hulaish* or *Hula'ch-un*

One ethnographer records that the Luiseño call the PLEIADES Hearts of the First People. In the following brief Luiseño story, Coyote follows *Chehaiyam,* seven women (PLEIADES), into the sky. The tale parallels a Cahuilla story in which Isilihnup (Coyote) follows *Chehaum,* three little girls (also PLEIADES), into the sky.

DuBois notes that *Hulaish* (also called *Hula'ch-un,* ORION's belt) appears in the sky with *Chehaiyam* and says that the two are always mentioned together.

At the time when the stars went up in the sky to escape death, the Pleiades, *Chehaiyam,* were seven young women, sisters; and when they went up, a rope was let down for them to climb on.

Coyote came along, and as there was no man with them he said, "I will go with you, girls." They did not answer him, but he took hold of the rope and kept on going up after them. But when they were safely up, they cut the rope and Coyote fell backwards. There is always a star following them, Aldebaran, and this is Coyote.

Hulaish went up at the same time.

(Reprinted from DuBois, The Religion of the Luiseño Indians of Southern California, 164.)

Milky Way: *Piwish*

The Luiseño describe *Piwish* (the Milky Way) as rising with the Star Chief *Nükülish* (Antares). *Piwish* is associated with the impermanence of life. In ground paintings made during puberty rites, the Milky Way is drawn as a circle that surrounds blood, the night, a panther, a raven, a rattlesnake, a spider, a stick, and other objects. (A more detailed description of ground paintings follows in the section on the Tipai.)

DuBois remarked on the beauty of the Milky Way when she was with the Luiseño. "The Milky Way glows brilliantly in the clear atmosphere of Southern California. It is there a much more imposing spectacle than it ever appears to the dwellers in the east. The ethereal quality of it, its vague outline and uncertain luminosity, make it easily an object of veneration."

Ipai and Tipai

Two closely related groups, the Ipai and Tipai, lived in the southernmost tip of California and in northern Baja. There were many bands within this area, and each band had its own chief, director of the hunt, leader of dances, and so on. In the summer, people sought shade under windbreaks and trees, while in the winter they used pole structures covered with brush, grass, and earth. Each village had a flat-roofed ceremonial structure with a dance ground.

The women had primary responsibility for collecting acorns, seeds, wild onion, stalks and roots from yucca, berries from elderberry and manzanita, and fruit from cactus, plum, and cherry. Men hunted, mostly small game like rabbits and birds, plus lizards, wood rats, and insects. Bands near water also fished.

Like the Luiseño, the Ipai and Tipai were adherents of the Chingichngish religion. Some rituals, including the boys' initiation rite, included the creation of ground paintings that portray various aspects of the universe. These ground paintings are doubtless an extension of the art form from the Navajo and Pueblo-dwellers, who create elaborate sandpaintings during ceremonials.

In 1910, the anthropologist T. T. Waterman described several Tipai ground paintings. They were fifteen to eighteen feet in diameter and were made in the village ceremonial house. The paintings were composed on the ground using powdered soapstone and iron oxide as the paint. The outer limit of the circle represented the horizon. Paintings included constellations, topographical features (these varied from village to village), and creatures associated with the Chingichngish cult, such as Rattlesnake, Tarantula, Bear, and Raven. Chingichngish had created these beings to make sure that people obeyed his laws and to punish those who broke them.

Stars and Constellations

The Ipai recognize SCORPIUS as a boy with a bow and arrow. The stars that move around the North Star, presumably the BIG DIPPER and the LITTLE DIPPER, are Wildcat's Rump. In some accounts ORION is a left-handed hunter with an arrow and in others *Mu* or *Emu* (Mountain Sheep); the tiny stars to one side of the constellation (perhaps ORION's shield?) are the horns of a mountain sheep. Aldebaran is identified as Coyote in ground paintings.

The stars represented in the Tipai ground paintings include Cross Star, NORTHERN CROSS; *Amai XatatkURL* (Sky-Its-Backbone), Milky Way; *Xatca*, PLEIADES; *Amu* (Mountain

Sheep), ORION's belt; *Saīr* (Buzzard), Altair; and *Watun* (Shooting), probably SCORPIUS. PLEIADES and Mountain Sheep travel together in the sky, and Shooting always points at Buzzard.

The Tipai recognize a constellation called the Hand, a group of five stars. This constellation appears to be the same as the Havasupai Hoop, which may be stars in the NORTHERN CROWN, the HERDSMAN, or the sickle of LEO. Or perhaps these five stars are the Chumash constellation that represents the handprint of a lizard.

The Jealous Star has not been identified.

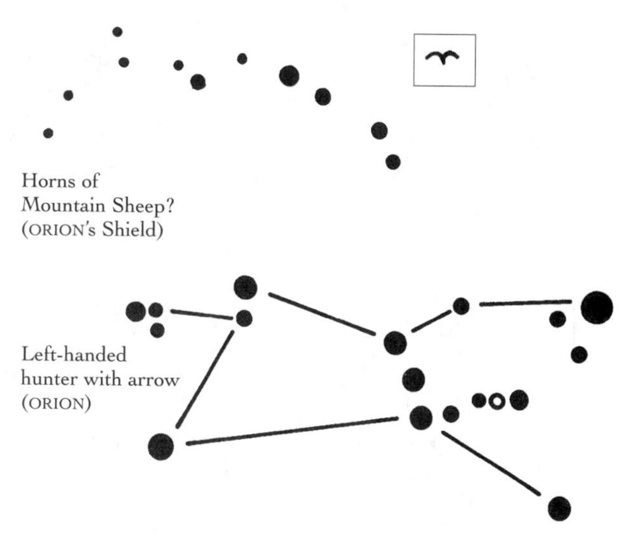

Fig. 9.9. The Ipai constellation representing the horns of a mountain sheep: "This apt name is given from the perfectly defined horns of the mountain sheep which can be traced on one side of the constellation [ORION] as we see it, in tiny stars." The stars of ORION's shield, with tiny stars that are not always visible, form two arcs above Betelgeuse and the belt when ORION is rising in the east. These arcs may be the mountain sheep's horns. One symbol used for mountain sheep in petroglyphs in southern California is shown in the inset. The Ipai also call ORION a left-handed hunter with an arrow.

PLEIADES: *Hecha* HYADES?: *Anyihai,* Girl Who is Planting

The Kamia are now considered a subgroup of the Tipai. Charles Beans and Narpai, two Kamia, told the following myths about *Hecha,* Chiyi, Chiyuk, and *Amuh* to E. W. Gifford, who published the information in 1931.

Anyihai (Girl Who is Planting) is a cluster of stars, most likely the HYADES, that rises in the east just before dawn in late June and early July. *Anyihai* is important because in the beginning of the world Wildcat told the people that when it appeared at dawn, it was time for them to plant their crops. The constellation *Hecha* (PLEIADES) rises before *Anyihai.*

Altair and Two Stars: Chiyi

In Tipai mythology Chiyuk and Chiyi are two supernaturals. Chiyuk makes the constellation *Anyihai* and then protects her. He tells the people to plant after *Anyihai* appears just before dawn. Chiyi then orders Chiyuk to go underground. "You will not stay underground forever," he says. Chiyi and Chiyuk alternate in the sky. When one is invisible, the other can be seen at night.

Chiyi is thought to be Altair and the two other stars that form the head of the EAGLE. Chiyuk is not clearly identified.

ORION: *Amuh*, Mountain Sheep

Various southwestern California tribes identify the three stars of ORION's belt as *Amuh* (Mountain Sheep). Narpai drew a picture of the constellation, specifying that the stars of the belt were a pronghorn antelope, a deer, and a mountain sheep. He also labeled

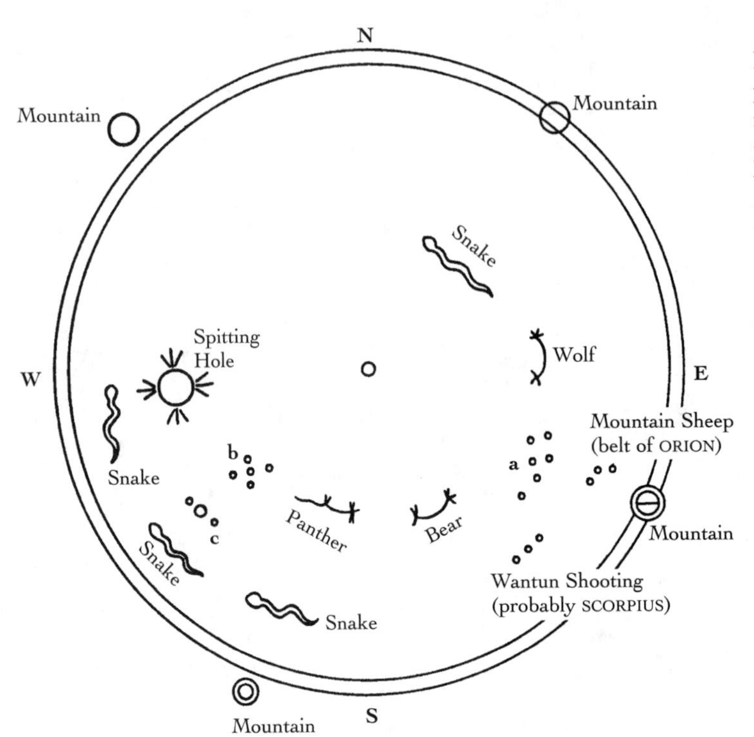

Fig. 9.10. The Milky Way, sun, and phases of the moon are not included in this illustration of a Tipai ground painting because Manuel Lachuso, who made the drawing, had forgotten their exact locations. The Milky Way, Amai *Xatatk*URL (Sky-Its-Backbone), was usually shown as a broad line from east to west. (Illustration from Waterman, The Religious Practices of the Diegueño Indians, 351.) *a. Xatca* (PLEIADES); *b.* Cross Star (NORTHERN CROSS); *c. Saĩr,* Buzzard star (ALTAIR).

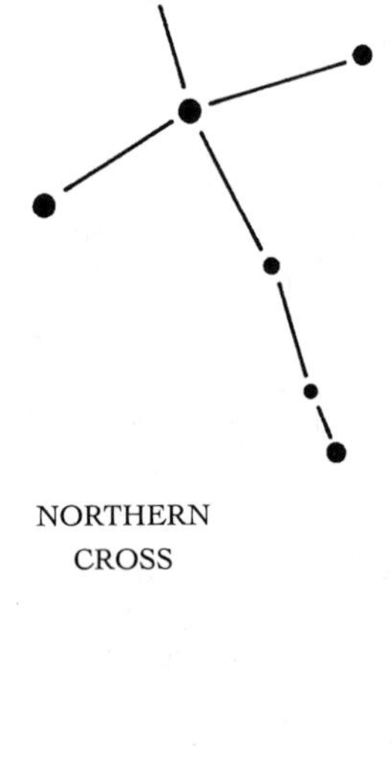

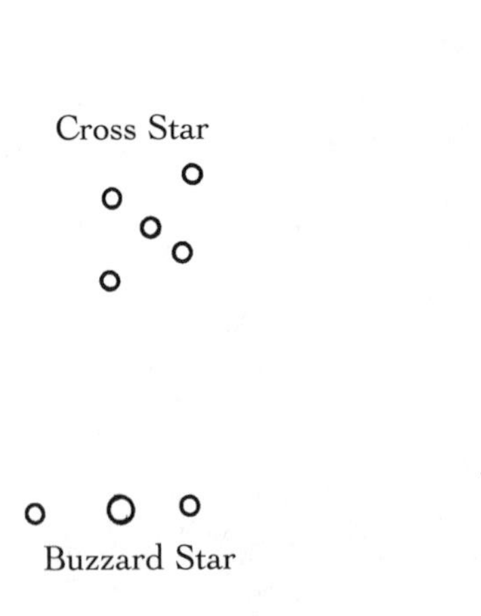

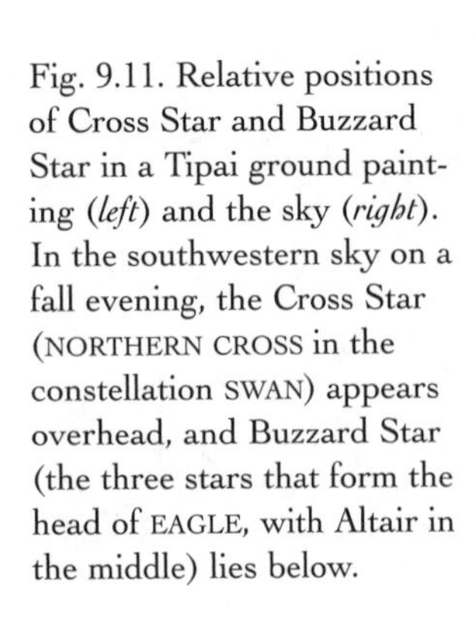

Fig. 9.11. Relative positions of Cross Star and Buzzard Star in a Tipai ground painting (*left*) and the sky (*right*). In the southwestern sky on a fall evening, the Cross Star (NORTHERN CROSS in the constellation SWAN) appears overhead, and Buzzard Star (the three stars that form the head of EAGLE, with Altair in the middle) lies below.

two hunters below the belt and an arrow point above it.

A Tipai myth recounts that in the beginning of the world two archers killed an antelope, a deer, and a sheep, and the souls of all five beings—as well as the arrow—went to the sky and became stars. The celestial hunter holds the arrow and the bow with his right hand and pulls the arrow back with his left. The antelope and deer have been pierced by the arrow.

Narpai explains that Chemehuevi and Yavapai, the two hunters, are skilled hunters, for they know well the mountains where the antelope, deer, and mountain sheep live. (The Chemehuevi and Yavapai are two tribes north and east of the Tipai, and Narpai appears to be saying that men in these tribes were better hunters of large game than were hunters in his own tribe.)

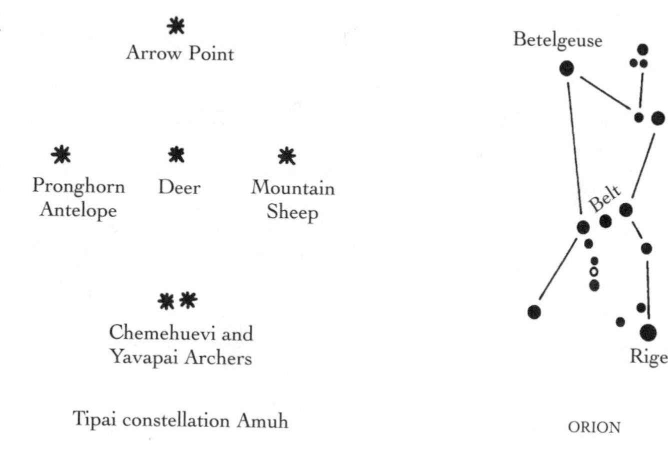

Fig. 9.12. Tipai drawing of the constellation *Amuh* (*left*), and the classical constellation ORION (*right*). The antelope (*kamul*), deer (*akwak*), and mountain sheep (*amuh*) form the belt of ORION. The arrow point may be Betelgeuse, and the two archers may be Rigel. (Tipai drawing from Gifford, *The Kamia of Imperial Valley,* 68.)

The arrow point could be reddish Betelgeuse, which matches the color of a bloodied arrow and is in the right place for an arrowhead that has pierced the antelope and deer. The two hunters could be Sirius, Rigel (which lies in a line with the antelope/deer and Betelgeuse), or another star altogether.

Capella?: *Kapwatai*

Kapwatai is a large star that is high in the winter sky at sundown and rises before the sun in the summer. The star Capella in CHARIOTEER matches this description. Capella is the sixth brightest star in the sky.

CONSTELLATIONS OF CALIFORNIA

Culture Group	Star or Constellation	Classical Equivalent
Northwest California		
Yurok	Many	PLEIADES
Karok	Road to Land-of-the-Dead	Milky Way
Karok	Little Stars	PLEIADES
Wiyot	Ocean (foam?)	Milky Way
Wiyot	Six People in a Boat	PLEIADES
Shasta	Path of the Soul	Milky Way
Shasta	Raccoon's Children and Coyote's Youngest Son	PLEIADES
Northeast California		
Atsugewi	Coyote's Cane	BIG DIPPER
Atsugewi	Devil's Trail; Dead Person's Trail	Milky Way
Atsugewi	Seven Sisters at Puberty Dance	PLEIADES
Sacramento Valley Region		
Northern Sierras	Cottontail Man	AQUARIUS, PEGASUS
Northern Sierras	Paw Print	DOLPHIN
Northern Sierras	Grizzly Sisters	PLEIADES?
Patwin	Knocking off Acorns	BIG DIPPER
Patwin	Seining in the Slough; West Star Now Comes	Evening Star
Patwin	Antelope's Road; Ashes	Milky Way
Patwin	Coyote Carries on Head	ORION
Yana	Coyote's Arrow	ORION's belt
Maidu	Looking Round	BIG DIPPER
Maidu	To Roast	DOLPHIN
Maidu	Morning Star's Trail	Milky Way
North Coast Range		
Pomo	Stick with a Hook	BIG DIPPER
Pomo	Night Woman	Evening Star
Pomo	Track of a Bear	Milky Way
Pomo	Day Woman; Morning Eye Fire; Big Star	Morning Star
Pomo	Eye of the Creator	North Star
Pomo	Strong Sucker	ORION's belt?
Pomo	Buckeyes Bunched Up	PLEIADES
Yuki	Young Women Who Announces the Sun	Morning Star

CONSTELLATIONS OF CALIFORNIA (*cont.*)

Culture Group	Star or Constellation	Classical Equivalent
San Joaquin Valley		
Yokut	Coyote	Aldebaran
Yokut and Monache	Buzzard	Altair
Monache	Seven Boys	BIG DIPPER
Yokut and Monache	Six Young Men	HYADES or ORION's belt?
Yokut	Racetrack	Milky Way
Yokut	Sky Coyote	North Star?
Yokut	Crane and Her Sons; Bear Chasing Antelope; Men (or Coyote) Chasing PLEIADES; Three Babies	ORION's belt
Yokut and Monache	Six Young Women	PLEIADES
Yokut	Eagle	Unidentified
Yokut and Monache	Blackbird	Unidentified
Yokut and Monache	Raven	Unidentified
Yokut and Monache	Roadrunner	Unidentified
Yokut and Monache	Dog	Unidentified
Yokut and Monache	Duck	Unidentified
Yokut and Monache	Hummingbird	Unidentified
Yokut and Monache	Y-Frame Cradle	Unidentified
South Coast Range		
Costanoan	Bunch of Little Ones Shaking	PLEIADES
Salinan	Star That Does Not Move	North Star
Salinan	Three in a Row	ORION's belt
Salinan	Group of Maidens	PLEIADES
Southeast California		
Cahuilla	Coyote; Suitor	Aldebaran
Cahuilla	Wildcat's Rump	BIG and LITTLE DIPPERS
Cahuilla	Kunvachmal (Coyote's stepbrother)	Canopus?
Cahuilla	Dust Kicked Up by Coyote and Wildcat	Milky Way
Cahuilla	Pretty Woman	North Star; other stars?
Cahuilla	Three Mountain Sheep	ORION's belt
Cahuilla	Arrow	ORION's sword
Cahuilla	Three Little Girls and Their Jewels	PLEIADES
Cahuilla	Hunter Who Shot Arrow	Rigel
Cahuilla	Funny Star	Unidentified
Cahuilla	Three Sisters (friends of Múkat)	Unidentified
Southern Coast		
Chumash	Coyote	Aldebaran
Chumash	Three Steady Persons	Altair and two stars

CONSTELLATIONS OF CALIFORNIA (*cont.*)

Culture Group	Star or Constellation	Classical Equivalent
Southern Coast *(continued)*		
Chumash	Seven Boys Who Became Geese	BIG DIPPER?
Chumash	Land-of-the-Widows	CASSIOPEIA?
Chumash	Sun's Female Cousins	Castor and Pollux?
Chumash	Ravine after Land-of-the-Widows	CEPHEUS and LIZARD?
Chumash	Wise Woman; Fox	LITTLE DIPPER's bowl
Chumash	Xolxol (Condor?)	Mars?
Chumash	Journey of the Piñon Gatherers; Night's Backbone; Road to the Land-of-the-Dead	Milky Way
Chumash	Teammate of Sky Coyote	Morning Star
Chumash	Star That Never Moves; also Sky Coyote?	North Star
Chumash	Bear	ORION's belt?
Chumash	Eight Wise Men	PLEIADES
Chumash	*Mech*	Sirius and HARE?
Chumash	Scorpion Woman	SWAN and LYRE?
Chumash	Monsters	SWAN, ARROW, DOLPHIN?
Chumash	Scorpion Woman's Stinger	Vega?
Chumash	Eagle; also Chief of the Land-of-the-Dead?	Venus as Evening Star?
Chumash	Two Thunderers	Venus?
Chumash	Hand	Unidentified
Luiseño	Coyote	Aldebaran
Luiseño	Buzzard (a sky chief)	Altair
Luiseño	*Nükülish* (a sky chief)	Antares
Luiseño	Nükülish His-Hand	Arcturus
Luiseño	*Nawiwit Chawachwish* (a sky chief)	Fomalhaut
Luiseño	*Tukmishwut* (a sky chief); Heart of the Night Wolves	North Star
Luiseño	*Piwish*	Milky Way
Luiseño	*Hulaish*	ORION
Luiseño	Seven Young Women; Hearts of the First People	PLEIADES
Luiseño	*Waonesh* (a sky chief)	Spica
Luiseño	Buzzard's Right Hand	Vega
Luiseño	That Which Is Left Over from Evening Until Morning	Venus
Ipai	Coyote	Aldebaran
Ipai	Wildcat's Rump	BIG and LITTLE DIPPERS
Ipai	Boy with Bow and Arrow	SCORPIUS

CONSTELLATIONS OF CALIFORNIA (*cont.*)

Culture Group	Star or Constellation	Classical Equivalent
Southern Coast *(continued)*		
Ipai	Mountain Sheep; Left-Handed Hunter with Arrow	ORION
Ipai	Horns of Mountain Sheep	ORION's shield?
Tipai	Buzzard	Altair
Tipai	Sky-Its-Backbone	Milky Way
Tipai	Cross Star	NORTHERN CROSS in SWAN
Tipai	Mountain Sheep	ORION's belt
Tipai	Hand	NORTHERN CROWN, HERDSMAN, or LEO?
Tipai	*Xatca*	PLEIADES
Tipai	Shooting	SCORPIUS?
Tipai	Jealous Star	Unidentified
Tipai (Kamia)	Chiyi (a supernatural being)	Altair and two stars
Tipai (Kamia)	Arrow Point	Betelgeuse?
Tipai (Kamia)	*Kapwatai*	Capella?
Tipai (Kamia)	Girl Who Is Planting	HYADES?
Tipai (Kamia)	Mountain Sheep	ORION's belt
Tipai (Kamia)	*Hecha*	PLEIADES
Tipai (Kamia)	Two Hunters	Rigel or Sirius?
Tipai (Kamia)	Chiyuk (a supernatural being)	Unidentified

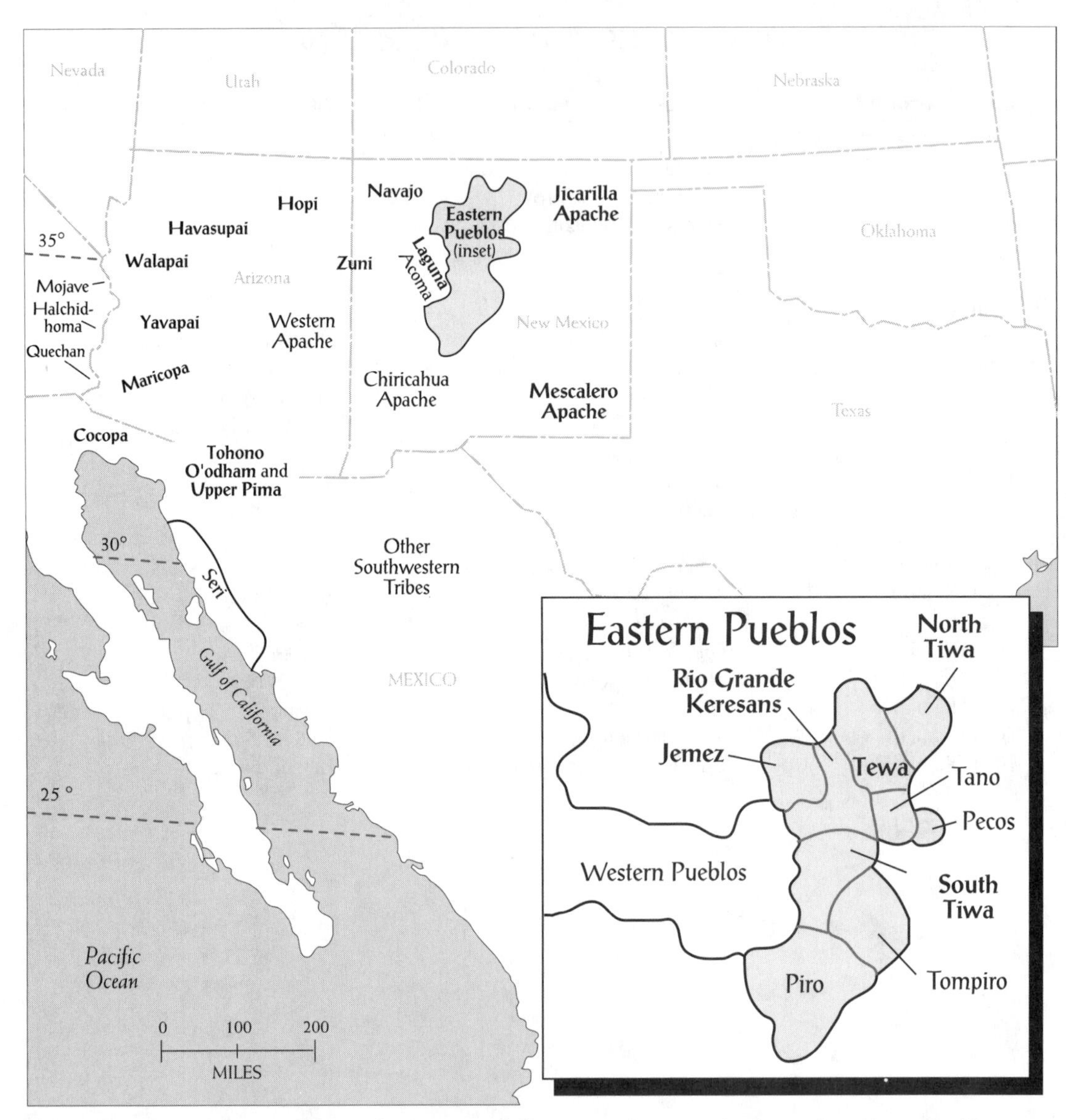

Fig. 10.1. Tribes of the Southwest culture area and their approximate territories at these times: Pueblos, 1500s; southwestern region of map, 1600s; the Navajo and Apache, 1700s. The tribes indicated in **boldface type** are included in this chapter. Not all tribes that inhabited this area are shown. (After Ortiz, *Handbook of North American Indians* Vol. 9, ix, and *Handbook of North American Indians* Vol. 10, ix.)

Chapter ten

Chief of the Night

Stars of the Southwest

The Southwest is a land of sharp contrasts. There are hot days and cool nights, high mountains and flat mesas. The sculpted, barren sandstone is at once harsh and beautiful. Drought-adapted plants burst into luminescent display with the advent of rain. The sky dominates all, with spellbinding sunrises and sunsets. At night, the stars spread over the sky in numbers too many to count.

People have lived in the Southwest for thousands of years, using irrigation and water management to grow food and cotton, which they wove into cloth. Prehistorically, the Anasazi built remarkable dwellings into cliffs, and the Mimbres Branch of the Mogollon made graceful black-and-white pottery. Their descendants include the people of the Eastern Pueblos (along the Rio Grande and its tributaries), and the Western Pueblos of the Zuni, Hopi, Acoma, Laguna, and the Tewa-speakers who live at Hano.

Several other groups made their homes in the Southwest. The Navajo and Apache, Athapaskan-speaking people who entered the Southwest relatively late, depended primarily on hunting and gathering. The Yuman-speaking tribes of the Western Uplands hunted, gathered, and engaged in some agriculture, while the Yuman- and Piman-speaking groups of the southern rivers and Sonoran Desert farmed or hunted, depending on their proximity to water.

The Southwest includes part of the Great Plains (in eastern New Mexico), the southern extension of the Rocky Mountains, through which runs the Rio Grande (north-central New Mexico), a swath of the Colorado Plateau, including the Western Uplands (northwestern New Mexico and northern Arizona) and an extensive section of basin and range territory in southern Arizona, New Mexico, and Mexico.

Star names and myths in the Southwest show great variation, so it is difficult to make generalizations. In both Pueblo and Navajo creation mythology, however, a key creation figure intends that the stars be placed in the sky in certain groupings, but the impatience or carelessness of another person ruins the plan and many stars are scattered without design. In both Navajo and Apache traditions, the grouped stars stand in the sky as examples of proper—or sometimes improper—behavior, and humans are to pattern their own behavior accordingly. Balance and harmony are important elements in constellations and in human life. In Zuni tradition there are various constellations, but the largest and most majestic is the Chief of the Night, whose figure is so large and encompassing that it cannot all be seen at once.

It is also interesting to note two parallels with California star myths. Tribes of the Southwest's southern rivers and desert describe ORION as a mountain sheep, hunter, and arrow, as do the Cahuilla, Ipai, and Tipai of southern California. The Walapai, Tohono O'odham, and Seri in the Southwest call the BIG DIPPER a cactus-gathering hook, a constellation that reflected their own food-procurement needs; in central California, the Patwin call the BIG DIPPER a stick for knocking off acorns, a staple in their diet.

PUEBLOS

Pueblo refers to a village of multistoried buildings made of mortared stone or adobe brick and also to the people who live in these villages. The buildings are constructed so that each level is connected to the next by a ladder that goes through the roof. This clustered, permanent housing was possible because the Native Americans of this area were adept farmers and did not need to travel far to gather food from a large area. Pueblo peoples grew corn, squash, and beans, which together composed more than half of their diet. They also hunted, fished, and gathered wild foods according to the abundance of the local environment. Each Pueblo had its own leaders and customs.

The people of this area developed a worldview in which the universe is both orderly and balanced. Good actions enhance this balance and evil actions destroy it. Each individual must strive toward balance, not only for himself or herself, but for the village as well. Pueblo cultures developed extensive ceremonies, some of which feature *kachinas,* representations of supernatural beings, who help people attain this balance. The rich culture of the Pueblos continues to the present, and Native peoples conduct a full complement of traditional ceremonies. Pueblo crafts, including pottery, weaving, silver work, and basketry, are highly prized.

The many illustrations of stars in their rock art and in murals in ceremonial structures called *kivas* suggest that the Pueblo peoples had a keen interest in the night sky and in the movements of the celestial bodies.

Eastern Pueblos

The Eastern Pueblos are situated in the southern Rocky Mountains, along the Rio Grande and its tributaries. For the most part, these rivers provided a steady source of water, although some of the tributaries ran only intermittently. The growing season is longer in these valleys than it is in the surrounding steppe and mountains.

Various Eastern Pueblos are affiliated by language, of which there are two major families. The Tanoan language family includes these branches: Tewa (San Juan, Santa Clara, San Ildefonso, and Pojoaque Pueblos); Tiwa (Taos, Picuris, Sandia, Isleta Pueblos); Towa (Jemez); and Piro (extinct in the Pueblos). An eastern form of the

Fig. 10.2. Star images. This image (*left*), on a rock wall in Chaco Canyon, New Mexico, may represent the supernova that appeared in the sky there near a crescent moon on July 4, 1054. A star *kachina* near Albuquerque has a headdress of eagle feathers (*center*). An animal with stars inside it (*right*) from San Cristóbal, Santa Fe County, New Mexico, probably dates from the 1300s to the 1700s. (Figures at center and right after Schaafsma, *Rock Art in New Mexico,* figs. 132, 142.)

Keresan language family includes Cochiti, Santo Domingo, San Felipe, Santa Ana, and Zia Pueblos.

Creation of the Stars

Pueblo creation stories describe how the First People lived in a world far below this one. In an epic voyage, they ascended to the present world and found their place in it. The following passage from a Zia creation story describes the making of the stars. Ût´sĕt is the mother of Indians, and her sister Now´ûtsĕt is the mother of other people.

> The two mothers created the moon from a slightly black stone, many varieties of a yellow stone, turkis, and a red stone, that the world might be lighted at night, and that the moon might be a companion and a brother to the sun. But the moon traveled slowly, and did not always furnish light, and so they created the star people and made their eyes of beautiful sparkling white crystal, that they might twinkle and brighten the world at night. When the star people lived in the lower world they were gathered into groups, which were very beautiful; they were not scattered about as they are in the upper world. . . .
>
> When the people were ready to pass through to the upper world, Ût´sĕt called the *I-shits* (mole) and gave him the sack of stars, telling him to leave first with the sack. The little animal did not know what the sack contained, but he grew very tired carrying it, and wondered what could be in the sack. After entering the new world he was very tired, and laying the sack down he thought he would peep into it and see its contents. He cut only

a tiny hole, but immediately the stars began flying out and filling the heavens everywhere. The little animal was too tired to return to Ût´sĕt, who, however, soon joined him, followed by all her people.

When Ût´sĕt looked for her sack she was astonished to find it nearly empty, and she could not tell where the contents had gone. The little animal sat by, very scared and sad. "You are very bad and disobedient and from this time forth you shall be blind," said Ût´sĕt, who was angry with him. (And this is the reason the mole has no eyes, so the old ones say.) The little fellow, however, had saved a few of the stars by grabbing the sack and holding it fast. These Ût´sĕt distributed in the heavens. In one group she placed seven stars (the Great Bear), in another three (part of Orion), into another group she placed the Pleiades, and throwing the others far off into the heavens, exclaimed, "All is well!"

(Adapted from Stevenson, *The Sia,* 30, 37.)

Two Cochiti tales take up a similar theme. Originally, the stars were all to have different names and occupy special places, but a little girl's curiosity (or, Coyote's irresponsibility) allowed most of the stars to escape, leaving only a few with names.

In one story, a little girl carries a heavy sack and releases the stars when she opens the sack to find out what is inside. She is able to recover a few stars and places them in orderly groups: the Shield Stars (the BIG DIPPER), the Sling-Shot Stars (the DOLPHIN), the Pot-Rest Stars (unidentified), and a few others.

In another story, a man has hung only the Seven Stars (the BIG DIPPER), the Three Stars (the Pot-Rest Stars), and the Morning Star before Coyote lets the other stars out of a jar. The Pot-Rest Stars are "the sign that the Indians are to use three stones to support the pot for cooking," notes Ruth Benedict, the anthropologist who recorded this information in 1924. Presumably these stones elevate the pot above the coals, much like a Dutch oven has three feet that elevate the oven above the embers. Though the Pot-Rest Stars are unidentified, one possibility is that they are ORION's belt.

ORION: Long Sash

The Pueblo myth of Long Sash taking his people to a new land is particularly striking because it parallels the Judeo-Christian narration of Moses taking the Israelites from Egypt. Long Sash leads his people from where they were living when they emerged from the world below to their present location in the Pueblos.

Pablita Verlarde, an artist from the Tewa village San Ildefonso, recorded this myth in her book, *Old Father, The Story-Teller,* in 1960. Verlarde identifies the stars involved: Long Sash is ORION, the Endless Trail is the Milky Way, the Place of Decision is Castor and Pollux, the Place of Doubt is CANCER, and the Three Stars of Helpfulness are three stars in LEO.

Long Sash was a brave and wise warrior, and the people trusted him very much. They turned to him when, year after year, their villages were destroyed and warriors killed by neighboring tribes. They said to Long Sash, "Take us to a land where we may live in peace."

"The journey will be difficult," answered Long Sash. "We will have little food and we must travel a long way on the Endless Trail. We must go into a land that even I have not seen. Are you willing to endure the hardships?"

The people assured him that they wanted to go, and packed their few belongings. They began the long journey. Long Sash had to instruct them in how to hunt and make clothes, as they had little to eat or wear. Babies were born and old people died. Quarrels broke out and there was much fighting.

Long Sash halted and said, "We cannot continue. There is violence among you. You do each other more harm than your enemies ever did to you. Let us rest here. There are women whose time is due. Let them give birth and then you can choose whether you want to follow me or go another way." They stopped by two bright stars, now called the Place of Decision. Even today, people look to these stars when making important decisions.

After the people had rested, they said, "We want to continue with you." Long Sash watched them closely to see if what they said was true. He believed them and they again set out on the Endless Trail. But Long Sash was growing old, tired, and doubtful. He had never been in this land, and he was unsure of which way to go. He asked for a sign. He closed his eyes and appeared to sleep, but when he awoke he told the people that he had been given many promises. They had already completed the most difficult part of the trip, he said. "There are times in each person's life when there is a Place of Doubt. You must seek help from the Above Persons when this happens." He laid his headdress down to mark the Place of Doubt and to remind people where they can seek guidance when they need it.

With the rest and Long Sash's new strength, people stopped quarreling and were kinder to one another. Two young men, for example, made a drag on which to carry not only their own belongings but also those of an old woman. All of the people continued on the Trail and at last came to the end of their journey.

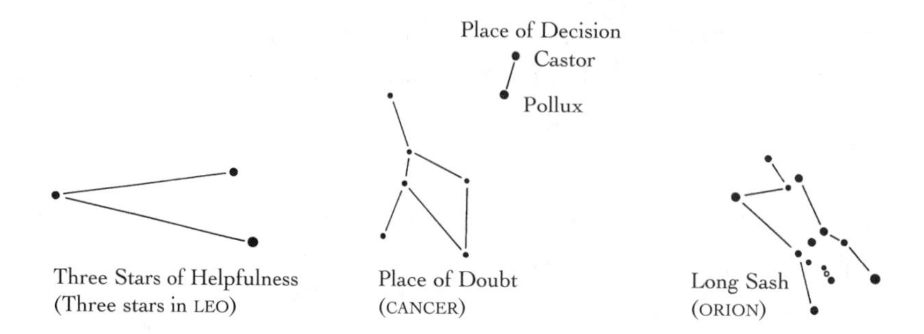

Fig. 10.3. The story of Long Sash (ORION) is recorded in the sky for all to see. At the Place of Decision (Castor and Pollux in GEMINI), Long Sash and his people decided to continue their journey. At the Place of Doubt (CANCER), Long Sash dealt with his own doubts about what the future would hold. Three Stars of Helpfulness (here tentatively identified as Deneb, Regulus, and another star in LEO) are a *travois* on which several men haul their own and an old woman's belongings.

They made their home in the Middle Place and have lived here forever.

(Adapted from Verlarde, *Old Father, The Story Teller*, 25–31; and Marriott and Rachlin, *American Indian Mythology*, 59–62.)

Stars and Constellations

Unless otherwise noted, the following constellations are from Tewa-speaking villages.

Seven Corner, Seven Tail, Long Tail and Dog Tail are all names for the BIG DIPPER. "Tail" refers to the handle of the DIPPER, which does indeed look like a tail. In her story of Long Sash, Pablita Verlarde wrote that the mole, the spider, the bear, the coyote, the cougar, the eagle, and the turtle each gave Long Sash and his people a sign about the exact place that they should live in the Middle World, and that the seven stars of the BIG DIPPER represent these seven signs. At Isleta the BIG DIPPER is Cradle.

In a Row is ORION's belt. At Isleta the belt is Fawns; at Jemez it is Rows; at Cochiti it may be the Three Stars or Pot-Rest Stars.

In a Bunch is the PLEIADES. At Taos, the PLEIADES are Deer; at Isleta they are Tumbled or Jumbled; at Jemez they are Seven Stars.

Yoke is the bowl of the LITTLE DIPPER.

Belly of a Sling, the wide leather piece attached to thongs, is the DOLPHIN.

Backbone of the Universe, also called Whitishness, is the Milky Way. At Taos and Jemez, the Milky Way is Sky Backbone, while at Cochiti it is Arch Across the Sky.

Meal-Drying Bowl is the NORTHERN CROWN. Women placed a meal-drying bowl, fired to a black finish, near a fire so that the meal it held would dry.

The Star in the South, also called Gray-Haired Old Woman, appears in the south at dawn in October. Canopus, a bright yellowish white star, fits this description.

Big Round Circle, a dance performed in October, is a large, irregular ring of stars near the NORTHERN CROWN. Some of the stars are very dim. The circle of stars in HERDSMAN (see the illustration of HERDSMAN in Chapter 2, page 19) is one possibility for Big Round Circle.

There are several other unidentified Tewa figures, including Star House, a big constellation that can be seen in the western sky after sunset in September; Sandy Corner, a large set of dim stars near ORION; Horned Star, a bright star; and Ladder. The Hand constellation, five stars representing the tips of fingers and one representing the wrist, is reminiscent of the Hand or Hoop constellations of the Western Uplands of the Southwest, except six rather than five stars are specified. At Isleta, there is an unidentified constellation called *Nadörna* (Wheel), which is a circle of stars with one in the center, and two unidentified stars called Bear.

The Tewa star Bull's Eye is unidentified, but it may be Aldebaran.

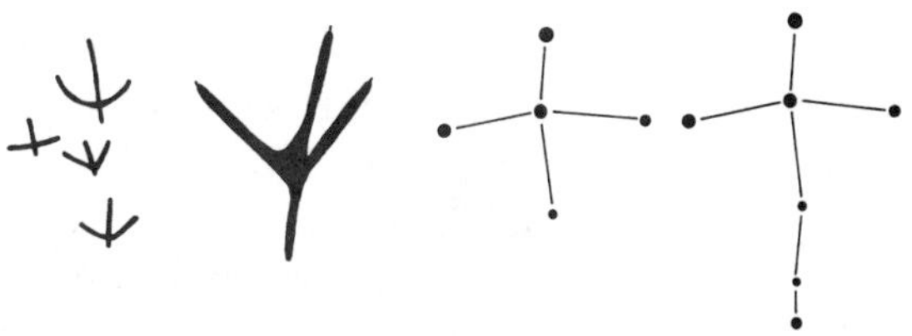

Fig. 10.4. The Tewa constellation Turkey Foot is unidentified, but the prints of the wild turkey are similar in shape to the NORTHERN CROSS. The NORTHERN CROSS has been suggested as the Pawnee constellation Bird's Foot. In the illustration are (*from left*): a petroglyph of bird tracks from a site along the Rio Grande; the footprint of a wild turkey; abbreviated form of the Northern Cross; the full NORTHERN CROSS. (Drawing of petroglyph after Schaafsma, *Rock Art in New Mexico*, fig. 108.)

Pueblo farmers kept turkeys, so it is not surprising that they should recognize a Turkey Foot constellation. John P. Harrington, an ethnologist who worked in the Southwest and published information about Tewa constellations in 1916, writes that the group is "an easily learned constellation of the exact form of a turkey's foot," but to the frustration of later readers he does not identify these stars. The shape of a turkey's foot is similar to that of the NORTHERN CROSS, which has also been suggested as the Pawnee Bird's Foot constellation (see page 225).

At Isleta, Bright Star is the Morning Star, and the Asking or Prayer Star is the Evening Star.

Morning Star: Big Star or Dark Star Man
Evening Star: Yellow-Going Old Woman
?: Star of San Juan Pueblo

To Tewa-speakers, the Morning Star is Big Star or Dark Star Man, a male divinity. In some contexts, Big Star is identified with the Twin War Gods, who are important in Pueblo mythology. The Evening Star is Yellow-Going Old Woman, a female deity whose gray hair (or yellowish hoary hair) falls over her face. In one story, presented below, the Morning Star continually chases the Evening Star in the sky; in other stories, it is a third star, Star of San Juan Pueblo, who chases the Evening Star.

The story of Big Star and Yellow-Going Old Woman elaborates on a widespread Native American theme of a man trying to follow his dead wife. People in Tewa-speaking villages believe that when a person dies, his or her soul wanders through the community for four days, asking for forgiveness. The spirit may appear as a puff of wind, a fire, or a dream. On the fourth day, relatives set out a bowl of food in a ceremony to speed the spirit on to the spirit world, for they do not want it to linger with them.

Here, Olivella Flower (called Deer Hunter in another version) and Yellow Corn Girl defy custom and remain together, but eventually Yellow Corn Girl's corpse becomes offensive and Olivella Flower tries to escape. Both Olivella Flower and Yellow Corn Girl ascend to the sky, standing as visible reminders to humans below to respect Pueblo traditions and customs. (This story is considerably more complicated, with more meaning, than is presented in this short retelling, which emphasizes the celestial aspects of the tale.)

Olivella Flower and Yellow Corn Girl were living nicely together, but Yellow Corn Girl died and was buried. Olivella Flower, however, went looking for her and found her. She was still around as a kind of wind. He wanted to stay with her and she agreed if he would continue to be a husband to her. He did, but was offended by her smell and ran away.

He ran to the house of T'owa'e, who agreed to hide him by putting him in a cane arrow and shooting him into the sky. He became Big Star. Yellow Corn Girl followed him, and she, too, was shot into the sky. She became Yellow-Going Old Woman. Big Star comes up first (in the morning) and Yellow-Going Old Woman comes up second (in the evening). But sometimes she passes him, for she has no heart.

(Adapted from Parsons, Tewa Tales, 22–23.)

?: Two Dove Maidens

When he was a boy, Rosendo Vargas (also called Feather-bunch Flying), heard the following myth from his grandfather. The myth offers a lesson to children who do not obey their parents and shows that even in mythic times, children wanted to be like their peers. Historically, women and girls ground corn on a flat stone called a *metate* using another stone, called a *mano*. Girls prayed to two stars for strength when grinding corn. The story ends with the phrase, "You have a tail," which means that it is someone else's turn to tell a story.

Two Dove Maidens and their grandmother lived at Picuris. The Dove Maidens always made baskets and never ground corn like the other girls in the pueblo. "We want to grind corn," they said to their grandmother. Their grandmother protested, but got out the big earthen jar and roasted some corn so that it could be ground in the morning.

The two girls set to work with the *metate*, singing a song, and as they sang they rose into the air. Their grandmother pleaded with them to stop, but they kept grinding and kept singing. Slowly they rose to the sky, where they became two stars above Jicarita Mountain.

You have a tail.

(Adapted from Harrington, *Picuris Children's Stories, with Text and Songs*, 351–352.)

Western Pueblos: Zuni

The Zuni, the only speakers of the Zunian language, have long lived on the Zuni River on the Colorado Plateau. In both Zuni and Hopi culture, supernatural beings called

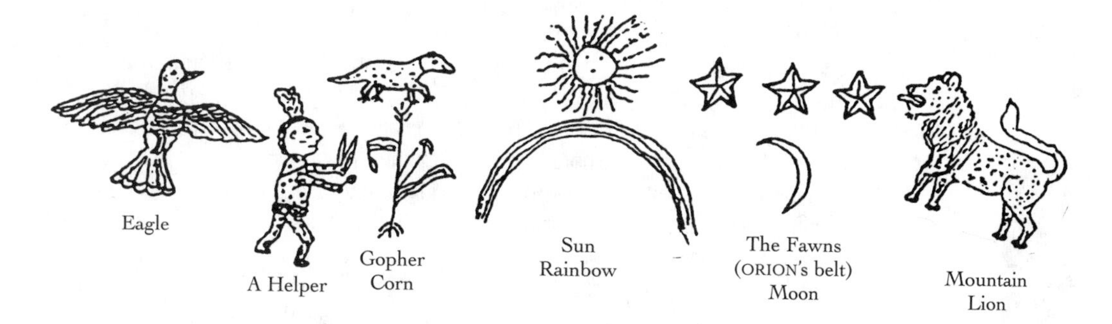

Fig. 10.5. Section of a design on a wall at Laguna Pueblo, a Western Pueblo, showing some of the Fathers. (Illustration from Parsons, *Isleta, New Mexico,* plate 17.)

kachinas were and are still particularly important, and individuals wear carved *kachina* masks in special ceremonies.

The Morning Star is the most revered heavenly body after the sun and moon and in past times served as the patron of warriors and hunters. The BIG DIPPER, LITTLE DIPPER, ORION, PLEIADES, Milky Way, North Star, and Evening Star are also among the pantheon of higher powers. Illustrations of stars are found on Zuni masks, headdresses, altars, murals, and other ceremonial objects.

The Zuni used the position of the moon, ORION, and the PLEIADES to mark time at night. The movement of these celestial bodies across the sky—seen through the overhead opening in the *kiva,* a ceremonial chamber built into the ground—signaled the start of various rituals and marked events in the agricultural cycle. The Zuni planted corn, for example, in the spring by the light of the "seven great stars" (the BIG DIPPER); these stars are in the shape of a gourd, with four stars as the main body and three as the gourd's handle. The rising of the Morning Star in August launched the rain ceremony, in which women wore headdresses adorned with images of the Morning and Evening Stars. Some of these ceremonies are still held today.

Morning Star and Evening Star: The *Ahayuta*

The *Ahayuta* are the twin warriors who figure prominently in the myths of the Zuni and other Southwest culture groups. The Twins guided the people from the Fourth Underworld into the present Fifth World and shaped them into finished humans. The brothers rid the Fifth World of monsters and gained the power to make rain. Together the Twins are called the *Ahayuta,* the name of the elder brother; the younger brother is Matsilema. Their father is the Sun and their mother is Water.

The elder brother is often depicted as the Morning Star (also called Big Star or Great Star) and the younger brother as the Evening Star (also called One Following the Sun): "Henceforth two stars at morning and evening will be seen, the one going before, the other following the Sun-father—the one Ahayuta, his herald; the other Matsilema, his guardian; warriors both, and fathers of men." In some stories the Twins create the Morning and Evening Stars rather than becoming these stars.

The *Ahayuta* appear in various Zuni stories, including ones in which they kill a monster and throw parts of its body into the sky, creating stars. In one story, the giant Cloud Swallower threatens the world by swallowing all the clouds and creating a drought. The Twins' grandmother tells them not to go near him, but they go anyway. With the help of Gopher, they kill Cloud Swallower and throw parts of his body into the sky. The liver becomes the Evening Star, the heart the Morning Star, the entrails the Milky Way, and the lungs the Seven Stars (the BIG DIPPER).

In another story, Bush Man captures and eats people who go out to collect firewood. The brothers again ignore their grandmother, who warns them to stay away, but find themselves in a stalemate—Bush Man cannot kill them, but neither can they capture him. Finally, they enlist the help of the Hopi and devise a plan in which they succeed. They kill Bush Man and throw his heart into the sky, where it becomes Star Liar, because Bush Man lied so often when he was alive. His liver becomes a star that moves through the sky as the corn ripens, and his head becomes the Seven Stars.

In yet another tale, the *Ahayuta* gamble with Takyeł´aci, who has magic powers. The Twins lose everything they have, including their clothes and bodies, and the gambler prepares to roast them at his fire. Their grandmother rescues them, however, by sending their sister with a special set of gambling sticks, and with these sticks she wins back not only her brothers and the gambler's house full of corn, but the gambler himself. The brothers cut out the gambler's eyes and throw them to the sky. One becomes the Lying Star and the other becomes the Morning Star. Lying Star is a bright "star"—probably a planet—that precedes the Morning Star.

A fourth story places the creation of the Morning Star outside the realm of the War Gods:

Fig. 10.6. A Zuni headdress worn by ceremonial impersonators of the Corn Maidens. The headdress is decorated with symbols of stars (two crosses), the sun, and the moon. (Illustration after Stevenson, *The Zuni Indians: Their Mythology, Esoteric Fraternities, and Ceremonies*, plate 38.)

There was once a woman living here in Zuni long ago who had a child born which she thought was a real child, but which was really a canteen water-jar. When it grew up, it walked like a turtle. He saw the other boys go out hunting, and he wanted to go along too, but he couldn't kill anything. And *K'yak'yalı* (the Eagle) saw him one day and said to him, "Poor thing! He can't kill anything, I will kill for him." So he killed some rabbits for him and the little fellow brought them home. And every time he went out, his friend Eagle hunted for him, and he always came home with meat.

When the snow melted away, he went down to the river every day and watched the other children play in the water, jumping up and down, and he wanted to play with them. At first he thought he wouldn't do it, but then decided that he might as well. So he jumped off a high place one day and hit on a hard spot and broke himself all to pieces.

His mother came along and picked up the pieces and brought them home and put them behind the fireplace. Pretty soon the little broken water-jar said, "Mother!" And she answered, "Yes, my child, are you all right now?" And he said,

"Yes, I am all right, but you must take the handles and the mouth of me and go out at daybreak and throw them to the east."

And she did as the little water-jar had told her, and took the handles and the mouth out early in the morning, and threw them to the east, and they became the Morning Star.

(Adapted from Handy, Zuni Tales, 464.)

?: Stick Game Ones

The Twins introduced a ritual game called *sho'liwe*, which is represented in the sky by an unidentified constellation, Stick Game Ones. The War Gods and members of the rain priesthood play this game for rain. In the version of the game that Pueblo people play, there are four sticks made of split reeds decorated with symbols indicating the parts of the day. The reeds are tossed against a basket or blanket and fall upon another blanket. The position of the reeds relative to one another determines whether the player wins a marker or not; twelve markers wins the game.

Other Stars and Constellations

The Seven Ones or The Seven (and, in one record, the Great White Bear of the Seven Stars) is the BIG DIPPER.

Star Zigzag is CASSIOPEIA; Those to the North is the LITTLE DIPPER; and Star Sling is DOLPHIN, as it is in the Tewa Pueblos. Seeds or Seed Stars are the PLEIADES. Because this star group rises at dawn in June, it likely marks an event in the agricultural cycle.

Red Star is the NORTHERN CROWN. The odd name—there is no red star within the NORTHERN CROWN—is not explained.

Four Big One or Square is the GREAT SQUARE OF PEGASUS.

Ashes Placed or Stretched Across is the Milky Way; in one instance, the Milky Way is described as a great snowdrift in the sky. The Twin War Gods made the Milky Way when they were searching for Corn Maidens and, in another context, when they killed Cloud Swallower. The Milky Way appears on Zuni masks and altars as checkered squares. (In Acoma sandpaintings the Milky Way appears as a ladder, for it is thought that these stars form a bridge to the heavens. In the Acoma creation myth, the roof beams of the first *kiva* represent the Milky Way.)

Chief of the Night

In his unpublished field notes of research with the Zuni, John P. Harrington describes a huge constellation called Chief of the Night. "This is the most majestic constellation ever reported from any tribe of Indians, as far as we know," he writes. "It is a gigantic human figure, even bigger than the whole visible sky, for its head had already set in the west or had been obscured by clouds lying near the horizon, while its heart was pointed out in mid-heaven, lying in the Milky Way, and its legs were said to extend beyond the horizon in the east." Harrington viewed this and other Zuni constellations with Nachapani, a Zuni worker in Chaco Canyon, early in the summer of 1929. His notes, unfortunately, do not fully identify the Chief of the Night.

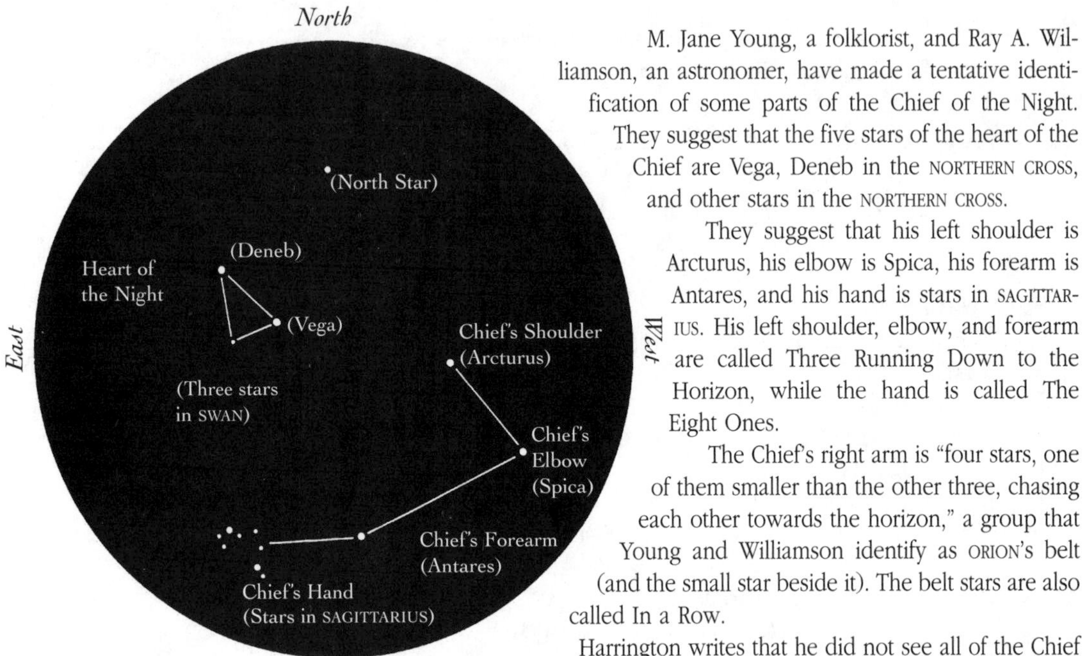

Fig. 10.7. The Zuni constellation Chief of the Night is so large that it cannot be seen all at one time. The heart and one arm are shown above; The North Star is included for reference. (This illustration shows the sky as seen from the latitude of the Zuni in the southern United States when the sky gets dark in the evening in August.)

M. Jane Young, a folklorist, and Ray A. Williamson, an astronomer, have made a tentative identification of some parts of the Chief of the Night. They suggest that the five stars of the heart of the Chief are Vega, Deneb in the NORTHERN CROSS, and other stars in the NORTHERN CROSS.

They suggest that his left shoulder is Arcturus, his elbow is Spica, his forearm is Antares, and his hand is stars in SAGITTARIUS. His left shoulder, elbow, and forearm are called Three Running Down to the Horizon, while the hand is called The Eight Ones.

The Chief's right arm is "four stars, one of them smaller than the other three, chasing each other towards the horizon," a group that Young and Williamson identify as ORION's belt (and the small star beside it). The belt stars are also called In a Row.

Harrington writes that he did not see all of the Chief of the Night when he watched the sky with Nachapani, so he never saw the Chief's huge legs. He does note, however, that "some part of the body of the Chief of the Night is perpetually visible, guarding the earth throughout the year."

Western Pueblos: Hopi

The Hopi, occupying the westernmost of the Pueblos, were farmers who developed a complex ceremonial life. Traditionally, the women made pottery and baskets, and the men hunted and wove cotton cloth. They call themselves *Hopitu-shinumu* (Peaceful People). Their language is part of the Uto-Aztecan language family. The Hopi now live on their reservation, which is surrounded by the Navajo Reservation, in Arizona.

Stars and Constellations

Alexander M. Stephen lived with the Hopi for several years in the 1890s, recording what he heard and saw in his journal. He noted that the appearances of PLEIADES and ORION were used in timing various rituals, and that these two constellations appeared on a star effigy on the War Chief's altar during the winter solstice ceremony. He also saw a design of the Milky Way drawn on the south side of the War Chief's house.

Stephen identified *Soñwuka* (untranslated) as the Milky Way; *Tala´shohü* as Big Star, the Morning Star; *Chavau´wutakamü* (Clustered) as the PLEIADES; *Wutom´kamü* or *Hotumkono* (Strung Together, Like Beads on a String) as ORION; and *Wu´yok sho´hü* as Broad Star, Aldebaran. The PLEIADES, the BIG DIPPER, and other stars were important in

Fig. 10.8. A Hopi mask used in the winter solstice ceremony. The small dots over the eyes represent the PLEIADES and the large dots on the cheeks represent the BIG DIPPER. (Illustration after Dorsey and Voth, *The Oraibi Soyal Ceremony,* plate 24.)

Fig. 10.9. A design from a Hopi *kiva* showing (*a*) ORION, (*b*) PLEIADES, (*c*) Morning Star, and (*d*) moon. (Illustration from Parsons, Hopi Journal of Alexander M. Stephen, fig. 143.) Reprinted with permission.

ritual and ceremony. They appeared on Hopi masks, *kachina* dolls, and other ceremonial objects.

Constellations do not occur often in recorded Hopi myths. In the creation myth, Coyote carries a heavy jar. When he becomes weary he opens it, releasing sparks and shining fragments that become the stars. There is also a story about how a *kachina* and his wife became stars.

Fig. 10.10. Hopi *kachinas:* (*left*) Coto, the Star *Kachina,* with three stars on top of the mask, a star and crescent moon on the face, and stars on the forearms and legs; (*right*) Yuña, the Cactus *Kachina,* with star and (possibly) the PLEIADES on the face and stars on breast and arms. Native Americans in the Pueblos and elsewhere often used crosses to represent stars. (Illustration after Fewkes, *Two Summers' Work in Pueblo Ruins,* plates 28, 49.)

MOUNTAINS, MESAS, AND PLAINS

The ancestors of the Navajo and Apache were Athapaskan-speaking people who are thought to have migrated from northwestern Canada into much of what is now Arizona and New Mexico, plus adjoining sections of Colorado, Oklahoma, and Texas, no earlier than A.D. 1000 and no later than the 1500s.

In contrast to the Pueblo farmers, who were concentrated along the Rio Grande and its tributaries, the Navajo and Apache were mainly hunters and gatherers who ventured over a wide area. The Navajo currently live on the Navajo Reservation centered in northeast Arizona and in several nearby communities. The Apache live on the Fort Apache Reservation in Arizona and on the Jicarilla, Mescalero, and Tonto-Apache Reservations in New Mexico.

The culture of both groups is rich in mythology. Coyote is prominent in many stories, sometimes performing a useful service like securing fire but often engaging in foolish or disgraceful behavior. Supernatural power resides in various natural forces, such as plants and animals. Illness comes about when individuals do not show proper regard for these forces. The extended family was and is still the basis for social organization.

Navajo

The Navajo call themselves the *Dineh* or *Diné* (the People). They settled in parts of what is now Utah, Colorado, Arizona, and New Mexico, living in small family groups rather than as a large tribe. By the late 1700s they had developed a culture of hunting, gathering, farming, and sheepherding. It was not until the 1900s, when the U.S. government mandated a tribal government, that the Navajo became more formally organized.

Ceremonies were and still are important to both the Navajo and the Pueblo Indians, but the two groups differ in several ways. The Navajo hold ceremonials (composed of a group of ceremonies) to cure individuals of disease or to secure blessings rather than to entreat the arrival of rain or enhance fertility. Thus, most Navajo ceremonials are held as they are needed rather than at set times of the year. These ceremonials are performed by chanters, also called singers, who have apprenticed with older and more experienced practitioners. This system contrasts with that of the Pueblos, where members of an established priesthood or religious society lead the rituals.

Despite these differences, there are similarities in the celestial motifs of these two groups. One astronomer believes that the Navajo learned some constellations as well as the use of stars in sandpaintings, ceremonial rattles, and rock art from the people of the Pueblos.

For the Navajo, the stars are living supernatural beings, and they embody the rules for living. First Woman used the stars as a tablet upon which to write the rules that would govern the behavior of humans. Water and sand were unsuitable because they constantly changed, but if the laws "are written in the stars they can be read and remembered forever." Indeed, each group of stars symbolizes a law by which to live and is a potent reminder of important values.

Pictographs on the ceilings of caves in Canyon de Chelly and other areas show stars or constellations; some of these pictographs were likely created between 1700 and 1864. Constellations also appear in sandpaintings or "dry paintings," murals that are created during Navajo ceremonials. Chanters and their assistants make these murals on the ground with colored sand and other objects to invoke Holy People. Because the goal of most ceremonials is to heal, the exact configuration of the constellations is of less importance than their healing power. Thus, individual chanters may create sandpainting constellations that are not exactly like their counterparts in the sky. In fact, there is no need for such accuracy.

The chanter Son-of-the-late-Cane, who drew a star map in the early 1900s, places summer and winter constellations in one illustration. He gives many constellations human forms, with arms, legs, a head, and other body parts, as well as a fire or "ignitor." His drawing, like those of constellations in ceremonials, does not show the actual configuration of stars in the sky.

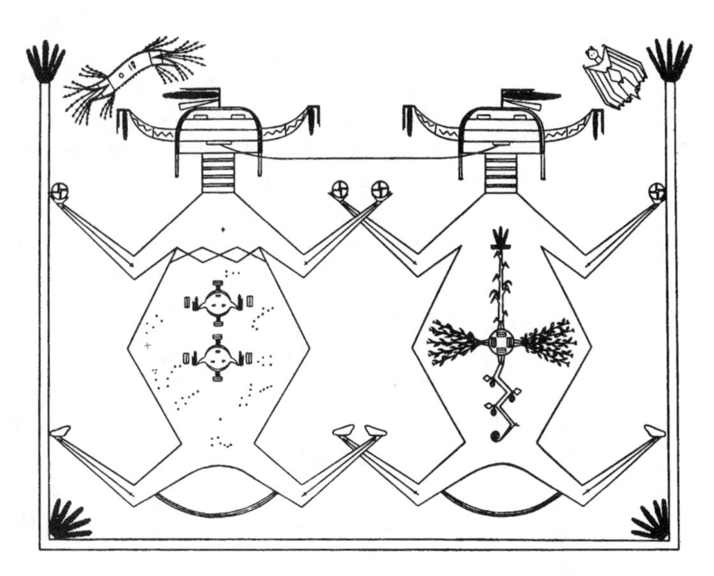

Fig. 10.11. A sandpainting from the Navajo Shooting Chant shows Father Sky (*left*), and Mother Earth (*right*). Father Sky has constellations on his body and the Milky Way (crosshatching) across his chest. The moon lies above the sun in the middle of his body. The constellations illustrated around the moon and sun on Father Sky's body are (*clockwise from upper right*): Rabbit Tracks (tail of SCORPIUS), above the moon; First Slim One (ORION), to the right of the moon; Man with Legs Ajar (CROW), to the right of the sun; His Soft Feather (unidentified), figures to the right of and below the sun; *Dilyéhé* (PLEIADES), left of and below the sun; Male One Who Revolves (BIG DIPPER), by the sun's horn; North Star on Top, a single cross; Female One Who Revolves (generally CASSIOPEIA, but some chanters identify it as the LITTLE DIPPER) above cross. (Illustration from Newcomb and Reichard, *Sandpaintings of the Navajo Shooting Chants*, fig. 5.) Reprinted with permission.

There are eight primary Navajo constellations and over three dozen secondary constellations and individual stars. The primary constellations include Male One Who Revolves (the BIG DIPPER and nearby stars); Female One Who Revolves (CASSIOPEIA); *Dilyéhé* (PLEIADES); First Slim One (ORION); Man with Legs Ajar (CROW); First Big One (the front section of SCORPIUS); Rabbit Tracks (the tail of SCORPIUS); and Awaits-the-Dawn (the Milky Way). These constellations are shown in a drawing of Father Sky, a powerful figure that appears in the ceremonials Male Shootingway, Nightway, Plumeway, Earthway, Big Starway, and Blessingway. In most instances, Father Sky is shown hand in hand with Mother Earth.

Stars were also important in farming, with ORION, the BIG DIPPER, and PLEIADES marking points in the agricultural cycle. The Navajo planted when ORION was setting at dusk (early May) and when the BIG DIPPER was horizontal in the evening sky (mid- to late May or early June). They stopped planting when they could see PLEIADES in the eastern sky at dawn (mid-June).

Two books are of particular interest with regard to Navajo star lore: *Earth Is My Mother, Sky Is My Father* written by Trudy Griffin-Pierce, an anthropologist and artist; and *Starlore Among the Navajo*, written by Father Berard Haile.

Creation Story

The Navajo creation myth is central to star beliefs as well as to Navajo beliefs in general. Although there are many versions of the creation story, they all involve the emergence of a primordial people onto the surface of the Earth. First Man and First Woman build a hogan (an earth-covered lodge) in which a group of creators fashions a world in miniature that becomes the Navajo world. Their actions set the world in order and provide every living thing with a proper place and a relationship to every other living thing.

Black God (or, in other stories, First Man and First Woman, or Fire Man and First Woman) places stars in the sky to represent supernatural beings as well as animals of the Earth. Coyote, in his usual role of trickster, intercedes in the orderly creation of the heavens. The following segment of the creation story explains why some stars are named and appear in patterns while the others are nameless and scattered.

Black God strode into the hogan of creation with the Pleiades on his ankle. He stamped his foot several times, and the star group jumped to a different location each time—his knee, his hip, his shoulder, and finally his left temple. "There it shall stay," he told the Creators. In this way he established his mastery over the stars and secured the right to arrange them in the sky.

Black God took mica crystals from the fawn-skin pouch that he carried and began placing them in the heavens. He made a group called Man with Legs Ajar in the eastern sky and placed Horned Rattler, Bear, Thunder, and First Big One in the southern sky. In the north, he created Male One Who Revolves and Female One Who Revolves, who whirl around the star that stands still. He made First Slim One, Pinching Stars, Rabbit Tracks, and *Dilyéhé*. He gave each group a star that would ignite it and provide light. Then, he placed stars across the sky to make the Milky Way.

He was ready to sit down when Coyote came near and exclaimed, "What have you done!" Coyote snatched the pouch from Black God, opened it up, and blew the rest of the stars into the sky in an unorganized array. "The sky looks nice that way," Coyote said. He held back one star, which he named Coyote Star, and placed it in the southern sky.

(Adapted from Haile, *Starlore Among the Navaho*, 1–3.)

BIG DIPPER: Male One Who Revolves
CASSIOPEIA: Female One Who Revolves

Náhookǫs Bikąʼii, Male One Who Revolves—also called Male Revolving One, Whirling Male, and Cold Man of the North—is the BIG DIPPER, and *Náhookǫs Baʼáadii*, Female One Who Revolves—also called Female Revolving One and Whirling Female—is CASSIOPEIA. The North Star, *Náhookǫs Bikąąʼdi* (North Star on Top), belongs to Male One Who Revolves but is considered the ignitor for both constellations and is called their campfire.

Another translation gives a more eloquent picture of these stars. The BIG DIPPER is Revolving Male Warrior with His Bows and Arrows; CASSIOPEIA is Who Carries the Fire in Her Basket; and the North Star is the Main Pole Which Holds All, that is, the main pole that holds up all of the sky.

Max Littlesalt, a Navajo who reported information that was generally known among the Navajo in the 1950s, says that Male One Who Revolves has feet and legs (stars in the BIG DIPPER's handle), a body (stars in the bowl), and head feathers (stars past the bowl of the BIG DIPPER). One of Griffin-Pierce's chanters says that Male One Who Revolves has a broken leg, the bend in the BIG DIPPER's handle.

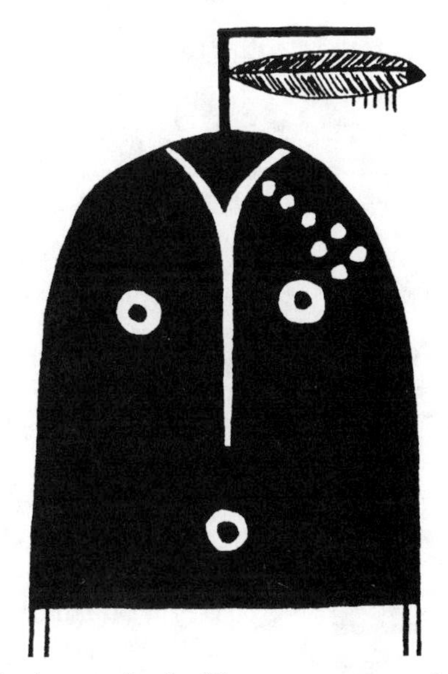

Fig. 10.12. The Navajo Black God with the PLEIADES on his temple. (Illustration after Haile, *Starlore Among the Navajo*.)

Navajo chanters have offered varying interpretations of Male One Who Revolves and Female One Who Revolves. Some chanters view them as wise leaders or First Man and First Woman. Others see them in terms of the hogan, the dwelling that is the focus of Navajo life; the two constellations are a married couple who stay at home and fulfill their responsibilities, people who value traditional family life. The North Star is their central fire.

These constellations embody values and practices. The image of just one couple using a fire that is written into the northern sky is linked with Navajo laws that proscribe two couples from living in the same hogan and from cooking over the same fire. Such laws help to maintain peace in the hogan and among families. Thus, the two carefully balanced constellations symbolize the *hózhǫ́* that is so important in Navajo life. *Hózhǫ́* cannot be translated directly but involves beauty, harmony, goodness, and well-being.

Other chanters interpret Male One Who Revolves and Female One Who Revolves as representations of the bows that the Navajo twins Monster Slayer and Born-for-Water use to slay monsters that once inhabited the Earth. Differences in interpretation arise because a chanter may specialize in a particular ceremonial, and each ceremonial has a certain theme, such as family life or harmony. The variations in interpretation provide different insights into how one is to live and are not seen as contradictory.

PLEIADES: *Dilyéhé*

Although there is no agreed-upon translation of *Dilyéhé*, one chanter calls it Sparkling Figure and another, Pin-like Sparkles. Both phrases well describe PLEIADES, with its tightly clustered pinpoints of light.

There are many interpretations of *Dilyéhé*. One of Griffin-Pierce's chanters describes the group as old men engaged in a game of dice; when one man loses, he becomes angry and tosses away the dice. The constellation also represents that man as he leads his wife and children from the event.

Dilyéhé is also associated with the seven Hard Flint Boys who assist Monster Slayer and use the Milky Way and rainbows as paths. These seven boys are running and dodging. "Before the Navahos [*sic*] had guns, they used to practice shooting

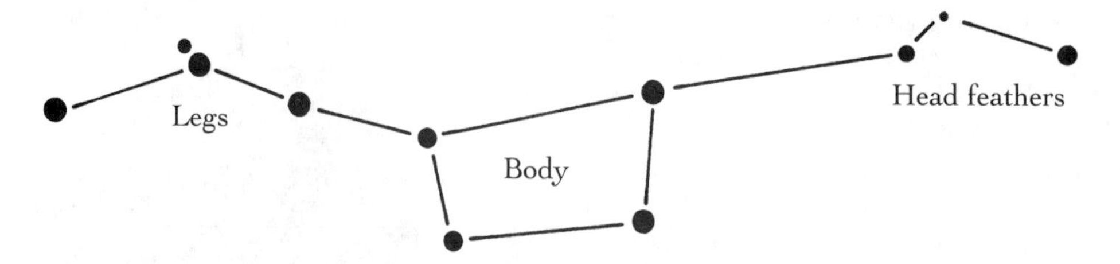

Fig. 10.13. The Navajo constellation Male One Who Revolves (BIG DIPPER). One chanter says that the man's leg is broken, thus producing the bend in the dipper's handle.

arrows and running and dodging at the same time; that way they would learn to put another arrow in their bow while they were dodging and making it hard for the enemy to hit them," explains Max Littlesalt.

Since *Dilyéhé* is the constellation that Black God uses to establish his mastery over the stars, this star group is also viewed as representing all of the stars in the sky.

Spider Woman taught the Navajo how to make string figures of *Dilyéhé* and other constellations in order to promote clear thought and concentration as well as to remind them of their relationship to the star beings.

HYADES: Hard Flint Women, Pinching or Doubtful Stars, Wrestlers

Dilyéhé is also a woman who marries First Slim One (or, Coyote Man) and their children are *Baalchini*, a pair of stars in the lower branch of HYADES known as Pinching or Doubtful Stars. In another story, *Dilyéhé* and First Slim One are not married, but *Baalchini* are twin daughters that they both claim. The twin associated with ORION is called Old Age That Goes About and the one associated with PLEIADES is called Happiness Caused by It. These two women are often mentioned in song and prayer.

The pair of stars in the HYADES is also called two Hard Flint Women—mothers of the Hard Flint Boys—who are pulling each other's hair and arguing over a gambling game.

In yet another instance, they are two boys who are wrestling in order to prevent the other from joining boys who are practicing dodging. Near the two boys is a red star, an old man who chides the boys because they are not paying attention to their practice. The star (likely Aldebaran) is "the fire of the twin stars," according to Littlesalt.

CROW: Man with Legs Ajar

Hastiin Sik'ai'í, Man with Legs Ajar, Man with Feet Ajar, or Old Man with Legs Spread, is an irregular square in CROW. Although Man with Legs Ajar does not contain bright stars, it stands out in an otherwise dark section of the sky.

The Navajo believe that when the Creators made the world, they made the months, too, assigning each month a feather or headdress that announces its arrival. Some of these feathers are natural occurrences like rain or wind while others are constellations. Man with Legs Ajar is the announcer or feather for the month called Parting of the Seasons (November), the second month in the Navajo year. When Man with Legs Ajar appears in the morning sky, it is time to get ready for winter.

SCORPIUS: *'Átsé'etsoh,* First Big One

'Átsé'etsoh, First Big One, is made up of stars in the forepart of SCORPIUS. Black God arranged this constellation in the southwest sky. First Big One is considered by one of Griffin-Pierce's chanters to be an old man who needs a cane to walk; Cane of the First Big One is also made up of stars in SCORPIUS. First Big One is the feather for the month Great Wind (December), and appears in the predawn sky at that time. During Great Wind, the Navajo made digging sticks and moccasins and began telling sacred stories.

Milky Way: *Yikáísdáhí,* Awaits-the-Dawn

Awaits-the-Dawn, the Milky Way, is the feather for Crusted Snow (January). When the Milky Way appeared in the morning sky, it was time for young men to learn sacred stories and to apprentice to chanters. It was also time to prepare for the impending growing season.

The Milky Way is considered a symbol of the cornmeal that First Woman scatters as she prays in the morning and a reminder that people should pray, too. The starry band is also mentioned as meal that Coyote spills as he takes a burro across the sky, and as ashes that Coyote drops as he steals some bread baked in an outdoor oven.

SCORPIUS's Tail: *Gah heet'e'ii,* Rabbit Tracks

The constellation *Gah heet'e'ii,* Rabbit Tracks, is composed of stars in the tail of SCORPIUS. Rabbit Tracks is the feather for the month called Baby Eagle (February), a time when the First Chief of the Winds wakens the plants and animals and the first plants appear. Rabbit Tracks does indeed appear in the morning sky in February.

This constellation shows the esteem in which the Navajo hold the rabbit, which was historically an important source of food. The rabbit in the sky symbolizes all game animals and honors the transfer of life from animals to humans. The Navajo did not hunt for rabbits when Rabbit Tracks appeared vertically in the sky, because the newly born animals were then too young to take care of themselves, but when the constellation moved into a more horizontal position after twilight (that is, in August), then it was permissible to hunt once more.

Von Del Chamberlain, a contemporary astronomer who has written extensively about Navajo and Pawnee constellations, points out that the constellation Rabbit Tracks looks like actual rabbit tracks. When a rabbit hops, it lands on one front foot and then the second front foot; then it swings the two hind legs forward and around the front feet, making a pair of prints beyond. The hind prints are a pair of stars in the tail of SCORPIUS. The prints of the two front feet are stars below or to the left, depending on when the constellation is being viewed. Chamberlain has also identified what he thinks are petroglyphs of rabbit tracks in rock art in New Mexico and Utah.

ORION: *'Átsé'ets'ózí,* First Slim One

'Átsé'ets'ózí, First Slim One or First Slender One, is ORION; sometimes just the belt and sword of ORION are named as the constellation. Max Littlesalt says that First Slim One has a stick, a curved line of stars to his right, and a basket, a circle of stars above him.

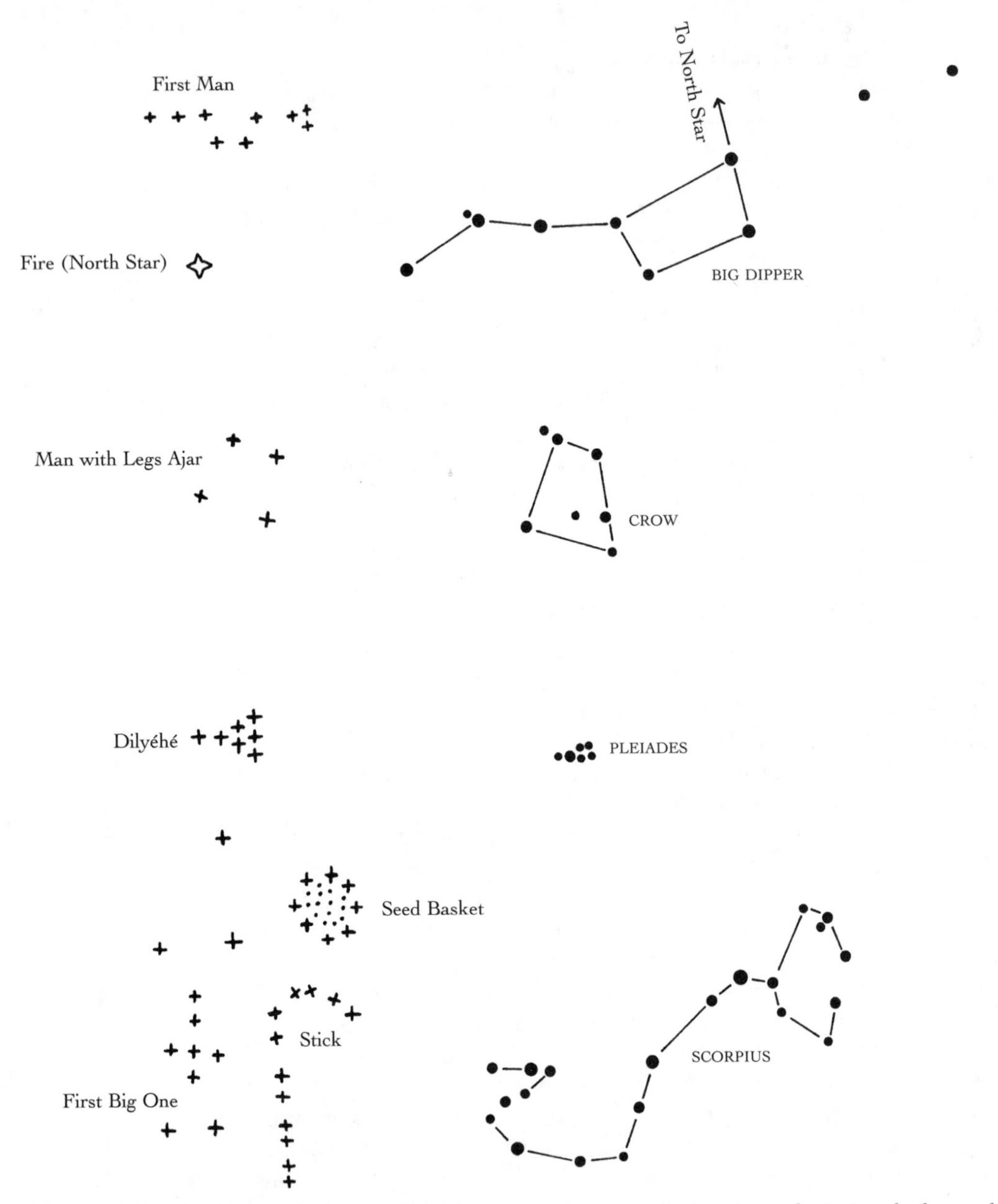

Fig. 10.14. Navajo chanters do not consider it necessary that constellations in sandpaintings look exactly like their counterparts in the sky. Compare the constellations from a Beadway Ceremonial sandpainting (*left*) with their celestial counterparts (*right*). In the Beadway depiction of First Man and the fire that he and First Woman share (*top*), the BIG DIPPER faces away from the North Star, rather than toward it. In Man with Legs Ajar (*second from top*) and *Dilyéhé* (*third from top*), the shapes are similar to their star counterparts. SCORPIUS is First Big One (*bottom*), shown here with his stick and seed basket, but the depiction of First Big One in the Beadway Ceremonial and the arrangement of stars in SCORPIUS are quite different. (Beadway drawings from Wyman, *The Sandpaintings of the Kayenta Navaho,* fig. 49.)

Stars and Planets

Sǫ'tsoh, Big Star, and *Sǫ'doo Ndízídí*, Morning Star, are Venus. *Sǫ'tsoh* is also translated as North Star and Evening Star. Coyote Star and No-Month Star—the one star that Coyote placed carefully in the sky—is likely Canopus or Antares. Red Star is Antares or Aldebaran.

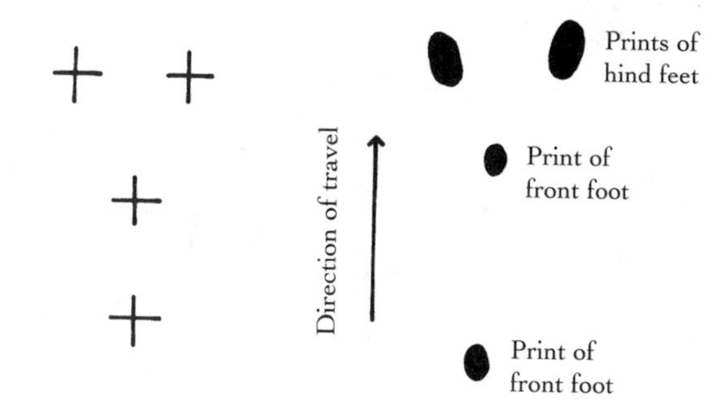

Fig. 10.15. The Navajo constellation Rabbit Tracks (*left*) as it appears in various ceremonials looks very much like the tracks of a rabbit (*right*).

Unidentified Constellations

Griffin-Pierce lists over two dozen unidentified constellations from various sources, including Butterfly, Porcupine, Bear, Red Bear, Thunder, Big Snake, Ram, Horned Rattler, the Crown, Two Stars Hooked Together, Mountain Sheep, Fire God, Monster Slayer, Born-for-Water, Corn Beetle, Turkey Tracks, Flash Lightning, the Ducks, Red Heavens, Lark Who Sang His Song to the Sun Every Morning, Wolf, Eagle, Lizard, Southern Cross, and Harvester. A figure shown in several sandpaintings, the Seed Basket, may not be an actual constellation. Other sources list Gila Monster, Turkey, Eagle, Swallow, Badger, Measuring Worm, Big Tail Feather, and Soft Feather. Deer Star, Dawn Feather, Feather of the Southern Blue, Feather of the Evening Twilight, Feather of Darkness, Horned Star, Pronged Star, Smoky Star, Out of Sight Star in the East, and Large Bright Stars Scattered Over the Sky also have not been identified.

Mescalero Apache

The Mescalero Apache lived in a large part of present-day southern New Mexico as well as sections of Texas and Mexico. Men stalked deer, elk, and bighorn sheep in the mountains and buffalo, antelope, and cottontail on the plains. Women collected a variety of wild plant foods, including agave, mescal, prickly pear cactus, mesquite, screw bean, various berries and greens, nuts (piñon, acorns, and walnuts), and grasses. The Mescalero farmed in some areas.

The Mescalero say that the Eternal Power or Creator made the Earth and the heavenly bodies, including the stars. The Creator then assigned to the stars the task of guiding the Mescalero. This guidance occurs in both the literal sense, by providing beacons for navigation at night, and the figurative sense, by embodying rules for proper living.

Milky Way: The Scattered Stars

The Milky Way is *Sųųs naatsdił* (Scattered Stars). According to myth, one of the Twin Warrior Gods carries a container of seeds. When the twins get into a fight, the container pops open and the seeds splash across the sky, becoming stars.

This story provides more than a simple explanation of how the Milky Way appeared in the sky. The Mescalero Apache believe that the Creator made a perfect, balanced universe. In this universe, certain groups—including those within the matrilineal

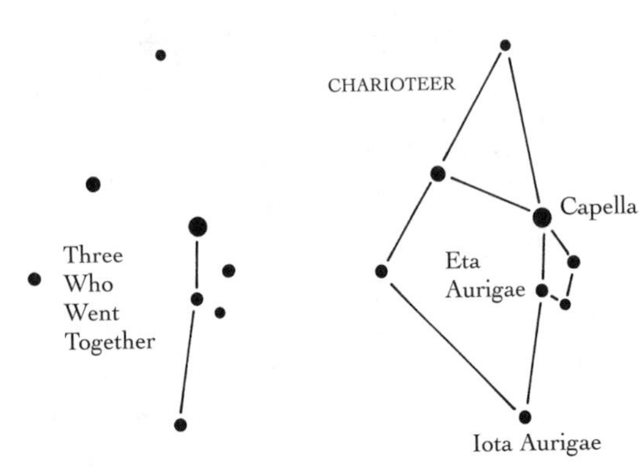

Fig. 10.16. The Mescalero Apache constellation Three Who Went Together (*left*), and CHARIOTEER (*right*).

family—are not supposed to fight with one another. By doing so, the Twins upset the balance, and their disruption is now visible in the sky. The Scattered Stars remind the Mescalero Apache that it is important to get along with one another, lest they contribute to disharmony in the world. For these people, as for the Navajo, the stars are a reminder of proper living.

CHARIOTEER: Three Who Went Together

Three Who Went Together—that is, who died at the same time—also offer a lesson for humans. Once, when two sisters loved the same man, they both married him and established a peaceful home. These sisters showed compassion within their household, to other members of the tribe, and to all creatures. When one sister was about to die, the Creator decided that the sisters should not be parted and placed all three humans in the sky as a constellation. This group of stars serves as an example of properly led lives. Three Who Went Together is composed of three stars in the CHARIOTEER (Capella and two nearby stars). Capella is *Sųųs biné*, also called Morning Star. Singers use it in the timing of the girls' puberty ceremony.

Other Stars and Constellations

Nahakus, which has two meanings (revolving around a central point, and falling) is the BIG DIPPER. Apache singers also use this constellation to time the girls' puberty ceremony.

Star-That-Does-Not-Move is the North Star.

Jicarilla Apache

The homeland of the Jicarilla Apache stretches from the southern Rocky Mountains to the Great Plains in what is now the northeast quarter of New Mexico and southeast Colorado. Although Native peoples farmed in some partly settled areas, hunting and gathering provided the bulk of the diet. Hunters secured mountain sheep, deer, elk, antelope and buffalo, supplementing this large game with prairie dog, porcupine, squirrel, chipmunk, rabbit, and beaver. The women secured a variety of wild plant foods.

Stars and Constellations

In the Jicarilla Apache creation myth, the union of Earth Mother and Black Sky produces *hą·šč̓ín* a supernatural being. Deep in the heart of the Earth, *hą·šč̓ín* makes Ancestral Man and Ancestral Woman, as well as animals. Eventually the descendants of this couple emerge as they search for light; the power of *hą·šč̓ín* then pervades many objects in the natural world, including stars. These objects (*hactin* in the following narrative) are very powerful and are not to be spoken of lightly.

Although the Two-Fighting Stars are not identified as the Twin Warrior Gods, whose fight with each other creates the Milky Way, it seems a logical association.

The *hactin* once lived on the earth in a large lodge, and at that time, all people were exactly alike. But when the *hactin* left the earth to put out what looked like a fire in the sky, they changed people so that each person was different. There are many *hactin* of different kinds. The star *hactin* include Big Dipper *Hactin,* North Star *Hactin,* Two-Fighting-Stars *Hactin* (Castor and Pollux), Three-Vertebrae-Stars *Hactin* (belt of Orion), two Morning Star *Hactin,* and two Evening Star *Hactin.*

When the Big Dipper *Hactin* went into the sky, he said, "When you look up at night, count the stars of the dipper first. Whenever you see only six stars there, something is going to happen, something bad. This dipper will stand as it is as long as the sky stands, as long as the earth stands. When you grow old, you must teach this ceremony to the young people."

(Adapted from Opler, Myths and Tales
of the Jicarilla Apache Indians, 147–161.)

PLEIADES: Six Hunters and a Little Girl

This story is told to children with the warning that they should not play at being a bear because the pretense can come true. The Jicarilla Apache also say that when the time comes that only six stars in the PLEIADES are visible, the bear will again be the enemy of humans. The Blackfoot and Kiowa of the plains tell similar bear-girl stories.

A girl played a game with other children in which she was a bear who would try to capture them. The first time she emerged from a hole in the arroyo, she was a girl, but each time after that she began changing into a bear. Finally, she was a bear in shape and thought and killed all the children except for her little sister, who escaped. The sister ran back to the camps to warn the others, and some people fled, but the bear killed those who didn't believe the little girl. The sister hid under a basket but the bear discovered her and took her back to the hole as a servant.

The sister's father and five other hunters had been away when all of this occurred, and the little sister kept watch for them to warn them. She provided the bear with food and water so the bear would not have to leave its hole. When the hunters finally returned, the little girl snuck out while the bear slept and told her father what had happened.

The father told the girl to bank the bear's hole with firewood and light the wood on fire. To blind the bear, the hunters threw deer fat into the bear's face when it came to the opening. Not knowing whether the bear lived or not, the hunters and the sister ran as far and as fast as they could. They prayed for help and a cloud took them to the sky, where they became seven stars that rise in the east and travel west, where their grandmothers live. They became the Pleiades.

(Adapted from Opler, Myths and Tales of the Jicarilla Apache Indians, 115–116.)

WESTERN UPLANDS

The Yuman-speaking tribes of the Southwest's Western Uplands were largely hunters and gatherers in an arid region dominated by sagebrush, creosote bush, and piñon-juniper woodland. The Havasupai, Walapai, and Yavapai who lived there developed a culture that was in many ways similar to that of the Paiute directly north of them across the Colorado River.

Today the Havasupai live on the Havasupai Reservation in Havasu (or Cataract) Canyon in Arizona; the Walapai live on the Hualapi Reservation in Arizona, and the Yavapai live with other tribes on the Yavapai Reservation and several other reservations in Arizona.

Havasupai, Walapai, and Yavapai

The Walapai relied on available game in the basins (deer, antelope, rabbit, and rodents) and the higher elevations (mountain sheep and birds) as well as on the edible portions of cacti, mesquite, paloverde, creosote bush and other vegetation.

The Havasupai, who split off from the Walapai, were both farmers and hunter-gatherers. In the summer they raised corn, beans, and squash along Cataract Creek, a tributary of the Colorado River that flows into the Grand Canyon. By mid-October they moved to the Coconino Plateau in search of game and wild plant foods.

The Yavapai, whose territory included desert and mountains, had access to a wider range of foods, including wild greens, mulberry, wild grape, acorn, piñon, wild onion, grasses, and yucca. They hunted antelope, deer, rabbit, and other game and grew some crops.

The people of these tribes generally lived in dome-shaped structures covered with thatch, bark, or earth in the winter and used rectangular structures and brush-covered shades in the summer.

Although star material for the individual tribes is sketchy, when grouped together the material reflects events throughout the year and reveals aspects of life on the Western Uplands.

Stars and Constellations

The dawn rising of the Hoop, a circular constellation of five stars, designates the beginning of the first month of the Havasupai year, which starts in mid-November. No classical constellation exactly fits the various descriptions of the Hoop, but the constellation could be the NORTHERN CROWN, part of the HERDSMAN, or (less likely) the Sickle in LEO. The Yavapai also name a Hoop constellation. The Hoop is part of the hoop and pole game, a favored pastime throughout Native North America. In the game, one person rolls a hoop onto a playing court, and then two players throw poles at it. The players try to position their poles so that the hoop falls onto them. The game requires skill.

The second Havasupai month (beginning in mid-December) starts with the rising of *I'pedja'ália*, Coyote Carrying a Pole (or piece of wood) on his shoulder. Coyote and his pole are SCORPIUS; perhaps Coyote is the bright star Antares and the stars of SCORPIUS's body are the pole. Coyote's travels mirror the movement of the constellation from the east to the west and its reappearance in the eastern sky. The brief story goes this way:

> Mountain Lion went to Coyote's camp and ate his food and Coyote was angry. Mountain Lion told Coyote to build a fire. Coyote did not want to burn his pole and so he "went to the west and came back from the east."
>
> (Adapted from Spier, Havasupai Ethnography, 167.)

The fourth Havasupai month (mid-February) begins with the appearance of Very Cold Star, Altair. Altair appears in the eastern sky before dawn from mid-January through February. The Walapai call this star Feces Interrupted (from the cold).

The Yavapai specify the dawn rising of Red Feathers, a large red star, as the beginning of the fourth month. The Walapai Eagle Feather Headdress is also a red, dawn-rising star in winter, and the Havasupai recognize the constellation *Tcĭ'ŭt,* Down Feather. These stars, whether the same or different, have not been identified. Eagle feathers have been important in ritual in many tribes, including the Yavapai, who believe that eagles carry messages to the spirit world.

The Yavapai call the PLEIADES *Hecha* and use its dawn rising to mark July. A constellation including either ORION's belt or the belt plus Betelgeuse and Rigel is called Mountain Sheep. The Yavapai describe a Hand constellation, but the identification is not known; it may be the same as the Hoop.

The Walapai call a group of stars *Hiné* (Giant Cactus-Gathering Crook). It seems likely that this constellation is the same as the Tohono O'odham constellation Cactus Hook and the Seri constellation Saguaro-Gathering Hook, both identified as the BIG DIPPER (see page 206 and page 208).

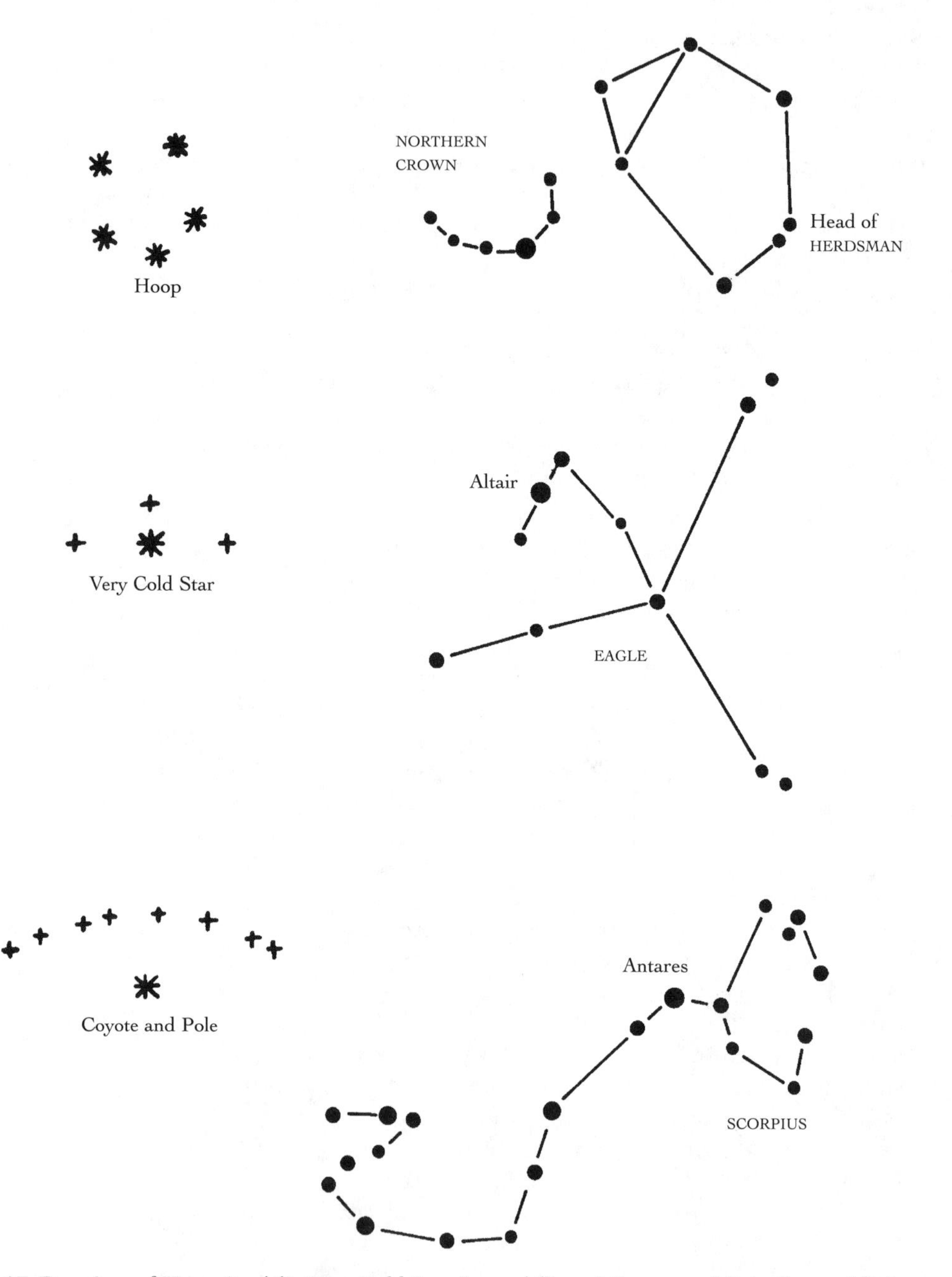

Fig. 10.17. Drawings of Hoop (*top left*), Very Cold Star (*center left*), and Coyote and Pole (*bottom left*) from Havasupai sketches. The Hoop is unidentified, but it may be the NORTHERN CROWN or the head of HERDSMAN. Very Cold Star is Altair, the middle star in the head of the EAGLE. Coyote is probably Antares, and the pole the line of stars forming SCORPIUS's body. (Havasupai drawings from Spier, Havasupai Ethnography, fig. 30.)

SOUTHERN RIVERS AND DESERT

Maricopa and Cocopa

The Maricopa and Cocopa, along with the Mohave, Quechan, and other Yumans of the Colorado and Gila Rivers, farmed along the rich floodplains and supplemented their diet with fish, game, and wild plant foods such as mesquite beans. They did not live concentrated in villages but in single-family structures separated by plots of land. Throughout much of the year they used brush-covered shades for shelter. The Maricopa now live in south-central Arizona and the Cocopa in southwestern Arizona, Baja California, and Sonora, Mexico.

Stars and Constellations

Kutŏ´x̣, who was then eighty-three years old, and Last Star, who was about sixty-five, shared the following Maricopa star lore with Leslie Spier, an anthropologist, in 1929 and 1930. The Maricopa include the movement of the stars in the annual calendar. Kutŏ´x̣ explains how the stars were of use to his people: "The stars (*xomace´*) change in the different seasons, so people will know when plants are ripe . . . They went for giant cactus fruit in the summer [mid-June] when the Pleiades (*Xĭtca´*) appear on the eastern horizon." He also said that when Cipa's Hand appears in the east at dawn, "the first crops will be frozen."

Cipa, the creator, placed his hand on the sky to push it upward and left the imprint of his fingers there. This constellation appears on the horizon in December. Cipa's Hand may be the NORTHERN CROWN or a circle of stars in the HERDSMAN.

The Milky Way is the track along which a deer and an antelope race. The deer tracks are widely spaced, but the antelope tracks are crowded. (See a similar Yokut story on page 149.)

Coyote's Fishing Net is the BIG DIPPER. Coyote holds the net at one end. When he hears fish jumping, he turns around because he is afraid that they will swim by. As he turns, he pulls the net behind him and makes a bend in it (the slight bend in the handle of the BIG DIPPER).

The constellation *Mani´c* (Scorpion) is SCORPIUS.

The Star That Travels Through the Sun is the Evening Star.

Big Star, the Morning Star, is an old woman who changes herself into a star. "When an old person is dying, he always says, 'I am not going to die until the Morning Star appears,'" said one of Spier's storytellers.

The North Star is called *Captain* (Chief), a Spanish name. In the Maricopa/Pima creation story, it forms the pole at one end of Coyote's net.

Two stars—Spier suggests Vega and Antares—are people. One is a Yavapai who wears a cap that is peaked forward (a bright star with a V of stars above it), and the other is a Pima who wears a strip of cloth around his head (a bright star with a circlet of dim stars around it).

The Quail is an unidentified constellation that appears near the Milky Way.

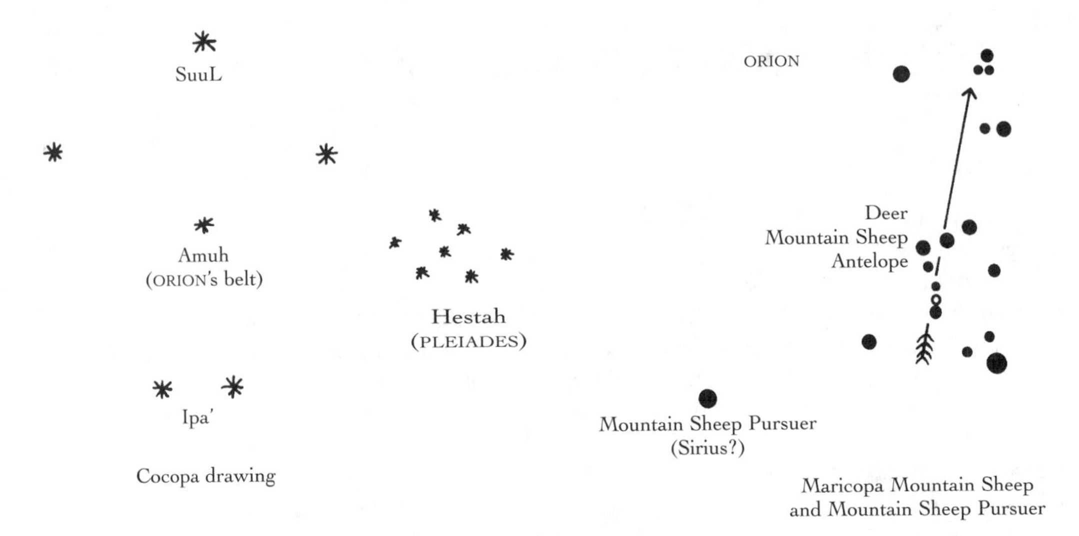

Fig. 10.18. In a Cocopa drawing of ORION and PLEIADES (*left*), *Amuh* (ORION's belt) is a man, and Hestah (PLEIADES) is the woman he wants to marry. Soba^r, who shoots the arrow at Amuh, is not shown. *Suu*L, the tip of Soba^r's arrow, is three stars above the man, and the arrow's shaft is Ipa', likely ORION's sword. The stars of the Maricopa constellations Mountain Sheep and Mountain Sheep Pursuer (*right*) also involve a hunter and an arrow. An antelope, a mountain sheep, and a deer are the three stars from left to right in ORION's belt. A hunter called Mountain Sheep Pursuer (perhaps Sirius) shoots an arrow toward the sheep. The apex of the arrow is three stars above ORION's belt, and the shaft may be ORION's sword. (Cocopa drawing from Gifford, The Cocopa, fig. 7.)

ORION: *Amŏ´s,* Mountain Sheep and Arrow
Sirius: Mountain Sheep Pursuer

For the Maricopa, *Amŏ´s* (Mountain Sheep) is ORION. The three stars in the belt are an antelope, a mountain sheep, and a deer. Three stars above the belt represent the head of the arrow and ORION's sword may represent the shaft. (The Tipai, a Yuman-speaking tribe in California, also recognize ORION's belt as these three animals; see page 165.) Mountain Sheep Pursuer, Sirius, is stalking and has shot Mountain Sheep.

The Cocopa, who, historically, lived downstream from the Maricopa on the Colorado River, describe ORION in a similar way but fold in the PLEIADES and present the love triangle as the root of jealousy. "Star people set an example, in the beginning of the world, for human jealousy. [They were] on the earth first, but ran up into the sky so no one could catch them," says a Cocopa storyteller about these stars.

Hestah, a young woman [Pleiades] plans to marry Amuh, a young man [Orion's belt]. Soba^r [unspecified, but likely Sirius] also wants to marry Hestah. Soba^r lies in ambush and shoots Amuh with an arrow, which, in the sky, has both a shaft and point.

(Adapted from Gifford, The Cocopa, 286.)

Altair: Cold's Cottonwood

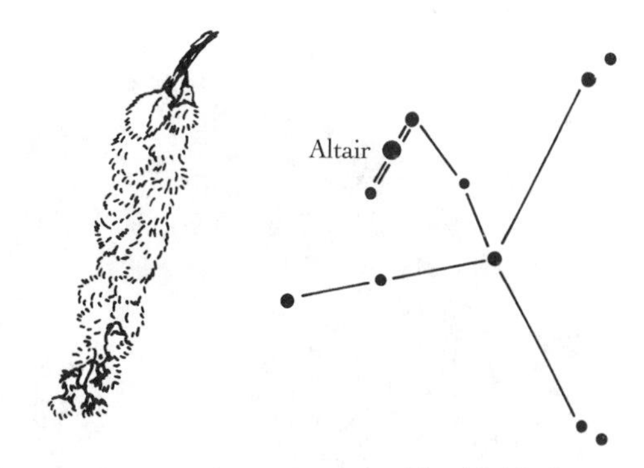

Fig. 10.19. Cottonwood catkin with seeds covered in white fuzz (*left*) and Altair and two flanking stars in EAGLE (*right*).

The Maricopa constellation Cold's Cottonwood (Altair) appears at dawn in January "when the yellow blossoms appear on the cottonwoods," says Last Star. The Maricopan Cold is a character in stories; the star is thought to look like a cottonwood. In fact, the bright star Altair, with a star on each side (forming the head of the EAGLE), does look like the flowering part of a cottonwood, called the *catkin*. The catkin is a narrow clusters of flowers several inches long; the mature seeds are enveloped in a white fuzz that looks like cotton and gives the tree its name.

Kutŏ´x̣ called Cold's Cottonwood a Yavapai, and said, "That is why the Yavapai can stand cold better than anyone else." The Yavapai's upland territory would have seemed cold in comparison to the Maricopa bottomland.

Maricopa/Pima Creation Myth

Although there are creation myths from both the Pima (see page 205) and the Maricopa, the combined story presented below, published in 1907, contains much more celestial information than do the individual stories.

The myth begins before the Earth exists. A whirlwind appears, and Sky and Earth become man and woman. Earth bears twin boys, Earth Doctor and his younger brother. These brothers create many of the stars in the beginning of the world. Then:

> At length the elder changed his brother into a spider and sent him to stretch webs north and south, east and west, and between points. Then a close web was woven outward from the center, where the lines crossed. On this web the earth was built of sediment deposited by the water. The elder brother then shaped the earth. The sky was so close the sun soon dried and cracked the earth into mountain ridges and deep canyons. So he put up his hand and pushed the sky away to its present position. There are five stars where his fingers touched the sky. They are called the hand of God. Then he went about making green things grow, shaping what came forth after subsequent whirlwinds into living things and men and women, teaching them how to build houses, and making the earth fit for them to live upon. So his Pima name is Earth Doctor.
>
> The Brother, ceasing to be a spider, followed and imitated Earth Doctor. Using common clay, he bungled so that misshapen animals were all that he could make. The man he formed had the palm of his hand extending out to the end of his fingers. Earth Doctor rebuked him, so he threw it down hard against the

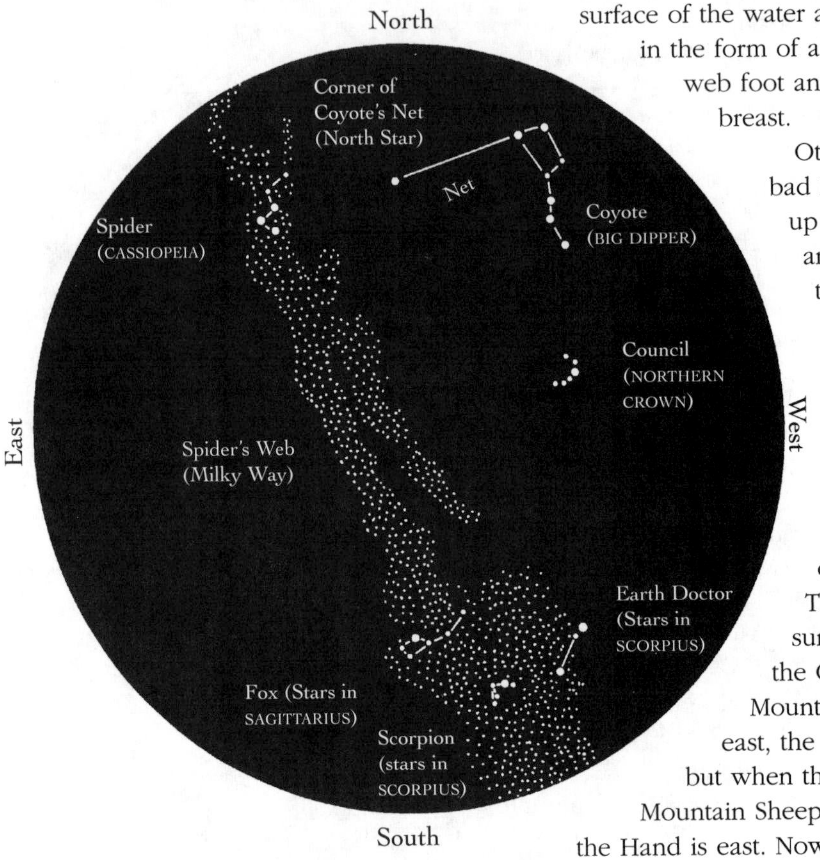

surface of the water and it swam off in the form of a duck, with a web foot and a very flat breast.

Others were so bad he threw them up against the sky, and they remain there. One of these is Gopher (Pleiades); one is Mountain Sheep (Orion), farther east; and one is the Scorpion (part of Scorpius). These go in the sun's path. When the Gopher and the Mountain Sheep are east, the Scorpion is west; but when the Gopher and Mountain Sheep are in the west, the Hand is east. Now all the things that were made then were of the first generation. The first flood came because the Brother made so much trouble and claimed to have more power than Earth Doctor, who at length drove him off the earth. Changing again to a spider, the Brother took refuge in the sky, across which he spun the web of the Milky Way.

North

Corner of Coyote's Net (North Star)

Net

Coyote (BIG DIPPER)

Spider (CASSIOPEIA)

Council (NORTHERN CROWN)

East

West

Spider's Web (Milky Way)

Earth Doctor (Stars in SCORPIUS)

Fox (Stars in SAGITTARIUS)

Scorpion (stars in SCORPIUS)

South

Fig. 10.20. These constellations featured in the combined Maricopa and Pima creation story are visible at the latitude of the Maricopa and Pima in the southern United States at 6 to 7 P.M. in late September and early October. Mountain Sheep (ORION) and Gopher (PLEIADES) are not in the sky at this time of year. Possible depictions of Earth Doctor and Scorpion are shown here; the exact stars in these constellations have not been identified.

Spider tried his own power and called upon it to rise up and wash away the earth. The waters rose, washing away all except the mountains and the representative races and animals that took refuge there. A truce was called; it was agreed that Earth Doctor should have power over the earth, the Brother over the water.

Meanwhile, Coyote and his twin brother Fox were born to the Sun and Moon. Coyote was as inept as Earth Doctor's brother, and soon there was dissension. Another flood was sent to destroy Coyote and all the Earth's inhabitants, but Coyote and some people survived. Because there was no war and very little disease, the Earth soon became overcrowded. Then:

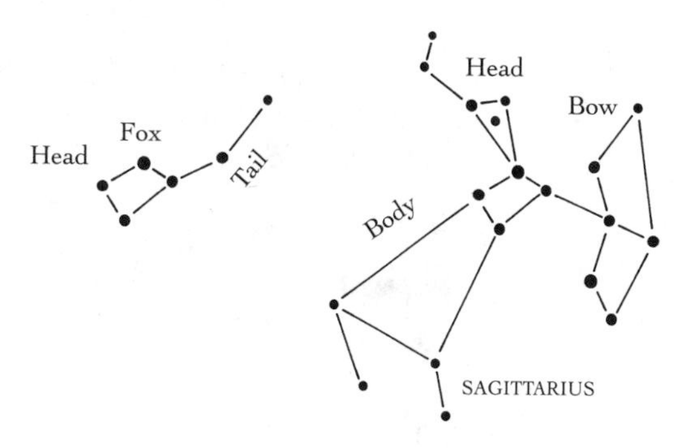

Fig. 10.21. The Maricopa/Pima constellation Fox (*left*) is composed of stars in SAGITTARIUS (*right*).

> A council was held in the skies. The seats of those who were there are in a circle (Northern Crown). They agreed to have the great flood, so there would not be too many people.
>
> The flood that followed continued for four years. The Brother, as Spider (Cassiopeia), sat on the northern end of the Milky Way opposite Coyote (the Big Dipper), who tended his fish net, fastened to the immovable star (the North Star). Fox, Coyote's brother, intent upon some prank, ran along the Milky Way toward the south and fell off, where he may be seen as six stars (in Sagittarius?) arranged like the seven stars that represent Coyote. He is generally seen with his head lower than his tail. But when the Moon is full, she takes him in her lap, and we can see him there as Rabbit (man in the moon).
>
> Earth Doctor took his seat at the end of the Milky Way opposite Fox. Only his head may be seen. It is very large and grand. His face is looking toward the west. The lower end of his long braid of hair is in the Milky Way. When "the moon is dead" and stars are thick, two eagle feathers may be seen in his hair, each composed of three very small stars in a row.
>
> (Reprinted from Culin, *Games of the North American Indians*, 201–203.)

The positions of the story's characters are consistent with the patterns in the sky. Spider (CASSIOPEIA) is opposite Coyote (the BIG DIPPER). At the other end of the Milky Way lies Earth Doctor, the Pima creator (part of SCORPIUS), and Fox (part of SAGITTARIUS). The myth says that only the Earth Doctor's head is visible, and that he faces west; therefore, his braid must be to the east. This description suggests that the Earth Doctor's head is Antares and his braid is stars in SCORPIUS that reach into the Milky Way. The feathers (stars above SCORPIUS?) are visible on nights when the moon is "dead," that is, when the moon is new and thus invisible.

The head of Fox, a small but recognizable rectangle in SAGITTARIUS called the Milk Dipper, is about a forth as large as the BIG DIPPER. The Milk Dipper is closer to the horizon than the tail, so Fox appears upside down, as is appropriate, because he has fallen off the Milky Way. The Moon passes near Fox and "takes him into her lap," meaning that the constellation disappears—dimmed by the Moon's light—and Fox (in the form of a rabbit) reappears on the face of the moon. She is, after all, his mother, and holds him as a mother would hold a child.

Scorpion (part of SCORPIUS), Mountain Sheep (ORION's belt), and Gopher (PLEIADES) are said to "go in the sun's path," meaning that they are seen in the part of the sky through which the sun travels. SCORPIUS and the PLEIADES in TAURUS are part of the classical zodiac, the belt of constellations that lies within the ecliptic. ORION is located nearby.

Pima

The Upper Pima, who call themselves the River People, lived along the Gila River and other waterways and still live in south-central Arizona. In the past, they secured more than half of their food from farming. To the Pima and their neighbors, the Tohono O'odham, the harvest of the fruit of the saguaro, or giant cactus, from June to mid-July marked an important time of year. The cactus grows twenty to forty feet high, and the fruit is located at the top, so some implement is needed for harvest. A long pole, such as the woody skeleton of a dead cactus with a hook or crosspiece tied to the end, is used.

The bright red fruit is eaten raw or dried. It is also used to produce a syrup that can be stored for up to a year. In the past, most of the syrup was kept in small containers in individual homes and was used in a nonalcoholic beverage. The rest was placed in large containers and fermented to make wine for a ceremony and dance that marked the beginning of the year. In the ceremony, the Pima entreated the rain to fall abundantly.

PLEIADES: Running Women

For the Pima, the appearance of PLEIADES on the eastern horizon at dawn meant it was time to harvest the saguaro and prepare for the wine festival. This song, translated by José Lewis Brennan, a Tohono O'odham, describes the morning sky at the time of the ceremony:

Singing to the gods in supplication;
Singing to the gods in supplication;
Thus my magic power is uplifted.
My magic power is uplifted as I sing.

Prostitutes hither running come;
Prostitutes hither running come,
Holding blue flowers as they run.
Talking in whispers they file along.

Along the crooked trail I'm going,
Along the crooked trail going west.
To the land of rainbows I'm going,
Swinging my arms as I journey on.

The bright dawn appears in the heavens;
The bright dawn appears in the heavens,
And the paling Pleiades grow dim.
The Moon is lost in the rising Sun.

With the women Bluebird came running;
With the women Bluebird came running;
All came carrying clouds on their heads,
And these were seen shaking as they danced.

See there the Gray Spider Magician;
See there the Gray Spider Magician
Who ties the Sun while the Moon rolls on.
Turn back, the green staff rising higher.
(From Russell, *The Pima Indians*, 283–284.)

One scholar, in an article about the saguaro wine ceremony, writes that the prostitutes are a metaphor for the PLEIADES. The Crooked Trail is the Milky Way, and the bluebird who runs with the women is Venus. Bluebird is a supernatural being who is able to change his appearance and sex; Bluebird might well appear in the sky as a "star" that does not follow the path of other stars because it is a planet. The song may describe a particular harvest when a new moon ("the Moon is lost in the rising Sun") and Venus (Bluebird) appeared in June at the same time.

Other Stars and Constellations

To the Pima, the Morning Star is *Su´mas Ho´-o* (Visible Star), the daughter of a magician. The North Star is Not-Walking Star. The Milky Way is created when a mule bucks off a load, spilling flour. Coyote eats some of it, but the rest is the Milky Way. In another story, the singularly Piman telling of creation, the Milky Way comes into existence in a much more lyric manner:

Earth Doctor saw that while the moon was yet above the horizon there was sufficient light, but when it disappeared the darkness was intense, so he took some of the water in his mouth and blew it into the sky in a spray, which formed the stars, but the night was still dark. Then he took his magic crystal and, after breaking it, threw it also into the sky to form the larger stars, so the darkness was less intense. Then he sang:

I have made the stars!
  I have made the stars!
Above the earth I threw them.
  All things above I've made
And placed them to illumine.

Next he took his walking stick, and placing ashes on the end he drew it across the sky to form the Milky Way.

(Reprinted from Russell, *The Pima Indians*, 208.)

Tohono O'odham

The Tohono O'odham, or Desert People (formerly called the Papago) occupied a large territory east and south of the Maricopa in what is now Arizona and Sonora, Mexico. This area is composed of a series of basins and ranges, and almost all of it is in the Sonoran Desert. The Desert People, who did not live on a river with permanently flowing water, took advantage of the brief summer rains to grow some crops in washes and otherwise gathered wild foods and hunted. Today the Desert People live in Arizona on a portion of their ancestral land.

Like the Seri to the south, the Tohono O'odham call the BIG DIPPER Cactus Hook.

Creation of the Stars

Coyote is an important and powerful figure. He is also impatient, irritating, and unable to follow simple directions. In the following story, he inadvertently scatters cornmeal across the sky, making the stars.

Coyote made so much noise on earth that Eagle took Coyote's wife to the sky. Buzzard told Coyote what had happened, and said that if Coyote would share what he caught when he hunted, then he, Buzzard, would help Coyote. Coyote agreed. Buzzard took Coyote to the sky, though Coyote almost died several times because he kept looking back at his home, contrary to Buzzard's instructions. Once he was in the sky, no one would give Coyote anything to eat and he thought he might starve. He found a sack of cornmeal in an empty house, and just as he was about to help himself, someone yelled at him to stop. He grabbed the sack in his teeth and the cornmeal spilled out across the sky, becoming the stars.

(Adapted from Saxton and Saxton, *O'othham Hoho'ok A'agitha: Legends and Lore of the Papago and Pima Indians*, 67–73.)

Milky Way: White Beans

The story of the Milky Way concerns a grandfather who hates his grandson and the grandson's selfless response. The Milky Way, say the Tohono O'odham, is made up of white beans.

There was an old man who disliked his daughter's son so much that he belittled him and scolded him and beat him when he did something wrong. A child should not grow up this way, and one day the boy left and did not return. He went to the sky and lay down there. When the grandfather became contrite and grieved over the loss of the child, the boy returned with a present to please him.

"Take this," he said. "It will multiply and you can eat it. When you want to see me, go outside and look into the sky, where I now live." The boy gave the old man some white beans and told him how to plant them.

The boy returned to the sky. He is the Milky Way. The old man planted the seeds and raised white beans. The white bean is called the child of the Tohono O'odham, the Desert People, and it grows even when there is little rain. That is why the Desert People can always live in this place.

(Adapted from Saxton and Saxton, *O'othham Hoho'ok A'githa: Legends and Lore of the Papago and Pima Indians*, 20–23.)

PLEIADES: Homeless Women

The Desert People call the PLEIADES Homeless Women:

Long ago, the Tohono O'odham did not mark the time when a girl reaches puberty. A wise man who lived in a cave taught the people beautiful songs to celebrate this event, and everyone enjoyed the ceremony. But some young women liked it so much that they sang and danced all of the time, wandering throughout the countryside without a home. They finally went to a powerful woman and asked her to help them. The woman turned them into stone and threw them into the sky, where they became the Pleiades. She did this so that everyone on earth would regard them, the Homeless Women, and understand that although it is all right to celebrate, it is not all right to celebrate endlessly.

(Adapted from Saxton and Saxton, *O'otham Hoho'ok A'agitha: Legends and Lore of the Papago and Pima Indians*, 24–25.)

Morning Star: A Beautiful Woman

This myth about a woman who becomes Morning Star addresses the plight of two young people who do not conform to the mores of their people and flee to the sky to find happiness. The Ceremonial Clown refers to one of the many masked clowns who take part in some Tohono O'odham ceremonies. These clowns and *kachina* dancers, who also appear in ceremonies, are probably of Pueblo origin.

There was a young woman and her brother who had been orphaned at an early age and were raised by their grandmother. When their grandmother died, they lived on in her house. Neither were married, and the man was a good hunter.

A great warrior decided that he would marry the woman and sent Coyote with presents for her. Each time she rebuffed Coyote and said that she was happy as she was. Finally the warrior decided to seize her, and approached the house with a gang of his friends. The brother came outside, but each time that they shot arrows at him, he jumped away. Then he flew into the sky. They were frightened, so they left.

But people were not content to leave the brother and sister alone. They hired a medicine man to intercede. Wind Man turned himself into a whirlwind and transported the woman to the top of a desolate mountain. With the help of Buzzard, the brother finally located her, but neither of them could get her down. They secured the assistance of a Ceremonial Clown who planted a vine, the gourd, that grew up to the woman and allowed her to escape. But the brother and sister knew that they could not be happy living where they were anymore.

The woman went into the eastern sky and became the Morning Star, while the man became a falling star that made earthquakes when he hit the earth.

(Adapted from Saxton and Saxton, *O´othham Hoho´ok A´agitha: Legends and Lore of the Papago and Pima Indians*, 11–19.)

Seri

The Seri call themselves *Comcáac* (the People). At the time the Spanish arrived they lived along the Gulf of California, south of the Upper Pima and Tohono O'odham, and ranged far inland for food. Although the Seri had access to shellfish, fish, and sea turtles, there was always an acute shortage of fresh water. They did not grow crops, but hunted, fished, and collected wild foods, using cane-balsa boats when they ventured into the Gulf. The harsh desert conditions did not allow for large settlements. People lived and searched for food in small groups, organizing around the extended family rather than a larger tribe. The Seri now live along the Gulf of California in a small portion of their original territory.

Stars and Constellations

Saguaro-Gathering Hook, a long stick with a bar set diagonally across the tip, is the BIG DIPPER.

A woman and her children who go outside to urinate are the PLEIADES.

Mountain Sheep is ORION's belt, Arrow is ORION's sword, and Antelope is Betelgeuse. A dog that follows Mountain Sheep is Sirius.

Pole-Carrier is SCORPIUS, as it is with the Havasupai and Walapai.

Queeto is Aldebaran. It has charge of other stars and lifts them into the sky. The month of October is *queeto yaao* (Aldebaran Its-Path). Aldebaran appears in the eastern sky in late October and early November.

Hee (Jackrabbit), is an unidentified star that is in the sky in November and provides the month name, *hee yaao* (Jackrabbit Its-Path).

Naapxa (Turkey Vulture or Buzzard) is in the night sky in December, the month of *naapxa yaao* (Turkey-Vulture Its-Path). The identity of this star is unknown, but the Luiseño and Tipai in southern California identify Buzzard as Altair, and Altair is indeed in the western sky in December evenings.

The constellation Quail is unidentified but is likely the same as the Maricopa constellation Quail.

CONSTELLATIONS OF THE SOUTHWEST

Culture Group	Star or Constellation	Classical Equivalent
EASTERN PUEBLOS		
Keresan-Speaking Pueblo		
Cochiti	Shield Stars; Seven Stars	BIG DIPPER
Cochiti	Sling-Shot Stars	DOLPHIN
Cochiti	Arch Across the Sky	Milky Way
Cochiti	Pot-Rest Stars	Unidentified
Tewa-Speaking Pueblos		
Various	Bull's Eye	Aldebaran?
Various	Seven Corner; Seven Tail; Long Tail; Dog Tail	BIG DIPPER
Various	Star in South; Gray-Haired Old Woman	Canopus?
Various	Belly of a Sling	DOLPHIN
Various	Yellow-Going Old Woman	Evening Star
Various	Yoke	LITTLE DIPPER's bowl
Various	Big Star; Dark Star Man	Morning Star
Various	Backbone of the Universe; Whitishness	Milky Way

CONSTELLATIONS OF THE SOUTHWEST (*cont.*)

Culture Group	Star or Constellation	Classical Equivalent
***Tewa-Speaking Pueblos** (continued)*		
Various	Turkey Foot	NORTHERN CROSS?
Various	Meal-Drying Bowl	NORTHERN CROWN
Various	In a Row	ORION'S belt
Various	In a Bunch	PLEIADES
Various	Ladder	Unidentified
Various	Star House	Unidentified
Various	Big Round Circle	Unidentified
Various	Hand	Unidentified
Various	Sandy Corner	Unidentified
Various	Horned Star	Unidentified
San Ildefonso	Place of Doubt	CANCER
San Ildefonso	Place of Decision	Castor and Pollux
San Ildefonso	Three Stars of Helpfulness	LEO
San Ildefonso	Endless Trail	Milky Way
San Ildefonso	Long Sash	ORION
San Juan	Star of San Juan Pueblo	Planet?
Tiwa-Speaking Pueblos		
Isleta	Cradle	BIG DIPPER
Isleta	Asking Star; Prayer Star	Evening Star
Isleta	Bright Star	Morning Star
Isleta	Fawns	ORION'S belt
Isleta	Tumbled or Jumbled	PLEIADES
Isleta	Bear	Unidentified
Isleta	Wheel	Unidentified
Taos	Sky Backbone	Milky Way
Taos	Deer	PLEIADES
Picuris	Two Dove Maidens	Unidentified
Towa-Speaking Pueblo		
Jemez	Sky Backbone	Milky Way
Jemez	Seven Stars	PLEIADES
Jemez	Rows	ORION
WESTERN PUEBLOS		
Laguna	Fawns	ORION'S belt
Zuni	Seven Ones; The Seven	BIG DIPPER
Zuni	Star Zigzag	CASSIOPEIA
Zuni	Star Sling	DOLPHIN
Zuni	One Following the Sun	Evening Star

CONSTELLATIONS OF THE SOUTHWEST (*cont.*)

Culture Group	Star or Constellation	Classical Equivalent
WESTERN PUEBLOS *(continued)*		
Zuni	Those to the North	LITTLE DIPPER
Zuni	Ashes Placed or Stretched Across	Milky Way
Zuni	Big Star; Great Star; Broken Water Jar	Morning Star
Zuni	Red Star	NORTHERN CROWN
Zuni	In a Row	ORION's belt
Zuni	Four Big One; Square	PEGASUS
Zuni	Seeds; Seed Stars	PLEIADES
Zuni* Ahayuta *Stories:		
Zuni	Lungs of Cloud Swallower; Head of Bush Man	BIG DIPPER
Zuni	Liver of Cloud Swallower	Evening Star
Zuni	Entrails of Cloud Swallower	Milky Way
Zuni	Heart of Cloud Swallower; Gambler's Eyeball	Morning Star
Zuni	Lying Star; Gambler's Eyeball; Heart of Bush Man	Unidentified
Zuni	Liver of Bush Man	Unidentified
Zuni	Stick Game Ones	Unidentified
Zuni Chief of the Night:		
Zuni	Chief of the Night	Many Constellations
Zuni	Left forearm of the Chief	Antares?
Zuni	Left shoulder of the Chief	Arcturus?
Zuni	Right arm of the Chief	ORION's belt?
Zuni	Left hand of the Chief; The Eight Ones	SAGITTARIUS?
Zuni	Left elbow of the Chief	Spica?
Zuni	Heart of the Night	Vega, Deneb, other stars in NORTHERN CROSS?
Hopi	Broad Star	Aldebaran
Hopi	*Soñwuka*	Milky Way
Hopi	Big Star	Morning Star
Hopi	Clustered	PLEIADES
Hopi	Strung Together; Like Beads on a String	ORION's belt
MOUNTAINS, MESAS AND PLAINS		
Navajo	Red Star	Antares or Aldebaran
Navajo	Male One Who Revolves	BIG DIPPER

CONSTELLATIONS OF THE SOUTHWEST (*cont.*)

Culture Group	Star or Constellation	Classical Equivalent
MOUNTAINS, MESAS AND PLAINS *(continued)*		
Navajo	Coyote's Star; No-Month Star	Canopus or Antares?
Navajo	Female One Who Revolves	CASSIOPEIA
Navajo	Man with Legs Ajar	CROW
Navajo	Big Star	Evening Star
Navajo	Doubtful or Pinching Stars; Hard Flint Women; Wrestlers	Two stars in HYADES
Navajo	Awaits-the-Dawn	Milky Way
Navajo	North Star on Top	North Star
Navajo	First Slim One; First Slender One	ORION
Navajo	*Dilyéhé*	PLEIADES
Navajo	First Big One	SCORPIUS's forepart
Navajo	Walking Stick of First Big One	SCORPIUS
Navajo	Rabbit Tracks	SCORPIUS's tail
Navajo	Big Star; Morning Star	Venus

(See page 193 for listing of unidentified Navajo constellations.)

Culture Group	Star or Constellation	Classical Equivalent
Mescalero Apache	Revolving Around a Central Point; Falling	BIG DIPPER
Mescalero Apache	Three Who Went Together	CHARIOTEER
Mescalero Apache	The Scattered Stars	Milky Way
Mescalero Apache	Star-That-Does-Not-Move	North Star
Jicarilla Apache	Two-Fighting Stars	Castor and Pollux
Jicarilla Apache	Three-Vertebrae Stars	ORION's belt
Jicarilla Apache	Six Hunters and Girl	PLEIADES
WESTERN UPLAND		
Havasupai	Very Cold Star	Altair
Havasupai	Hoop	NORTHERN CROWN or HERDSMAN?
Havasupai	Coyote Carrying a Pole	SCORPIUS
Havasupai	Down Feather	Unidentified
Walapai	Feces Interrupted (from the cold)	Altair
Walapai	Giant Cactus-Gathering Crook	BIG DIPPER
Walapai	Coyote Carrying a Pole	SCORPIUS
Walapai	Eagle Feather Headdress	Unidentified
Yavapai	Hoop	NORTHERN CROWN or HERDSMAN?
Yavapai	Mountain Sheep	ORION
Yavapai	*Hecha*	PLEIADES
Yavapai	Red Feathers	Unidentified
Yavapai	Hand	Unidentified

CONSTELLATIONS OF THE SOUTHWEST (*cont.*)

Culture Group	Star or Constellation	Classical Equivalent
SOUTHERN RIVER AND DESERT		
Cocopa	Amuh (a young man)	ORION's belt
Cocopa	*SuuL* (tip of arrow)	Stars in ORION
Cocopa	*Ipa' (*shaft of arrow)	ORION's sword?
Cocopa	Hestah, a young woman	PLEIADES
Cocopa	Soba^{r} (another man)	Sirius?
Maricopa	Cold's Cottonwood	Altair
Maricopa	Coyote's Fishing Net	BIG DIPPER
Maricopa	Star that Travels Through the Sun	Evening Star
Maricopa	Deer and Antelope Tracks	Milky Way
Maricopa	Big Star	Morning Star
Maricopa	*Captain*	North Star
Maricopa	Cipa's Hand	NORTHERN CROWN or HERDSMAN?
Maricopa	Mountain Sheep and Arrow	ORION
Maricopa	Scorpion	SCORPIUS
Maricopa	Mountain Sheep Pursuer	Sirius
Maricopa	Two Men with Hats	Vega and Antares?
Maricopa	Quail	Unidentified
Maricopa/Pima	Coyote	BIG DIPPER
Maricopa/Pima	Spider	CASSIOPEIA
Maricopa/Pima	Spider's Web	Milky Way
Maricopa/Pima	Pole	North Star
Maricopa/Pima	Council	NORTHERN CROWN
Maricopa/Pima	Mountain Sheep	ORION
Maricopa/Pima	Gopher	PLEIADES
Maricopa/Pima	Fox (Coyote's brother)	SAGITTARIUS
Maricopa/Pima	Earth Doctor	Stars in SCORPIUS
Maricopa/Pima	Scorpion	Stars in SCORPIUS
Maricopa/Pima	Creator's Hand	Unidentified
Maricopa/Pima	Duck	Unidentified
Pima	Flour; Ashes	Milky Way
Pima	Not-Walking Star	North Star
Pima	Running Women	PLEIADES
Pima	Bluebird; Visible Star	Venus as Morning Star
Tohono O'odham	Cactus Hook	BIG DIPPER
Tohono O'odham	White Beans	Milky Way
Tohono O'odham	Woman	Morning Star
Tohono O'odham	Homeless Women	PLEIADES

CONSTELLATIONS OF THE SOUTHWEST (*cont.*)

Culture Group	Star or Constellation	Classical Equivalent
SOUTHERN RIVER AND DESERT *(continued)*		
Seri	*Queeto*	Aldebaran
Seri	Turkey Vulture	Altair?
Seri	Antelope	Betelgeuse
Seri	Saguaro-Gathering Hook	BIG DIPPER
Seri	Mountain Sheep	ORION's belt
Seri	Arrow	ORION's sword
Seri	Women and Children	PLEIADES
Seri	Pole-Carrier	SCORPIUS
Seri	Dog	Sirius
Seri	Jackrabbit	Unidentified
Seri	Quail	Unidentified

Chapter eleven

A Bright Light Across the Heavens

Stars of the Great Plains

The Great Plains culture area stretches from the Mississippi River on the east to the Rocky Mountains on the west, and from southern Alberta, Saskatchewan, and Manitoba south to Texas. Two distinct cultures developed in this area. In the east, largely on the tall-grass prairie, lived semiagricultural groups that engaged in both farming and hunting bison. These tribes grew corn, squash, and beans in the summer, depending on a higher annual rainfall than was available on the Great Plains itself. They also hunted bison, often during a summer and a winter hunt. The people of this area used earth-covered lodges (north of Kansas) or lodges covered with skins, mats, or grass (to the south), which were clustered in permanent villages, generally along the Missouri and other rivers. Some of the settlements, such as those of the Wichita, Hidatsa, Arikara, and Mandan, became trade centers. The village-oriented tribes included the Wichita, Pawnee, and Arikara, who spoke dialects of the Caddoan language family, and the Dakota, Mandan, Hidatsa, Omaha, Osage, and other groups, who spoke dialects of the Siouan language family.

The second culture developed on the Great Plains and centered solely on the bison. Here, hunting groups lived a nomadic life following the herds upon which they depended. They traveled at first on foot and then on horse. They lived in tipis, structures made of poles and bison hide that could be dismantled, packed, and then reassembled at a new site. These tribes include the Blackfoot, Plains Cree, Gros Ventre, Cheyenne, and Arapaho, who spoke Algonquian dialects; the Assiniboin, Crow, and Lakota, who spoke Siouan dialects; the Kiowa and Comanche, who spoke Uto-Aztecan dialects; and the Sarsi and Plains Apache, who spoke Athapaskan dialects.

This village-nomad division is somewhat arbitrary. The Cheyenne, for instance, were once village-oriented people who migrated to the western plains and became bison hunters. Native Americans of the plains moved around a great deal—the Siouan and Algonquian tribes migrated from the east, the Caddoan tribes from the south—and culture evolved as location and resources changed.

The introduction of the horse in the late 1500s and 1600s dramatically changed life on the Great Plains. Securing food became much easier. Hunters could reach game

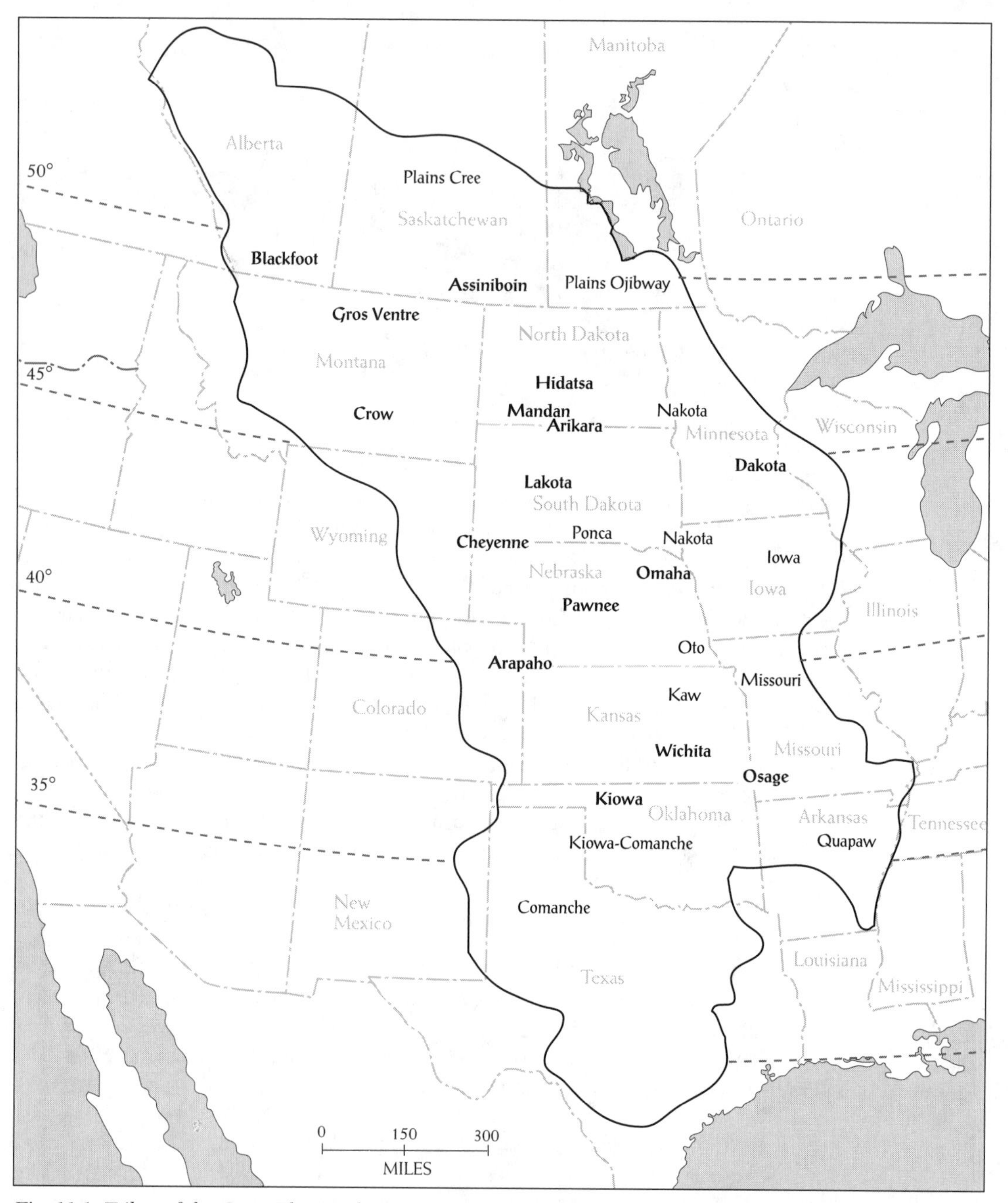

Fig. 11.1. Tribes of the Great Plains culture area and their approximate territories in the 1820s. The tribes indicated in **boldface type** are included in this chapter. Not all tribes that inhabited this area are shown. (After Waldman, *Encyclopedia of Native American Tribes*, 188.)

quicker and with less effort, and could roam farther in search of it. Plains tribes became renowned for their skill in riding and breeding horses. The villages were more easily moved, too, for horses could pull a larger *travois* (a drag made of two long poles with a platform or net) than could dogs. With horses, people were able to transport more food and personal possessions.

Bison were the primary source of food. Native people ate the meat raw or roasted, dried it for later use, and mixed dried meat with berries and fat to make a staple called *pemmican*. They used the hide, horns, bones, and other parts of the bison in many ways, ranging from tipi coverings (hide) to ladles and spoons (horn) to hoes (shoulder blades) to knives (bone).

Fig. 11.2. This Arapaho painting of a vision includes a humanlike spirit who appeared to the painter and said that the Morning Star (the Maltese cross in the painting) would protect him. The turtle is the individual's supernatural helper. The moon's points are almost horizontal, which is a sign of prosperity. (Illustration from Kroeber, *The Arapaho,* Part 4, fig. 165.)

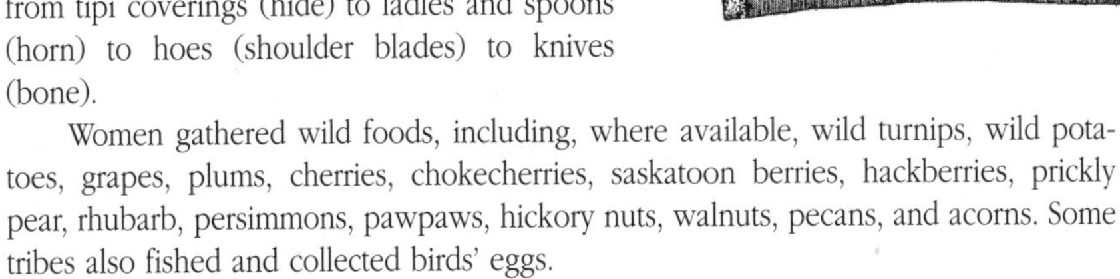

Women gathered wild foods, including, where available, wild turnips, wild potatoes, grapes, plums, cherries, chokecherries, saskatoon berries, hackberries, prickly pear, rhubarb, persimmons, pawpaws, hickory nuts, walnuts, pecans, and acorns. Some tribes also fished and collected birds' eggs.

Visions were and still are important to many Native American groups as a way of making contact with the spirit world and securing power or assistance from a guardian spirit. Native people of the plains, like some other groups, engaged in this process through a vision quest. After a purifying sweat bath, an individual would set out without clothes or food and seek a vision through fasting and physical challenge. This vision might present itself in the form of an animal or some other aspect of nature. When the individual returned, a medicine man helped interpret the experience. The individual would then put together a personal medicine bundle with items of particular importance, such as stones, herbs, feathers, or parts of animals. There were also medicine bundles of sacred items that were kept for the entire tribe.

Many plains tribes, along with several Plateau and eastern Great Basin tribes, practiced some form of the Sun Dance, a ceremony meant to renew life, protect the bison as a supply of food, and ensure success in tribal efforts. The ceremony lasted from several to ten days and usually included extensive singing, dancing, and drumming. Specially prepared dancers performed around a center pole in the Sun Dance lodge to fulfill vows while other community members gathered and gave support. In some tribes, dancers lacerated themselves or tore away strips of flesh as part of the ceremony, and many had visions at that time. The Sun Dance—also called the Mystery Dance, New Life Lodge, Medicine Lodge, and other names—played an important part in the spiritual life of the tribes that embraced it.

Of the tribes of the Great Plains culture area, the star lore of the Pawnee has been the most extensively recorded and researched. To the people of this tribe, the stars are all-important. Stars created the Earth and gave it life. One star controls the buffalo, another the remaining animals; stars serve as the patrons of individual villages. Stars even serve as models for behavior. Throughout most of the plains area, the BIG DIPPER is considered a stretcher or litter upon which lies a sick person, and this star group shows Native people how to attend to their own sick. The Pawnee also believe that their own chiefs should sit in a circle, as does the Council of Chiefs in the sky (the NORTHERN CROWN). It is not surprising that on the vast plains, where the sky stretches unimpeded from horizon to horizon, an ancient people should deem the stars to be creators of and active interveners in life on the Earth.

CADDOAN-SPEAKING TRIBES

The Wichita, Pawnee, and Arikara are thought to have migrated onto the Great Plains from the south or southwest. The Wichita remained on the southern plains while the Skidi Pawnee and Arikara moved to the Platte River region (in present-day Nebraska) by the mid-1600s; other Pawnee bands soon followed. The Arikara continued north and settled along the Missouri River (in what is now North Dakota) in the late 1600s. Thus, these tribes with a common heritage came to live in widely separated areas in the central part of the continent.

Today, the Wichita and Pawnee live in Oklahoma, and the Arikara are located on the Fort Berthold Reservation in North Dakota.

Pawnee

The Skidi, one of the four independent bands of the Pawnee, have the best preserved celestial cosmology of the plains tribes. The supreme power, Tirawahat (or Tirawa), created the stars. The stars then made the Earth and sent a girl-child to populate it. They also sent sacred items to the Skidi for use in ceremonies honoring the Evening Star and the North Star, as well as necessities like corn, implements, and weapons. Each member of the tribe comes under the supervision of a star and when a person dies, he or she becomes a star.

There were more than a dozen Skidi villages, each said to have been created when a star or constellation gave the founder of that village a sacred bundle. These bundles generally included ritual and symbolic objects such as a pipe, tobacco, sweet grass, corn, parts of birds and mammals, and in certain cases a scalp or skull, all wrapped in a bison hide. When the Skidi assembled for certain ceremonies, they arranged their villages according to the spatial relationships of their patron stars in the sky. People married within the village so that the bundle—and its power—would not leave. Each village had chiefs, who advised and made decisions through consensus; priests, who cared for the bundles, led ceremonies, and attended to the well-being of the village; and shamans, who cured the sick. Skidi society was divided into commoners and an upper class of chiefs, priests, and other esteemed people. Although some positions were inherited, individuals could move into the upper class by hard work, skill, and bravery.

The Skidi lived in these small villages until the 1600s and 1700s, when they began forming somewhat larger settlements for safety. They built circular, domed earth lodges that could hold one or several families. On the summer bison hunt they used a dwelling of saplings and skins, whereas on the winter hunt they lived in tipis. Corn was important in their daily and ceremonial lives.

Some Pawnee tales could only be told during the winter months, when snakes were not in evidence. Coyote, the trickster, does not like to be talked about, and Coyote Star tells Snake Star to bite anyone who tells tales about Coyote during the summer, when both stars are visible. (Coyote Star is Sirius and Snake Star is a star in SCORPIUS.)

Although the star lore of many North American tribes received scant attention, James R. Murie, whose mother was a Skidi, devoted himself to collecting cultural material—including celestial information—from his tribe. Murie's Pawnee name was Sakuruta (Coming Sun), and he was born in 1862 in what is now Nebraska. Although he lamented that he was able to record only the shreds of a disappearing culture, he nonetheless preserved a great deal of information.

The cosmology of the Skidi Pawnee is considered in depth in the book *When Stars Came Down to Earth: Cosmology of the Skidi Pawnee Indians of North America*, by astronomer Von Del Chamberlain.

Creation

The creation story is central to Skidi beliefs about the stars. Running Scout (also called Roaming Scout) was an influential and prestigious priest of the Skidi band whose account was published in 1904. James Murie also recorded several versions of the creation. Although the tellings differ somewhat in detail, the main narrative is similar. The following narration combines elements from both Running Scout and Murie.

At first, there was Tirawahat (the universe and everything that it contains), who is all-powerful. Tirawahat spoke to the assembled gods through Bright Star. "I am going to put each one of you in the heavens," said Tirawahat. Tirawahat used the Pathway of Departed Spirits to separate the eastern portion of the sky, which was male, from the western portion of the sky, which was female.

"You, Great Star, shall be in the east and shall push the other stars toward the west. You shall be the greatest power in the east. And you, Bright Star, shall be in the west." Bright Star was very beautiful.

Tirawahat then placed gods in the four semicardinal directions: Big Black Meteoric Star in the northeast, Red Star in the southeast, Yellow Star in the northwest, and White Star in the southwest. He assigned the powers Thunder, Lightning, Wind, and Cloud to Bright Star.

Tirawahat placed a circle of stars in the sky to represent chiefs. He put two beings, Wind Ready to Give and Breathe, in the northern sky and assigned Wolf Star to the east-southeast.

He positioned the Star That Does Not Move in the north, directing it to serve as chief of the other gods in the sky. Spirit Star he assigned to the south, saying that it would appear only at an appointed time of the year. He set Black Star in the east-northeast.

Then, Tirawahat put Sun and Moon in the sky. "You, the Sun, shall be in the east and give light and warmth. You, the Moon, shall be in the west and provide light when there is no sun."

Great Star called a council in the east, and all the heavenly gods attended. Here they discussed a plan to create the earth. Bright Star opposed the idea, for she would have to marry Great Star and bear the child that would people the earth. She had denied previous suitors and did not want to marry Great Star.

Still, the council favored the plan, and Great Star began moving from the eastern sky to the western sky to capture Bright Star, who had returned there. As Great Star started his quest, he asked his younger brother to accompany him and carry a sacred bundle from Tirawahat.

Bright Star lived by a beautiful garden with ripening grain, bison, and streams. She was guarded by four powers, Mountain Lion, Wolf, Black Bear, and Wildcat. As Great Star advanced, she created ten challenges—a serpent, thorns, monsters, cactus, and others—to stop him, but each time he produced a ball of fire from his pouch and conquered the obstacle. Even the four powers who waited near Bright Star's garden could not stop Great Star.

When he at last reached her, she required him to secure first a cradle-board, then a mat for the child that was to come, and then water to wash the child. Great Star did all of these things; the cradle-board was covered with a wildcat (bobcat) skin, whose speckles portray the stars in the heavens. At last she yielded to him and they merged their powers for the good of the people who were to come from the union. Great Star's strength came from a bed of flint and Bright Star's power came from storms. When they joined, Great Star gave Bright Star a fire drill [with which to make fire], which she later gave to the people of the earth.

After the union of Great Star and Bright Star, Big Black Meteoric Star presided over a second council of the gods in which they all talked about the creation. They gave Bright Star major responsibility for providing for the people who would come. Thunder, Lightning, Wind, and Cloud were to help her.

Bright Star presented Great Star with a pebble that he dropped into water. The pebble became the earth, but it was covered with water. Thunder, Lightning, Wind, and Cloud sang songs and hit the water with clubs, creating land.

When her child was born, Bright Star gave her daughter Mother-Corn seeds, put her on a cloud, and sent her to earth, where she fell as rain. The daughter found a boy, the offspring of Sun and Moon, and from these two came the people that inhabit the earth. In addition to corn, Bright Star gave them bison and other necessities for living.

Many years later, Great Star told the people that he wanted a human sacrifice to honor his ordeal in securing Bright Star. Bright Star told Errand Man to gather the various bands, and all but two came together. At this time, the people performed the four-pole ceremony.

The stars sometimes send storms to punish people who are evil, but Tirawahat himself does not get angry. He is changeless.

(Adapted from Dorsey, Traditions of the Skidi Pawnee, 3–6; Murie in Curtis, *The Indians' Book*, 99–104; Murie, *Ceremonies of the Pawnee*, 38–39; and Murie, Pawnee Societies, 552–554.)

Venus and Mars: Bright Star and Great Star

Bright Star (Evening Star, Female White Star) is Venus. She is the leading power in the western sky, a beautiful woman who has rebuffed many suitors. Tirawahat tells her, "To you I give power to create all things. It is through you that many people shall come." Although initially unwilling, she is responsible for giving birth to a daughter and sending her to the Earth; Bright Star also gives the Skidi corn, bison, and a sacred bundle that bears her own name. This bundle is more important than other bundles given the band by various stars and is the first to be unwrapped at the annual Thunder Ceremony held in the spring.

Great Star (Mighty Star of Fire, Big Star, Morning Star) is Mars (with Jupiter or Venus substituting when Mars is not visible). Great Star is "a great warrior, painted red, carrying a club in his folded arms, and having on his head a downy feather, painted red." The downy feather symbolizes breath and life. In an important ritual, Skidi priests

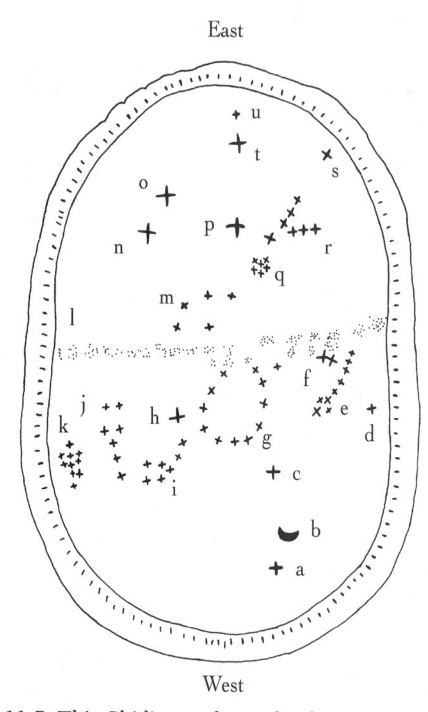

Fig. 11.3. This Skidi star chart of unknown age was given to James Murie, who presented it to the Field Museum of Natural History in Chicago in 1906. The buckskin chart was not meant to portray all the stars in the sky with precision but rather to remind the Skidi of the importance of particular stars that guided them in their lives. The larger stars and those in patterns are shown here; not shown are many smaller stars seemingly scattered throughout the chart. The PLEIADES are shown twice—once in winter as ten stars and once in summer as six stars. (Illustration after Chamberlain, *When Stars Came Down to Earth: Cosmology of the Skidi Pawnee Indians of North America,* fig. 48.) Starting from lower right: *a.* Evening Star (Venus); *b.* moon; *c.* White Star (Sirius?); *d.* unidentified, but may be South Star (Canopus); *e.* Real Snake (SCORPIUS); *f.* Duck Stars (SCORPIUS's tail); *g.* Circle Stars, Chiefs (NORTHERN CROWN); *h.* North Star; *i.* Little Stretcher (LITTLE DIPPER); *j.* Big Stretcher (BIG DIPPER); *k.* Seven Stars, wintertime (PLEIADES); *l.* Milky Way; *m.* Bow (DOLPHIN); *n.* Black Star (Deneb?); *o.* Big Black Meteoric Star (Vega?); *p.* Fools Wolves (Sirius); *q.* Seven Stars (PLEIADES); *r.* Deer Stars (stars in ORION); *s.* Red Star (Antares?); *t.* Morning Star (Mars); *u.* Morning Star's Brother

sang, "Oh Morning Star, your form we see! Clad in shining garments do you come, your plume touched with rosy light. Morning Star, you now are vanishing." The Skidi observed that this being was sometimes brighter, sometimes fainter, and sometimes disappeared altogether, as is indeed true for Mars.

The courtship may relate to actual events in the sky. Von Del Chamberlain, the astronomer, suggests that the creation story describes the movement of Mars from the east to the west and its subsequent conjunction with Venus.

Stars of the Four Quarters

Tirawahat set four stars in the semicardinal directions and then assigned Thunder, Lightning, Wind, and Cloud to Bright Star's garden. (In one version of the creation story the four semicardinal stars are these four elements.) Tirawahat says to these stars, "You four shall be known as the ones who shall uphold the heavens. There you shall stand as long as the heavens last, and, although your place is to hold the heavens up, I also give you the power to create people. You shall give them the different bundles, which shall be holy bundles. Your powers will be known by the people, for you shall touch the heavens with your hands, and your feet shall touch the earth." Each of these four stars subsequently gave a Skidi village a holy (medicine) bundle and became the patron power of that village.

Big Black Meteoric Star in the northeast controls bison and other animals as well as the arrival of night. This star is the patron god of medicine men. Chamberlain suggests that this star is Vega.

Red Star in the southeast controls certain animals as well as the arrival of the day. Chamberlain proposes that Red Star is Antares, a red star in SCORPIUS that rises in the southeastern sky.

Yellow Star was assigned to the northwest, where the sun shines gold. Chamberlain suggests that the yellow star Capella in CHARIOTEER is Yellow Star.

White Star is in the southwest and faces north, from where the snow comes. Chamberlain suggests that Sirius is White Star.

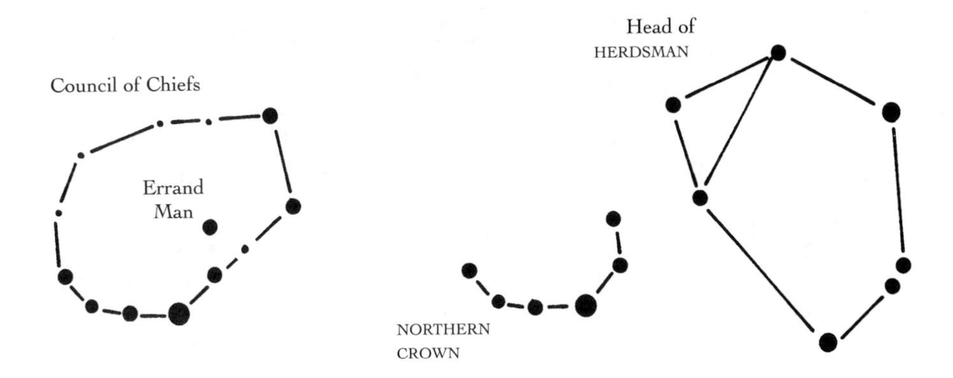

Fig. 11.4. The Pawnee constellation Council of Chiefs includes six stars in the NORTHERN CROWN, faint stars nearby, and two stars in HERDSMAN. One star of the NORTHERN CROWN is the Errand Man, who sits in the center of the circle to tend the fire and run errands for the chiefs. (Illustration after Chamberlain, *When Stars Came Down to Earth: Cosmology of the Skidi Pawnee Indians of North America*, 234.)

North Star: The Star That Does Not Move

The North Star is the Star That Does Not Move. Tirawahat told this star to stay in one place, lest the other stars become confused in their movements. The Star That Does Not Move is chief of the stars and leads them when they sit in council, an event represented by the NORTHERN CROWN and nearby stars. Certain Skidi rituals are dedicated to the North Star, and during them, participants sit in a circle like the Council of Chiefs. The star that is located in the middle of the circle is the Errand Man. In Skidi society, the Errand Man tended the fire, cooked food, and took care of other duties during ceremonies.

Canopus?: The South Star

The South Star (also Spirit Star, Death Star, or Midway Star) is thought to be Canopus. This star is the father of the North Star and appears only for a short time to see whether his son is still at his station. The South Star is in charge of the land of the dead and taught the people the Buffalo Dance. The South Star appears in the fall, low in the sky, and marks the beginning of winter, considered the second half of the year.

Canopus rises above the horizon at thirty degrees north latitude (the latitude of New Orleans) but is not seen at forty degrees north latitude (the Kansas/Nebraska line). Therefore, the Skidi could not see Canopus when they lived along the Loup River in what is now Nebraska. Perhaps the cosmology concerning the South Star developed before the Pawnee moved north to the Loup; the Wichita in the south, who are related to the Pawnee by language, also identify a Great South Star. Or, the South Star may be some other heavenly body, such as Sirius.

According to Pawnee belief, the North Star and the South Star will bring the world to an end. Young Bull, a prominent medicine man in one of the Pawnee South Bands in the early 1900s, describes these events:

When the time approaches for the world to end, the South Star will come higher, until at last it captures the people who are carrying the two people upon the stretchers [the Big and Little Dippers]; as soon as the South Star captures these two people upon the stretchers they will die. The North Star will then disappear and move away and the South Star will take possession of the earth and of the people.

The command for the ending of all things will be given by the North Star, and the South Star will carry out the commands. Our people were made by the stars. When the time comes for all things to end our people will turn into small stars and will fly to the South Star, where they belong. When the time comes for the ending of the world the stars will again fall to the earth. They will mix among the people, for it will be a message to the people to get ready to be turned into stars.

(Reprinted from Dorsey, The Pawnee Mythology, 135–136.)

Milky Way: Bright Light in a Long Stretch Across the Heavens

A Skidi named Fox tells the following story about the Milky Way and a star that receives spirits in the north, which Chamberlain hypothesizes may be a star in CEPHEUS. In another source, the Milky Way is described as a "bright light" that is a "long stretch across the heavens." Spirits move along the Milky Way after death; wind pushes them toward the Spirit Star, where they will dwell. Warriors who die in battle travel on a short fork of the Milky Way, and other people travel on the long fork.

The star that receives the Pawnee after they are dead stands at the end of the Milky Way, in the north. He receives them upon the earth, takes them with him on a long journey to the north, and after he gets to this place he places them upon the Milky Way. If the dead man had been a warrior, he was put on the dim Milky Way; if he died of old age, or if it was a woman, they were put upon the wide traveled road. Then they started on the journey toward the south. There, at the end of the Milky Way, in the south, stands another star, who receives the spirits of the departed, and there they make their home.

(Reprinted from Dorsey, Traditions of the Skidi Pawnee, 57.)

A second story about the Milky Way is more anecdotal. A horse and a bison were running a race:

The horse jumped farther than the buffalo, making dust here and there, so that he beat the buffalo; the buffalo, taking

> quicker steps, made more dust;
> their course is the Milky Way.
> (Reprinted from Dorsey, Traditions of the Skidi Pawnee, 57.)

BIG and LITTLE DIPPERS: Two Stretchers

When it was being decided where each star would stand in the sky, two beings, one old and one young, fell ill. Other stars placed them on litters and carried them. The four stars in the bowl of the BIG DIPPER carry the old man, and the three stars in the handle—the medicine man, the wife and dog, and the errand man—follow. The wife is the bright star Mizar, and the dog is the faint star Alcor. The LITTLE DIPPER is the litter for the baby. "The people took their way of living from the stars, so they must carry their sick or their dead as shown, the mourners following."

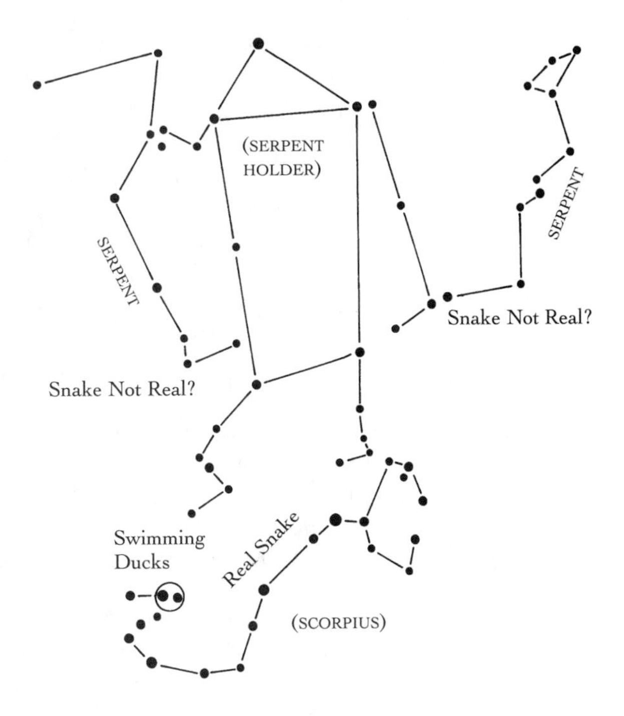

Fig. 11.5. The Pawnee constellation Swimming Ducks is composed of two stars in the tail of SCORPIUS. Other stars in SCORPIUS are Real Snake, which tries to protect Bright Star. Stars in SERPENT may form the Pawnee constellation Snake Not Real.

Animal Stars

The two stars in the tail of SCORPIUS are the Swimming Ducks. When the Ducks could be seen through the smoke-hole of the lodge, the priests began to listen for prolonged rumbling thunder throughout the heavens. This spring thunder provided the signal for the Thunder (or Creation) Ritual, performed in honor of the Evening Star bundle. This ritual marked the beginning of spring. The arrival of the Swimming Ducks also meant that animals would return. "It is believed that at their appearance the animals in the water revive, or come out and break holes in the ice. This is spoken of as the taking of new breath of life. When the first thunders come, it is said that they are calling the water animals." (The Swimming Ducks may also be the Loons, a constellation mentioned but not identified.)

The Skidi identify the other stars in SCORPIUS as a constellation, Real Snake. Chamberlain suggests that Real Snake is the serpent that tried to protect Bright Star in the creation story. Mars passes through SCORPIUS, which is part of the zodiac, and thus Great Star could be seen as slaying the serpent.

A second snake constellation, called Snake Not Real, precedes Real Snake into the sky; Chamberlain suggests that Snake Not Real is stars in the classical constellation SERPENT.

CASSIOPEIA is Rabbit. Chamberlain suggests that it is a rabbit seen from behind, with the bent legs of the animal forming the shape of the constellation.

A set of stars—likely ORION's belt, or the belt, sword, and Betelgeuse—is Deer.

Bird's Foot (sometimes called Turkey's Foot) is a constellation in the Milky Way; it may be stars in the NORTHERN CROSS. This constellation was painted on the forehead of a warrior in certain rituals. "Each warrior places on his head a ball of down feathers, and transversely in the scalplock, an eagle feather. Their faces are painted red, streaked down the sides, and on the forehead is a bird's foot mark like the constellation in the Milky Way."

Murie writes that Bright Star is protected by four powers—Bear, Panther, Wildcat, and Wolf—who are also stars. The various accounts differ here, for although Running Scout calls Bear the Black Star, Panther or Mountain Lion the Yellow Star, Wildcat the White Star, and Wolf the Red Star, Murie does not make these associations. Chamberlain suggests that the Bear is SAGITTARIUS, the Panther is CHARIOTEER (which contains Capella, proposed as Yellow Star), the Wildcat is Procyon, and the Wolf is Sirius. (Chamberlain previously linked White Star with Sirius, believing that the Skidi may have called one star different names according to the occasion.)

Wildcat, or Bobcat, has special significance. In addition to being a star, its coat represents all the stars in the sky. The cradle-board that Great Star gives to Evening Star is covered by a wildcat skin, just as the first covering used for upper-class Skidi infants was also a wildcat skin. To the Skidi, wrapping children in this covering is like wrapping them in the stars themselves. Skidi cradle-boards were sometimes decorated with a celestial design representing the Morning Star.

Wolf Star (also called Wolf Got Fooled, Fool Wolf, Fool Coyote, or Coyote Star) is Sirius. It is interesting to note that, perhaps coincidentally, Sirius is also the classical Dog Star in the constellation BIG DOG—in both cultures the star is a canine.

Roaming Scout narrated the following story about Wolf Star, which explains where the Skidi get their name:

Wolf Star became jealous of Bright Star and decided that he, too, could place a being on the earth. He sent a wolf to follow Lightning (one of the powers who protected Bright Star), who was visiting the earth. Bright Star had suggested that Lightning take a sack full of stars with him so that he would have company. He let the stars out of the bag from time to time, and they set up a village. But Wolf stole the sack and let out the stars, thinking they were food, and they killed him by mistake. In this way, death appeared on the earth.

The people prepared the hide of the wolf and carried it with them, and for this reason they are called Skidi (Carry Wolf's Hide), the Wolf Band of the Pawnee.

(Adapted from Dorsey, *Traditions of the Skidi Pawnee*, 17–18.)

PLEIADES: *Chaka,* The Seven Stars

The Seven Stars—the PLEIADES—figure prominently in a Skidi ritual called the *Hako* or *Calumet* (pipe) Ceremony. When these stars appear on the horizon during the ceremony, the following stanza is sung:

Look as they rise, up rise
Over the line where sky meets the earth;
Pleiades!
Lo! They ascending, come to guide us,
Leading us safely, keeping us one;
Pleiades,
Teach us to be, like you, united.

Tahirŭssawichi, a member of one of the Pawnee South Bands, explains:

This song to the Pleiades is to remind the people that Tirawa has appointed the stars to guide their steps. It is very old and belongs to the time when this ceremony was being made. This is the story to explain its meaning, which has been handed down from our fathers:

A man set out upon a journey; he traveled far; then he thought he would return to his own country, so he turned about. He traveled long, yet at night he was always in the same place. He lay down and slept and a vision came. A man spoke to him; he was the leader of the seven stars. He said, "Tirawa made these seven stars to remain together, and he fixed a path from east to west for them to travel over. He named the seven stars *Chaka*. If the people will look at these stars they will be guided aright."

When the man awoke he saw the Pleiades rising; he was glad, and he watched the stars travel. Then he turned to the north and reached his own country.

The stars have many things to teach us, and the Pleiades can guide us and teach us how to keep together.

(Reprinted from Fletcher, *The Hako: A Pawnee Ceremony*, 152, 330.)

Another story describes the moon and "the seven stars" but does not name them as the PLEIADES. Woman Cleanse the People, a Skidi, explains how Tirawahat gave knowledge in the form of songs and a dance to the first humans by sending stars to Earth. Although the plot differs from the previous myth, the point—that the stars "guide us"—is the same.

This story also describes the origin of the basket dice game, a favorite pastime of Skidi women. The game involves a basket, plum seeds decorated with symbols (the dice), and twelve sticks used as counters. The seeds are placed in the basket, then flipped into the air. Bets are placed on which marks will fall face-up.

The four star daughters of Big Black Meteoric Star are not identified. Perhaps, speculates Chamberlain, if Big Black Meteoric Star is Vega, the brightest star in LYRE, then other stars in the constellation are his daughters.

The use of a grass lodge rather than an earth lodge indicates a southern setting for the story.

One evening as the first man and the first woman sat in their grass lodge, they heard singing. The man investigated the next day and found a lodge with a field of corn next to it. He and his wife both went to the lodge. There, they found an old woman (the Moon), Moon's daughters, Evening Star, four old men (Wind, Cloud, Lightning, and Thunder), and the four daughters of Big Black Meteoric Star. Evening Star carried a basket while she danced in front of the old men; Big Black Meteoric Star's daughters also danced and placed sacred objects in the basket. The celestial beings taught the man songs, dances, and ceremonies so that he could pass along this knowledge to others.

These heavenly beings had come to the earth in a basket (the moon) to teach the man and woman things that they needed to know. After they had done so, they gave the couple a basket to remind them of the event and taught them a game. The woman was told to find plum seeds and make marks on them to represent the stars. There were also twelve counting sticks to remind the earth people of the twelve stars in the circle of chiefs in the sky.

When they saw that the two humans understood what they were to do on earth, the heavenly beings leaped into their basket and returned to the sky.

(Adapted from Dorsey, The Pawnee Mythology, 44–46.)

Skidi priests watched the PLEIADES to decide when to begin planting; myth also associated the star group with the harvest. In one story, a girl is pursued by a rattling skull that rolls along the ground. A group of brothers saves the girl from being killed by the skull, and she in turn gives them food by planting and harvesting corn, beans, and squash. She joins them as a sister as they travel nightly through the sky.

Other Stars and Constellations

Great Star's younger brother, who follows with a sacred bundle while his sibling pursues Bright Star, may be Mercury or Jupiter. The Star of Sickness may be Saturn.

Pahukatawa (Kneeprint by the Water) was a great prophet who sent the Skidi bison, warned them of enemy attack, and helped them in other ways. Chamberlain suggests that the Kneeprint Star is one in PERSEUS or CEPHEUS.

Wind Ready to Give and Breathe are two other stars (Chamberlain again suggests stars in PERSEUS) that Tirawahat placed in the north. "By their breaths they send snowstorms upon the land and drive buffalo to the people."

Lone-Chief, a Skidi, narrated a story published in 1904 about Wind Ready to Give, who was also called Pahukatawa. This man, Lone-Chief said, went out to do battle with the enemy, but he was killed and the parts of his body were scattered over the land. The gods put him together, substituting feathers for his missing brain. Thereafter, Wind Ready to Give advised the people of his village, teaching them ceremonies and helping them gain what they desired. At last, however, they no longer listened to him, and he left to become a star near the North Star.

The Pawnee constellation Bow is DOLPHIN, a small constellation just outside the Milky Way. The Bow represents the bow that Great Star gave the Skidi so that they could secure game.

Chamberlain suggests that the Feather Headdress Star is the Great Nebula in ORION's sword or the sword itself, with the hazy nebula representing the downy feather. He also proposes that Black Star (different from Big Black Meteoric Star) is Deneb.

Wichita

The Wichita lived, often along rivers, in the southern plains in present-day Kansas, Oklahoma, and Texas. They built grass-thatched structures; grew crops; hunted deer, bear, and antelope; and went on an annual bison hunt. If their agricultural efforts yielded a surplus, they traded produce with tribes that did not farm. Kinnikasus (the creator), the Sun, the Moon, Morning Star, and other beings were important deities.

At the time of creation, the land floated on a huge sea. The first man and first woman appeared on Earth with corn, arrows, and a bow. All was darkness until three deer were shot, and then there was light, and stars appeared. Eventually there were animals and villages. First Man and First Woman went from village to village, teaching the people and giving them power. Finally, he became the Morning Star and she became the Moon.

Stars and Constellations

There are few recorded Wichita star stories. In one, Bear Woman tries to kill her brothers and sister, but they escape and become The Seven, the BIG DIPPER. In another, two boys slay monsters and join their father, a star, in the sky; the three stars are unidentified.

This Wichita story about the South Star, probably Canopus, and a star known as Flint Stone Lying Down Above illustrates the importance both of corn (by the name of the village) and flint (by the name of the star). Hunters used flint in knives and arrowheads and also to make fire; flint figures in several stories of Great Plains tribes. In the Skidi Pawnee creation story, Great Star's strength comes from a bed of flint, and in the Blackfoot story of Blood-Clot, the culture hero uses a flint knife to kill monsters.

In a village called Stone Corn Mills Lying on the Hillside or Where Large Ears of Corn Grow, there lived a great man called Protector of Warriors. This man was always victorious in war, and persons who accompanied him knew they would be safe. But one day he told his wife, who was pregnant, that he would be traveling far to the south so that he would be closer to the enemy; he would not ever return. Although he took four men with him, they each gave up before reaching the destination. Protector of Warriors traveled on alone and built a lodge in the south, where he continued to kill his foes.

His son was born and, years later, decided to try to find his father. After a long, long journey, during which he gained advice from the men who could not keep pace with Protector of Warriors, he reached his father's lodge. At first the elder man did not believe it was his own son, but when the young man put some cornmeal in his father's mouth, his father said, "Ah, it tastes just like the kind I used to eat at home." Young Flint Stone lived with his father and helped him slay the enemy, but eventually his father said, "It is time for you to go home. I will never return there, but shall become the South Star. Those who offer a smoke to me will be victorious in battle and will meet with good fortune."

Young Flint Stone did as Protector of Warriors instructed and became a great man in his village. He offered a smoke to his father whenever the South Star appeared in the sky. At last he decided that he, too, should go to the sky. He is a star known as Flint Stone Lying Down Above.

(Adapted from Dorsey, Mythology of the Wichita, 58–62.)

Arikara

The tribal name *Arikara* means "horn" and refers to that tribe's practice of twining hair around two bones, making what look like horns. The Arikara lived in villages south of the Mandan and Hidatsa along the Missouri River in what is now North Dakota. All three tribes used small, round bull boats—made of a bison hide stretched over a willow frame—for navigating rivers. The Arikara constructed large earth lodges about forty feet in diameter that housed several families.

The Arikara are closely related to the Pawnee, and for several years in the early 1800s, many Arikara lived with the Pawnee.

Stars and Constellations

The constellation Invalid Being Carried is likely the BIG DIPPER; a girl and her brothers who escape a bear are the PLEIADES. The Winding Trail is an unidentified group of four stars close together in the southern sky.

There are other stories that involve the heavens, but they are not generally tied to a specific star. Two Holy Men, also called Star Fathers, are related to the bison. These men visit the Earth and help in a time of starvation. Star Boy, son of the Morning Star, leaves his celestial home to go to Earth. There he slays the enemy in a series of encounters, then marries the chief's daughter. After their first child is born, he returns to the sky. Star Boy himself is not identified.

The Arikara, like many tribes across North America, tell a story about a woman who falls in love with a star and goes to the sky to live with him. She is unhappy there and tries to escape to the Earth on a bison-sinew rope, but her husband kills her as she is lowering herself to the Earth. Their son, sometimes called Old Woman's Grandchild, survives the fall and is the hero of many Arikara myths.

PLEIADES: A Girl and Her Brothers

The following story about the PLEIADES describes the creation of Devil's Tower, a dramatic volcanic pillar in northeast Wyoming. The 865-foot monolith, which is visible for miles, has huge columnar joints from top to bottom that were created when the molten rock cooled. Many tribes of the northern plains call the formation Bear Lodge and consider the area sacred. The Kiowa, who lived in the vicinity of the Black Hills and Devil's Tower before moving to the southern plains, tell a tale nearly identical to this one but say that the stars become the BIG DIPPER. Squash Blossom, the grandmother of Alfred Morsette, told him this story; Morsette calls it a holy story.

The children were playing one day, taking turns pretending to be a bear. They then said to one girl, who was fourteen winters, that she should be next. "No," she said. "I don't want to do this." But they asked and asked and at last she reluctantly agreed. "A bad thing will happen," she said, and told her younger sister to hide.

The other children kept on playing and were surprised to see that when the older sister came out of the bushes, she had become a real bear. The bear killed them all, then went on to the village and killed her own father, mother, sister, grandmother, and brother—she killed everyone there.

The bear returned to the place that her sister had hidden and demanded that she come out and accompany her. The two went into the woods and dug a hole for a home. The younger sister got food for them both, first dried meat from the village and then wild turnips, plums, and chokecherries. The younger sister kept watching for her four brothers, who had been away at war when this all happened.

The brothers returned one day and the younger sister told them what had gone on. They sent their sister back to the cave and instructed her to find out the bear's weakness. How could it be killed? The sister did as they said. "No one can kill me, except by shooting me in either of my two smallest toes. But if even a little blood should escape, then I will come back," said the bear.

The younger sister reported this information to her brothers and they formed a plan. They gathered the thorns of the cactus and thorn apple bush, which the younger sister placed outside the bear's den. She then yelled to the bear and got it to leave the cave. The bear stepped on the thorns, and while it was rolling on the ground in pain, the brothers slashed off its smallest toes, then burned the bear and everything around it to remove all traces of blood.

The boys and the girl ran far from that place, but somehow some blood had escaped the fire and the bear came to life again and followed them. The brothers threw an awl that became thorn apple bushes, a knife that became wide cracks in the ground, and a comb that became a cactus patch, but each time the bear got through the obstacles. Finally the youngest brother seized a grooved stone and asked it for help. "Please, Grandfather, we can do nothing more!" The stone grew and grew—higher and higher—with the children on top of it. When the bear arrived, it clawed at the rock but could not climb it or knock it over. The claw marks are the grooves in the rock. At last the bear gave up. "You are safe from harm," she told her brothers and sister. Meanwhile, they had become five stars bunched together in the sky.

(Adapted from Parks, *Traditional Narratives of the Arikara Indians*, 179–185.)

SIOUAN-SPEAKING TRIBES

Tribes that spoke dialects of the Siouan language came to occupy a vast area from Arkansas through North Dakota, from the Mississippi to the Rocky Mountains. There was a great deal of movement as tribes migrated onto the Great Plains, established their homes, and responded to encroachment from other tribes and eventually from Euro-American trappers, traders, and settlers.

Today, most Omaha and their descendants live in Nebraska; the Osage are centered in northern Oklahoma; most Dakota and Lakota live on reservations in Minnesota, Nebraska, Montana, and North and South Dakota; the Assiniboin live in Montana,

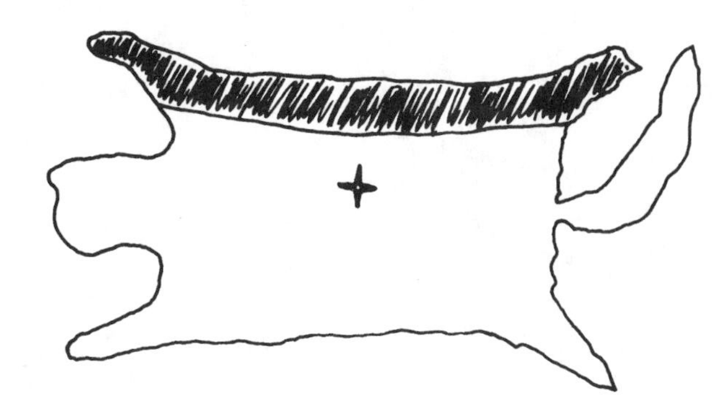

Fig. 11.6. Omaha tipi and robe with Morning Star designs. The darker part of both the tipi and the robe represents the night. The robe was worn with the dark strip over the shoulders. (Illustration after Dorsey, *A Study of Siouan Cults,* 398.)

Alberta, and Saskatchewan; the Mandan and Hidatsa live on the Fort Berthold Reservation with the Arikara in North Dakota; and the Crow live in eastern Montana.

Omaha and Osage

The Omaha and Osage are two closely related Siouan-speaking tribes of the lower Missouri River that moved onto the prairie through the Ohio Valley. Their languages, tribal organizations, and religious rites are similar, as are their constellations. Both tribes have two major divisions, the Sky People and the Earth or Land People. *Wakonda* is a supernatural primal life force who lives in all things in the Earth and in the sky.

For the Omaha, the stars are the children of the moon and the sun. Certain stars and constellations, the sun, the moon, Darkness, Thunder Being, and other powers are called *Wakonda*(s). These powers are the object of prayer for serious ventures, such as setting out on a long journey.

When an Omaha child was old enough to be considered part of the tribe, he or she was introduced to the cosmos. The chant began:

> Ho! Sun, Moon, Stars, and all that move
> in the heavens,
> I bid you hear me!
> Into your midst has come a new life.
> Consent, I implore!
> Make its path smooth, that it may reach
> the brow of the first hill!

Stars and Constellations

The Omaha say that the North Star is the Star That Does Not Walk and the BIG DIPPER is a litter. Big Star refers to both the Morning and Evening Stars. A group of brothers forms a circle of stars (probably the NORTHERN CROWN) near the handle of the BIG DIPPER. A path that the spirits use as they travel to the land of the dead is the Milky Way. Large Foot of Goose is ORION's belt. Deer's Head is the ancient, religious name for the PLEIADES, while Little Duck's Foot is the constellation's common name.

The Osage believe that they came from the stars and that the actions of the sun, moon, and stars show perfect order. They honor about a dozen chief stars or constellations. These celestial beings are addressed as grandmothers or grandfathers.

The Osage say that a stretcher or litter in a funeral bier is the BIG DIPPER. This constellation figures prominently in Osage ceremonies. Deer Head is the PLEIADES. Great Star or Red Star, a male being, is the Morning Star. A female being is the Evening Star.

The North Star is also called Red Star and serves as a guide when traveling.

Wolf that Hangs at the Side of the Heavens is the constellation BIG DOG or the star Sirius that lies within it. (The Skidi Pawnee Wolf Star is also thought to be the star Sirius.) One of the Osage tribal divisions used the Wolf and the North Star as war symbols.

Three Deer is ORION's belt, and the constellation Stars Strung Together is two stars in the sword. The stars are "strung together" by the Great Nebula in ORION, clouds of gas that appear as a hazy star.

Red Corn (William Matthews) drew a symbolic chart in the early 1880s that depicts events in the history of the Osage and includes stars and constellations. Before they had bodies or souls, the ancestors of the Osage came up through a tunnel and into the upper worlds. At the fourth upper world, the female Red Bird gave adults but not children bodies. Black Bear then went to the important stars and asked them for help, but they each gave the enigmatic answer, "I am not the only mysterious one; wait a while." Black Bear went back to Red Bird, who gave parts of her body to make the children. When these ancestors were fully formed, they emerged into the world and split into two groups or *gentes*, the war *gentes* and the peace *gentes*. Red Corn's chart belongs to the peace *gentes*, and there is a chant telling the story that goes with it.

Fig. 11.7. Osage symbolic drawing of the world. A cedar tree, representing the tree of life, grows next to a river. Beneath the river, at left, is the Red Star (Morning Star). To the right are six stars forming Three Deer (ORION's belt) and, presumably, Stars Strung Together (ORION's sword). To the right are the Evening Star and Small Star. Below, in an arc, are the Seven Stars or Deer Head (PLEIADES), with the moon on the left and the sun on the right. A pipe, a hovering bird, and a hatchet lie below. The four horizontal lines represent the four upper worlds of the Osage, which are resting in an oak tree. (Illustration from Dorsey, *Osage Traditions*, fig. 389.)

Dakota and Lakota (Sioux)

The Sioux, the largest of the Siouan tribes, is composed of the Santee, who speak the Dakota dialect; the Yankton and Yanktonai, who speak the Nakota dialect; and the Teton, who speak the Lakota dialect. These groups call themselves by their dialect names, the Dakota, Nakota, and Lakota.

The Sioux migrated up through the Ohio Valley and into the woodland area west of the Great Lakes. There, they fought often with the Chippewa, and in the early 1700s they began moving west onto the plains. The Lakota relocated far west into what is now South Dakota and adopted the life of a hunting tribe. The Nakota moved into what is now North and South Dakota. The Dakota lived in villages in the eastern shortgrass prairie and woodlands in present-day Wisconsin, Minnesota, and eastern South Dakota, where they continued to grow crops as well as hunt bison.

Dakota Stars and Constellations

Carrying the Dead Man is the BIG DIPPER. A hunter died suddenly and his friends (the four stars in the bowl) lifted him up. The man's wife (handle star nearest the bowl) and two children (other handle stars) follow.

A trail of bubbles from a diving contest is the Milky Way. Beaver and a bird argued about who could dive the deepest. All animals with fur and feathers assembled and dove into the water. The beaver won. The bubbles they created became the Milky Way.

Largest Star refers to any of four stars: the Morning Star, the Evening Star, a star in the south, and a star in the southeast.

A constellation called *Ta´maopa´* (untranslated), which resembles "a kite," appears overhead in winter and is not seen in the sky from midwinter until June. One likely candidate for this constellation is DOLPHIN, a small but distinctive group of stars that does indeed look like a kite. DOLPHIN can be seen in the evening sky from June through January.

The eight stars in the Sweat Tipi constellation represent a small tipi with two men inside, singing. The sweat lodge is a small structure used for a sweat bath—a time to pray, sing, and purify oneself. A "sweat" often precedes or follows other rites. This unidentified constellation is visible in the summer in the southwest sky, but it cannot be seen in winter.

Lakota Stars and Constellations

For the Lakota, the stars are *wakan* (something mysterious, holy, incomprehensible), a term that involves power or contact with the spiritual world. They are the "holy breath" of Wakan Tanka (Creator, Great Spirit) and represent sacred speech that is explained through myth and ritual. Humans should not try to learn about stars or talk about them.

Seven Stars (the BIG DIPPER) is a group that dances in a circle around the Star That Does Not Move (North Star). In honor of the Seven Stars in the sky, the Lakota make seven council fires.

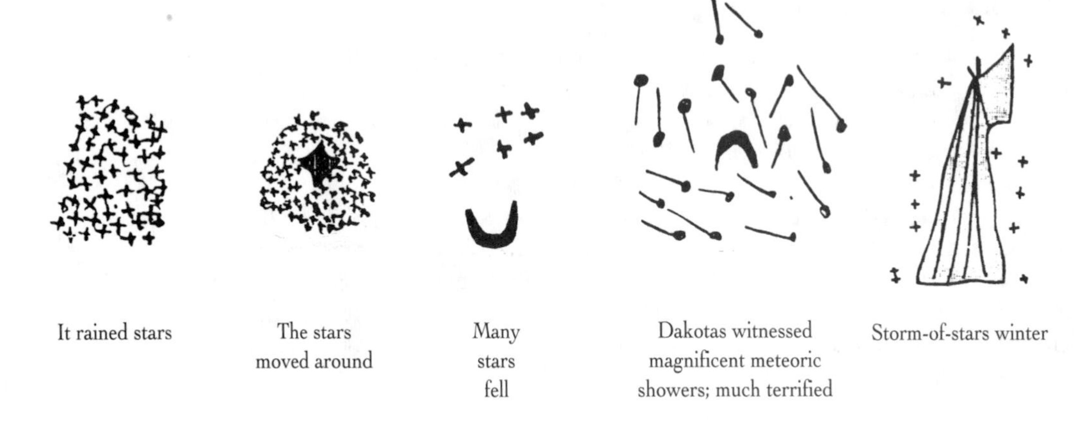

Fig. 11.8. On the night of November 12–13, 1883, a magnificent meteor shower lit the sky over the Great Plains. The shower was so memorable that some tribal historians, who kept a chronological record of the years by drawing pictures on tanned hide, used the shower to represent that year. This sequence of pictures on a hide is called a Winter Count. The Winter Counts mentioned here were all kept by Dakota or Lakota tribal historians. *From left:* "It rained stars," from Cloud-Shield's Winter Count; "The stars moved around," from American-Horse's Winter Count; "Many stars fell," from The-Flame's Winter Count; "Dakotas witnessed magnificent meteoric showers; much terrified," from The-Swan's Winter Count; "Storm-of-stars winter," from Battiste Good's Winter Count. (Illustrations from Mallery, *Picture-Writing of the American Indians*, figs. 1219–1223.)

The Spirit Way (Milky Way) moves around the sky so that bad spirits cannot find it. The wind shows good spirits how to find the Spirit Way at the edge of the world.

The constellation Bear's Lodge is eight stars around GEMINI.

The disappearance each spring of the Hand (stars in ORION) is a sign of the Earth's loss of fertility. A sacrifice is needed to renew the Earth, just as Inyan (Rock) created the Earth by shedding his own blood. This annual sacrifice was historically provided during a midsummer ceremony, the Sun Dance, when Lakota dancers mortified the flesh of their own bodies. Many dancers had visions at this time.

The Hand itself represents the chief who lost his arm because he refused to sacrifice; the Thunders ripped it from his body and hid it. A young man, Fallen Star, wished to marry the chief's daughter and could do so only after retrieving the arm. Fallen Star received gifts that gave him power and journeyed through the Black Hills and star villages in his quest. He secured the arm, returned it to the chief, married the daughter, had a son, and became the new chief. Though restored, the former chief no longer had power because he had gone against the nature of the universe. The Hand appears in the sky each winter as part of the renewal cycle and as a reminder to people on Earth that they must live according to the principles in the heavens (see Fig. 11.9).

Assiniboin

The Assiniboin lived in the Lake Winnipeg area in the 1600s but, according to tradition, were once part of the Yankton when that group lived in Minnesota. The Assiniboin led a typical plains life: hunting bison, living in tipis, and celebrating the annual Sun Dance. As in Siouan-speaking tribes to the south, *Wakonda* was the creator of all things.

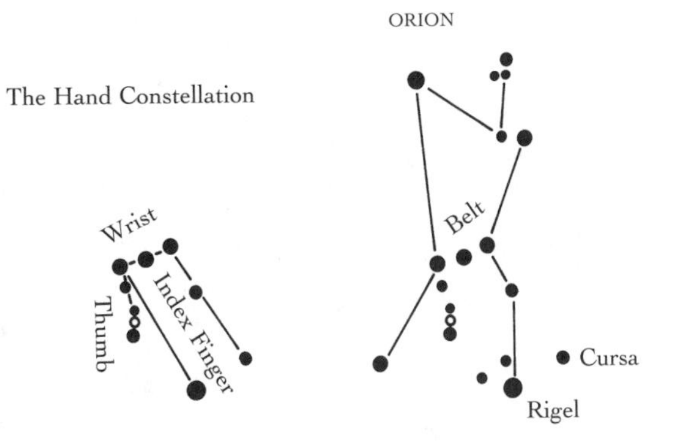

Fig. 11.9. The Lakota Hand constellation (*left*) is made up of stars in ORION (*right*). ORION's belt is the wrist, the sword is the thumb, Rigel is the index finger, and the star Cursa in the classical constellation RIVER ERIDANUS is the tip of the little finger. (RIVER ERIDANUS is a long, rather dim constellation that is not featured in this book.)

BIG DIPPER: Seven Stars

Wa-se-a-ure-chah-pe (North Star) is the North Star, and Backbone of the Sky is the Milky Way. The constellation Seven Stars is the BIG DIPPER. A story about the BIG DIPPER combines elements found in Pawnee and Arikara stories, including the rolling skull and magic flight from a pursuer.

A woman had illicit intercourse with a snake. Her husband once watched her as she was fetching firewood. She pounded a tree-stump, and from its hollow a snake came out. The man decided to kill his wife. He went on a hunt, killed a moose, and returned in a zigzag line. Then he told his wife to fetch the meat, following his track. The woman obeyed. In the meantime, the husband went to the tree, killed all the snakes, gathered their blood in a cup, and boiled soup from it. When his wife returned, he gave her the soup. After she had swallowed it, he told her it was the snakes' blood. She started towards the tree and found all the snakes dead. She was furious.

The couple had seven children, one of whom was a girl. The man bade them flee. They ran away while he stayed in the lodge, having fastened the door-flap. He heard the woman approaching with yells. When she stuck her head inside, he cut it off with a bone knife. The head began to roll after the children, caught them, and carried them to a tipi, where they lived together. One of the boys once killed a moose and was stretching the skin. The head went out to scrape it, warning the children not to look at it. One of the boys wanted to look out, though another boy warned him. He insisted, and his brother finally yielded, saying, "Well, then look at our mother." The head knew the boy was peering out, and said, "I will kill all my children." The children ran away, pursued by their mother's head.

The girl had an awl. She gave it to her brother, who threw it behind them. A great number of awls sprang up, and the head could not get across the points. After three days it managed to get through, and again pursued the children.

The girl gave her brother a little piece of flint. He threw it behind them, and a big fire started up. The head was burnt in the fire, all its hair being singed off. After a long while it got through the fire and continued the pursuit.

When it came near, the sister gave her brother a piece of rock. When he threw it behind them, it turned into a big mountain. The head could not get over it at first, but finally it passed across.

The children came to a deep river. Two cranes were standing there. The boys said, "Let us travel across your necks to the other side." The birds allowed them to cross. Then the children asked them not to allow their mother to cross. When the head came and had passed to the middle of the crane's neck, the bird threw it into the water.

After a while the head got out again and started after the children. The sister made a ball and said, "Let us play ball." They stood in a circle and threw the ball to one another. While doing so, they rose to the sky and became the Dipper. The head could not jump high enough to reach them.

(Reprinted from Lowie, The Assiniboine, 177–178.)

PLEIADES: Wise-One and his Brothers

The following Assiniboin story likely details the origin of the PLEIADES, because the previous story chronicles the BIG DIPPER and the PLEIADES are the other "seven stars" in the sky. This story is similar to one told by the Natchez in the Southeast (see page 279).

There were seven youths on this world. One of them was red-haired. They did not know whether they had any parents. They were having a hard time of it. "What shall we turn into?" they asked one another. One said, "Let us change into the earth." The one named Ksā'be (the Wise-One) said, "No, the earth is mortal. It gets caved in."

Then another one said, "Let us become rocks." "No, they are destructible. They all break asunder."

A third one said, "We must change into big trees, into very big ones." "No, they are perishable. When there is a storm, they are blown down."

Again one of them said, "Let us change into water." "No, it is destructible. It dries up completely."

The fifth said, "Let us change into the night." "No, the night is fleeting. Soon the light appears again."

The sixth boy said, "Let us be the day." "No, it is fleeting. When the sun disappears, it is dark once more."

The Wise-One said, "No, the blue sky above is never dead. It is always in existence. Shining things live there. Such we shall change into. In that region let us dwell."

Well, so they do. The smallest of them took them up, hoisting them by means of his spider-web. He set three on one side and three on the other, seating himself in the middle. When the last one had gotten up, he tore the web in the middle, threw it down, and gave it to the spider.

(Reprinted from Lowie, The Assiniboine, 177.)

Morning Star: Star Shining Daylight Chases

The Assiniboin word for Morning Star means Star Shining Daylight Chases. The story about the Morning Star, like that of the BIG DIPPER, involves snakes and a paramour.

A man and his wife were camping by themselves. She was pregnant. While her husband was away, another man would come and embrace her. Her lover wished to elope with her, but he did not like to take her with the baby in her womb. So he once entered her lodge and said, "I want to eat food from your belly." She asked, "How shall I sit?" "Lie down on your back and place the dish on your belly." She obeyed. When he was done eating, he stuck a knife into her, and took out the child, which he left in the lodge.

Then the lovers fled underground, entering the earth under the fireplace. When the woman's husband returned, he found the child's body, and saw that his wife was gone. He split trees and dried up the creeks where he thought she might have fled. When the lovers came above ground again, he tracked them. They turned into snakes and crawled into a hollow tree. He followed in pursuit, and saw the snakes, but did not recognize them as the fugitives. He thought the lovers had gone up the tree. He climbed up, but could not find them. At last he climbed higher still, reached the sky, and became the Morning Star.

(Reprinted from Lowie, The Assiniboine, 176.)

Mandan and Hidatsa

The Mandan and Hidatsa lived in villages along the middle Missouri in what is now North Dakota—growing corn, beans, and squash and hunting bison. It is thought that the Mandan moved into this region from the southeast and that the Hidatsa came from the Lake Winnipeg area. The Mandan called the Hidatsa *Minitaree* (They Crossed the River) and themselves *Numakiki* (People). Like other plains Indians, both tribes practiced the Sun Dance and had numerous religious and military societies.

The Hidatsa separate stories into the secular and the holy, but stories of any kind were traditionally told only from fall to spring. Coyote plays an important part in many of the narrations. Although he is sometimes foolish, he is an honored person in the ceremonial lodge.

Stars and Constellations

The Mandan believe that stars and people are one, for a person is a star in human form. When a child is born, a star descends from the sky to the Earth; when that person dies, the spirit returns to the sky as a star. The moon has several children, including Woman Who Wears a Plume (the Morning Star), Woman of the West (the Evening Star), and a "high star" near the North Star.

The Hidatsa BIG DIPPER is an ermine, a description reminiscent of Northeast tribes that call this constellation a fisher. Both the ermine and the fisher are members of the weasel family. The North Star is the Star That Does Not Move and the Milky Way is the Ashy Way.

Like the Lakota, the Hidatsa say that the Hand Star is stars in ORION. ORION's belt is the wrist, where the hand was cut from the arm, and the thumb and fingers hang below. Bear's Arm told this Hidatsa story, and Arthur Mandan interpreted it. The myth, which is different from the Lakota version, begins with a village in the sky. Long Arm is the chief and a holy man there, and Charred Body is a mighty hunter.

Charred Body decided to leave his village in the sky to follow the bison herd that he saw on earth. He founded thirteen lodges and brought down his friends and relatives to join him. Charred Body wanted to marry the chief's daughter from a nearby village, but she would not have him, and he killed her. To avenge this evil deed, the chief killed many people in Charred Body's village, but Charred Body escaped. Coyote, who had been around during the fight, went to Charred Body, and with his power he created a new lodge and food for Charred Body, the hunter's sister, and himself.

The sister, who was pregnant, was not supposed to let anyone in her lodge while Charred Body and Coyote were away, but she admitted a stranger who was a monster. The monster insisted on eating his supper on her belly, and as he did so, he killed her. Twins were born at that instant. Lodge-Boy remained in the dwelling, but the monster threw Spring-Boy into the spring.

When Charred Body returned, he mourned his sister and began raising Lodge-Boy. At first he did not realize that there was another child, for Spring-Boy was wild and would only appear when Lodge-Boy was alone. Eventually Charred Body tamed Spring-Boy, and the twins, who had supernatural powers, returned their mother to life. They also killed many monsters that lived in the area.

The people in the sky village began to worry that the twins would rise to the sky and kill them, too. They convinced Long Arm to steal Spring-Boy from the earth, then they beat Spring-Boy and secured him to a torture tree. Spring-Boy said to them, "I am one of your people so it is not right for you to do this. You should not be afraid of me." The people were silent, for they really had no cause to hurt him.

Lodge-Boy, meanwhile, awoke from the deep sleep that Long Arm had caused and realized that his brother was missing. He discovered where Spring-Boy had gone and followed him into the sky. He changed himself into a little boy so that no one would recognize him and visited the lodge where his brother was being held. A light appeared above Spring-Boy's head when he recognized his disguised brother. Seeing the light, the people knew that Lodge-Boy had arrived, but they could not find him.

Lodge-Boy could not get close to his brother, so he turned into a spider. He freed Spring-Boy from the tree and rubbed his wounds with bison fat. Both boys, in the form of spiders, escaped the lodge through a hole. The holy men knew what was happening but did not have enough power to stop the boys. Long Arm put his hand through the hole to try to catch them, and Spring-Boy cut off his hand at the wrist. "Your hand has now committed two crimes, and I will put it in the sky, as a sign to the people below."

The Hand Star is still visible today. It is a group of stars that the White people call Orion.

(Adapted from Beckwith, Mandan-Hidatsa Myths and Ceremonies, 22–42.)

Crow

Many years ago, the Crow and Hidatsa were one people living in villages along the Missouri. There was, reportedly, a dispute over the division of bison meat, and the Crow moved west, where they became nomadic bison hunters.

Individuals known as good storytellers were featured guests at Crow feasts on winter nights. The raconteur would begin, pausing after every sentence or two for the required "yes" from the audience. When the audience no longer responded, which meant that everyone was asleep, the storyteller quit for that evening.

Stories can be told only at night because that is when the stars come out. The Crow believe that the named stars used to live on Earth.

Stars and Constellations

The Crow tell a Bear Woman story much like that of the Arikara and Kiowa, except that the siblings become the BIG DIPPER rather than PLEIADES, and Devil's Tower is not part of the story. The siblings rather calmly sit on the side of a hill, smoking and talking about what they should do to escape. The sister suggests that they go to where they point the pipe—to the stars that form the BIG DIPPER—and they agree. "What shall we do there?" asks one brother. "At night we shall walk, in the daytime we'll go into the ground. We'll always come again on the other side of the earth. As the sun and the moon and the Morning Star do, so shall we do," says the little girl.

Another Crow tale about the BIG DIPPER substitutes an evil husband and a powerful but wicked old woman for the bear. A young woman runs to the tipi of seven brothers to escape her husband, who has tried to kill her. The youngest brother kills the husband, and the brothers adopt the woman. But she carelessly allows an old woman into the tipi, and this women threatens them all. The young woman and the brothers flee. The old woman kills most of the brothers, but the youngest brother kills the old woman and restores his siblings to life. Then, all of the brothers and their adopted sister become the "seven stars," the BIG DIPPER. The woman is the small star Alcor.

Old Woman's Grandchild, the child of the sun and a human woman who subsequently dies, is the North Star. The child is raised by his grandmother. He is the subject of many stories, including those in which he kills various monsters. At last he becomes the North Star and his grandmother becomes the moon. Old Woman's Grandchild is also identified with the Morning Star.

The Crow used tobacco as a medicine, and the plant was important in ritual. The Crow Tobacco Society held three ceremonies each year—at the time of initiation, when seeds were planted, and when crops were harvested. According to tradition, the tobacco plant was once a star. As the Transformer (Creator) walked around the recently made Earth, he saw someone and said to his companions, "Look, that person is a star from the stars above." When they neared the person, they saw instead a tobacco plant, the first plant on the Earth.

The Crow, like the Hidatsa and Lakota, call the stars of ORION the Hand Stars.

ALGONQUIAN-SPEAKING TRIBES

The Blackfoot, Cheyenne, and Arapaho moved onto the Great Plains from the east and developed a rich culture in which the sun, the stars, the bison, and the tipi played an important part. The Blackfoot now live in Montana and Alberta, the Cheyenne in Montana and Oklahoma, and the Arapaho in Oklahoma and Wyoming.

Blackfoot

By the time Euro-Americans had arrived in central North America, the Blackfoot had been on the northern plains, living in tipis and hunting bison, for many years. The Blackfoot confederacy includes three closely allied tribes, the Northern Blackfoot, the Blood, and the Piegan.

The Blackfoot believe that the Sun, Moon, Morning Star, and Thunder have special powers, and they revere other celestial beings, including Bunched Stars, Seven Stars, Poïa, Blood-Clot, and the woman who marries a star, all of whom are involved in stories that follow. The stories explain how these supernaturals convey ceremonies, songs, and sacred items to the Blackfoot.

As the following narrations show, individuals were taught how to decorate tipis and how to carry out ceremonies for each design. Although the information might come in a personal vision, the designs involved universal themes such as bison and stars and served to unite the members of the tribe.

The Blackfoot often featured the BIG DIPPER, PLEIADES, Morning Star, the puffball fungus, and various animals on their tipis. Puffballs are called "dusty stars," and each puffball is thought to be a meteor that has fallen to the earth. The puffball, a fungus that is in the same class as mushrooms, has a hollow spherical reproductive structure that is full of tiny spores. Pressure on the puffball expels a "puff" of spores, which is perhaps why puffballs are called dusty stars. The Blackfoot believe that when people die, their spirits go into the sky and become stars, thus ever increasing the number of stars in the heavens.

Of the planets, the Blackfoot name the Early Riser or Day Star (Venus); Big Fire Eater or Big Fire Star (Mars); and Young Morning Star or Mistake Morning Star (Jupiter).

Wolf-head, a Blackfoot, recounted some of the star myths presented here to anthropologists Clark Wissler and C. D. Duvall (who was half Blackfoot) in 1903. Wolf-head was born around 1840, so he had led a traditional Great Plains life before his tribe was forced onto a reservation. Brings-Down-the-Sun was also an old man when he recounted stories to Walter McClintock, an ethnologist, in the early 1900s. Given the widely acknowledged hostility the Blackfoot showed to hunters and trappers, it is remarkable that McClintock was able to live with the Piegan and record their history in his book *The Old North Trail*, published in 1910.

BIG DIPPER: The Seven Brothers and Their Sister

Both Wolf-head and Brings-Down-the-Sun independently describe how a woman changes into a bear and chases seven brothers and their sister into a tree. When one of the brothers waves his "medicine feather," says Brings-Down-the-Sun, they all escape to the sky and became the BIG DIPPER; the sister is the small star Alcor. In Wolf-head's version, the youngest brother becomes the North Star, and his sister, who used to carry him, is the pointer star at the lip of the bowl.

The star at the end of the handle is often called the Last Brother. The Blackfoot used the movement of the Last Brother relative to the bowl—whether the brother was right, left, below, or above the bowl—to mark the passage of time before the introduction of clocks.

PLEIADES: The Lost Children

Wolf-head and Brings-Down-the-Sun narrate similar accounts of the creation of the Bunched Stars or the Lost Children. The story explains why the PLEIADES are visible in the evening sky from fall through spring but not the rest of the year.

Plains hunters used several techniques to capture their quarry. Sometimes they pulled a bison skin over themselves in order to get close enough to a bison to kill it with a lance or bow and arrow. Or, when the situation presented itself, they stampeded a herd into or over a *piskun* (an impoundment, corral, or cliff for trapping animals). Both the disguising and the stampeding are mentioned in this story. The hunters are using a *piskun,* and the children are "playing buffalo" by pulling a skin over their heads and pursuing one another.

Children of any culture, it seems, are sensitive to the views of their peers and capable of taunting one another.

There is also a family of six stars we call the Lost Children (Pleiades). These children were lost a great many years ago from a large camp of Blackfeet during the moon when the Buffalo Calves are Yellow (spring). The hunters had been running buffalo over a *piskun* and had secured a large number, among them many buffalo calves. The little yellow hides were given to the children, who played with them a game of buffalo. There was a poor family of six children who were unable to secure any of the yellow skins and went naked.

One day, when many of the children were on the prairie playing buffalo together, putting the skins over their heads and running from each other, they made fun of the poor children, calling them "scabby old bulls" and shouting derisively that "their hair was old and black and coming out." The six children did not go home with the rest. They were ashamed because their parents gave them no yellow skins. They wandered off on the plains and were taken up to the sky. They are not seen during the moon When the Buffalo Calves are Yellow (spring, the time of their shame), but every year, when the Calves Turn

Brown (autumn), the lost children can be seen in the sky every night.

(Reprinted from McClintock, *The Old North Trail*, 490.)

Castor and Pollux: Ashes-Chief and Stuck-Behind

Many tribes (including Cahuilla, Zuni, Navajo, Mescalero Apache, Hidatsa, and Iroquois) tell stories about twin brothers who create the Earth, kill monsters, and give the people essential tools and food. In some cases one of the twins turns out to be evil while the other one performs good deeds.

The Blackfoot tale of twins describes the creation of the stars Castor and Pollux, found in GEMINI. The tale begins when Smart-Crow goes out to hunt, leaving his pregnant wife alone. He warns her not to admit anyone to the tipi, but she ignores his words and allows a man to enter. This unknown but powerful man cuts the wife open and pulls out two boys, who are named Ashes-Chief (because he is set down near ashes) and Stuck-Behind (because he is placed behind the lining of the tipi). Smart-Crow returns to find his wife dead and then begins searching for the stranger. The stranger assures him that Smart-Crow's wife will return to life and gives Smart-Crow a special tipi and instructions concerning the ritual to be used with it.

All unfolds as the stranger has said. The boys, who are subsequently raised by a rock and a beaver, eventually return to their father, revive their dead mother, and outwit a cannibal woman who tries to eat them all. When a crow hides the bison and the people are starving, Stuck-Behind finds the bison and returns them to the plains. Eventually, Ashes-Chief goes to the sky in the branches of a tree, and Stuck-Behind joins him there. They are large stars, side by side.

The following excerpt from Wolf-head's long story about Ashes-Chief and Stuck-Behind describes how a particular tipi style came into use and illustrates how concepts of the stars are woven into Blackfoot daily life.

Then the stranger took the bundle from his back and gave it to Smart Crow, saying, "I give you this lodge and the running-fisher skin." The stranger set up the lodge. There were four buffalo-tails hanging on its sides. Two of these were cow-tails, and two were bull-tails. One of each hung in front, and also behind. This lodge was called the Four-Tail Lodge. The stranger told Smart-Crow that the hanging of the buffalo-tails on the lodge would make the buffalo range near it, so that the people would always have meat. The stranger transferred this lodge to Smart-Crow. He sat down upon a stump, explained the ritual to him, and also taught him the songs.

Among the other things he said, "The punk which you use to make fires is made of bark, and does not kindle quickly; take puff-balls (fungus) instead, for they are much better. They are the Dusty Stars. You are to paint these stars around the bottom of the lodge. At the top of the lodge you are to paint the Seven Stars on one side and the Bunch Stars on the other. At the back of the lodge, near the top, you must make a cross to represent the Morning Star. Then around the bottom, above the Dusty Stars, you shall mark the mountains. Above the door, make four red stripes passing around the lodge. These are to represent the trails of the buffalo."

When Smart-Crow had received all of the instructions belonging to the new lodge and had learned all the songs, he went away with it and returned to his own lodge.

(Reprinted from Wissler and Duvall, Mythology of the Blackfoot Indians, 40–53.)

The Woman Who Marries a Star

Tribes across North America tell various stories about a woman who marries a star. In many cases the woman conveys important knowledge, sacred ceremonies, and holy articles from the Sky People to the people of the Earth. This Blackfoot account, for example, explains how Feather Woman takes a sacred Medicine Bonnet, a dress trimmed with elk teeth, a Turnip Digger, and Sweet Grass (incense) to the Blackfoot. "Ever since those days, these sacred articles have been used in the Sun Dance. . . . The Turnip Digger is always tied to the Medicine Case, containing the Medicine Bonnet, and it now hangs from the tripod behind my lodge," says Brings-Down-the-Sun.

Feather Woman was the Blackfoot's first medicine woman, and in some accounts it is she rather than her son who gives the Blackfoot the Sun Dance. Because of Feather Woman's central role in the Sun Dance, a woman sponsored the Blackfoot Sun Dance ceremony. This woman, known as a medicine woman, and her husband organized the event, saw to the construction of a Sun Lodge, and secured a medicine bundle.

In the following version of the woman-who-marries-a-star story, the North Star is the hole through which Feather Woman rejoins her homeland; it is also the hole through which the radiance of the Sun God shines. This star alone does not move and is called the Star That Stands Still. The other stars walk around it.

The Lodge of the Spider Man is the NORTHERN CROWN, and his five fingers, with which he spins a thread to let Feather Woman down from the sky, are stars in HERCULES. Brings-Down-the-Sun explains that in the ancient days, the Blackfoot used stones rather than stakes to hold down the edges of the tipis. When the tipi was removed, the stones were left in place. "Whenever you see the half-buried and overgrown circles," he said, "you will know why the half-circle of stars was called by our fathers, the Lodge of the Spider Man."

Brings-Down-the-Sun says that the events in this story happened "long ago, when the Blackfeet used dogs for beasts of burden." The story was related through many generations to his father, who told it to him.

It was a cloudless night and a warm wind blew over the prairie. Two young girls were sleeping in the long grass outside the lodge. Before daybreak, the eldest sister, So-at-sa-ki (Feather Woman), awoke. The Morning Star was just rising from the prairie. He was very beautiful, shining through the clear air of early morning. She lay gazing at this wonderful star until he seemed very close to her, and she imagined that he was her lover. Finally she awoke her sister, exclaiming, "Look at the Morning Star! He is beautiful and must be very wise. Many of the young men have wanted to marry me, but I love only the Morning Star."

When the leaves were turning yellow (autumn), Feather Woman became very unhappy, finding herself with child. She was a pure maiden, although not knowing the father of her child. When the people discovered her secret they taunted and ridiculed her, until she wanted to die. One day while the geese were flying southward, Feather Woman went alone to the river for water. As she was returning home, she beheld a young man standing before her in the trail. She modestly turned aside to pass, but he put forth his hand, as if to detain her, and she said angrily, "Stand aside! None of the young men have ever before dared to stop me."

He replied, "I am the Morning Star. One night, during the moon of flowers, I beheld you sleeping in the open and loved you. I have now come to ask you to return with me to the sky, to the lodge of my father, the Sun, where we will live together, and you will have no more trouble."

Then Feather Woman remembered the night in spring, when she slept outside the lodge, and now realized that Morning Star was her husband. She saw in his hair a yellow plume and in his hand a juniper branch with a spider web hanging from one end. He was tall and straight and his hair was long and shining. His beautiful clothes were of soft-tanned skins, and from them came a fragrance of pine and sweet grass.

Feather Woman replied hesitatingly, "I must first say farewell to my father and mother." But Morning Star allowed her to speak to no one. Fastening the feather in her hair and giving her the juniper branch to hold, he directed her to shut her eyes. She held the upper strand of the spider web in her hand and placed her feet upon the lower one. When he told her to open her eyes, she was in the sky. They were standing together before a large lodge. Morning Star said, "This is the home of my father and mother, the Sun and the Moon," and bade her enter. It was daytime and the Sun was away on his long journey, but the Moon was at home.

Morning Star addressed his mother, saying, "One night I beheld this girl sleeping on the prairie. I loved her and she is now my wife." The Moon welcomed Feather Woman to their home. In the evening, when the Sun Chief came home, he also gladly received her. The Moon clothed Feather Woman in a soft-tanned buckskin dress, trimmed with elk-teeth. She also presented her with wristlets of elk-teeth and an elk-skin robe, decorated with the sacred paint, saying, "I give you these because you have married our son."

Feather Woman lived happily in the sky with Morning Star and learned many wonderful things. When her child was born, they called him Star Boy. The Moon then gave Feather Woman a root digger, saying, "This should be used only by pure women. You can dig all kinds of roots with it, but I warn you not to dig up the large turnip growing near the home of the Spider Man. You have now a child and it would bring unhappiness to us all."

Everywhere Feather Woman went, she carried her baby and the root digger. She often saw the large turnip but was afraid to touch it. One day, while passing the wonderful turnip, she thought of the mysterious warning of the Moon, and became curious to see what might be underneath. Laying her baby on the ground, she dug until her root digger stuck fast. Two large cranes came flying from the east. Feather Woman besought them to help her. Thrice she called in vain, but upon the fourth call, they circled and lighted beside her. The chief crane sat upon one side of the turnip and his wife on the other. He took hold of the turnip with his long sharp bill and moved it backwards and forwards, singing the medicine song, "This root is sacred. Whenever I dig, my roots are sacred."

He repeated this song to the north, south, east, and west. After the fourth song he pulled up the turnip. Feather Woman looked through the hole and beheld the earth. Although she had not known it, the turnip had filled the same hole through which Morning Star had brought her into the sky. Looking down, she saw the camp of the Blackfeet, where she had lived. She sat for a long while gazing at the old familiar scenes. The young men were playing games. The women were tanning hides and making lodges, gathering berries on the hills, and crossing the meadows to the river for water. When she turned to go home, she was crying, for she felt lonely, and longed to be back again upon the green prairies with her own people.

When Feather Woman arrived at the lodge, Morning Star and his mother were waiting. As soon as Morning Star looked at his wife, he exclaimed, "You have dug up the sacred turnip!"

When she did not reply, the Moon said, "I warned you not to dig up the turnip, because I love Star Boy and do not wish to part with him."

Nothing more was said, because it was daytime and the great Sun Chief was still away on his long journey. In the evening, when he entered the lodge, he exclaimed, "What is the matter with my daughter? She looks sad and must be in trouble."

Feather Woman replied, "Yes, I am homesick, because I have today looked down upon my people."

Then the Sun Chief was angry and said to Morning Star, "If she has disobeyed, you must send her home." The Moon interceded for Feather Woman, but the Sun answered, "She can no longer be happy with us. It is better for her to return to her own people."

Morning Star led Feather Woman to the home of the Spider Man, whose web had drawn her up to the sky. He placed on her head the sacred Medicine Bonnet, which is worn only by pure women. He set Star Boy on her breast, and wrapping them both in the elk-skin robe, bade her farewell, saying, "We will let you down into the center of the camp and the people will behold you as you come from the sky." Spider Man then carefully let them down through the hole to the earth.

It was an evening in midsummer, during the moon when the berries are ripe, when Feather Woman was let down from the sky. Many of the people were outside their lodges when suddenly they beheld a bright light in the northern sky. They saw it pass across the heavens and watched until it sank to the ground. When they reached the place where the star had fallen, they saw a strange looking bundle. When the elk-skin cover was opened, they found a woman and her child. Feather Woman was recognized by her parents. She returned to their lodge and lived with them, but never was happy. She used to go with Star Boy to the summit of a high ridge, where she sat and mourned for her husband. One night she remained alone upon the ridge. Before daybreak, when Morning Star arose from the plains, she begged him to take her back. Then he spoke to her, "You disobeyed and therefore cannot return to the sky. Your sin is the cause of your sorrow and has brought trouble to you and your people."

Before Feather Woman died, she told all these things to her father and mother, just as I now tell them to you.

(Reprinted from McClintock, *The Old North Trail*, 491–497.)

Jupiter: Young Morning Star
Venus: The Morning Star
Milky Way: The Wolf Trail

Brings-Down-the-Sun continues his story, describing the fate of Star Boy, who is later called Poïa (Scarface). Poïa eventually becomes Young Morning Star, Jupiter; his father, as described above, is the Morning Star, Venus. Brings-Down-the-Sun notes that Morning Star and Young Morning Star sometimes travel together, but they may travel apart for many years. When he told this story to Walter McClintock in July 1905, Venus and Jupiter were in conjunction. "I see them in the eastern sky, rising together over the prairie before dawn," he said. "Poïa comes up first. His father, Morning Star, rises soon afterwards, and then his grandfather, the Sun."

Although born in the home of the Sun, Star Boy was very poor. He had no clothes, not even moccasins to wear. He was so timid and shy that he never played with other children. When the Blackfeet moved camp, he always followed barefoot, far behind the rest of the tribe. He feared to travel with the other people, because the other boys stoned and abused him. On his face was a mysterious scar, which became more marked as he grew older. He was ridiculed by everyone and in derision was called Poïa (Scarface).

When Poïa became a young man, he loved a maiden of his own tribe. She was very beautiful and the daughter of a leading chief. Many of the young men wanted to marry her, but she refused them all. Poïa sent this maiden a present, with the message that he wanted to marry her, but she was proud and disdained his love. She scornfully told him she would not accept him as her lover until he would remove the scar from his face. Scarface was deeply grieved by the reply. He consulted with an old medicine woman, his only friend. She revealed to him that the scar had been placed on his face by the Sun God, and that only the Sun himself could remove it. Poïa resolved to go to the home of the Sun God. The medicine woman made moccasins for him and gave him a supply of pemmican.

Poïa journeyed alone across the plains and through the

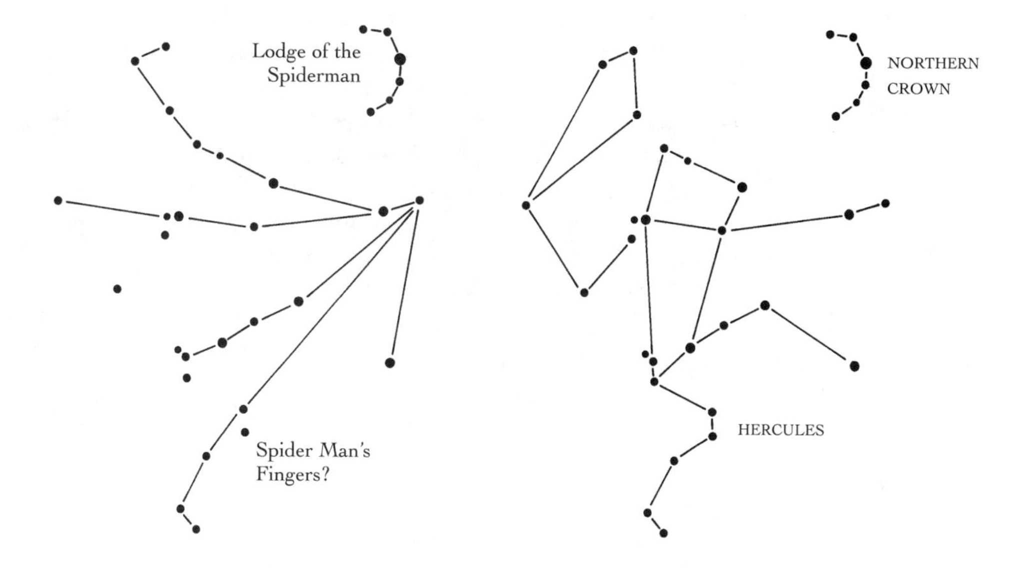

Fig. 11.10. Possible configurations of the Blackfoot constellations Lodge of the Spider Man and Spider Man's Fingers (*left*), consisting of stars in the NORTHERN CROWN and HERCULES (*right*).

mountains, enduring many hardships and great dangers. Finally he came to the Big Water. For three days and three nights he lay upon the shore, fasting and praying to the Sun God. On the evening of the fourth day he beheld a bright trail leading across the water. He travelled this path until he drew near the home of the Sun, when he hid himself and waited. In the morning, the great Sun Chief came from his lodge, ready for his daily journey. He did not recognize Poïa. Angered at beholding a creature from the earth, he said to the Moon, his wife, "I will kill him, for he comes from a good-for-nothing-race," but she interceded and saved his life.

Morning Star, their only son, a young man with a handsome face and beautifully dressed, came forth from the lodge. He brought with him dried sweet grass, which he burned as incense. He first placed Poïa in the sacred smoke, and then led him into the presence of his father and mother, the Sun and the Moon. Poïa related the story of his long journey because of the rejection by the girl he loved. Morning Star then saw how sad and worn he looked. He felt sorry for him and promised to help him.

Poïa lived in the lodge of the Sun and Moon with Morning Star. Once, when they were hunting together, Poïa killed seven enormous birds, which had threatened the life of Morning Star.

He presented four of the dead birds to the Sun and three to the Moon. The Sun rejoiced when he knew that the dangerous birds were killed, and the Moon felt so grateful that she besought her husband to repay him. On the intercession of Morning Star, the Sun God consented to remove the scar. He also appointed Poïa as his messenger to the Blackfeet, promising if they would give a festival (Sun Dance) in his honor, once every year, he would restore their sick to health. He taught Poïa the secrets of the Sun Dance, and instructed him in the prayers and songs to be used. He gave him two raven feathers to wear as a sign that he had come from the Sun, and a robe of soft-tanned elk-skin, with the warning that it must be worn only by a virtuous woman. She can then give the Sun Dance and the sick will recover. Morning Star gave him a magic flute and a wonderful song, with which he would be able to charm the heart of the girl he loved.

Poïa returned to the earth and his camp by the Wolf Trail (Milky Way), the short path to the earth. When he had fully instructed his people concerning the Sun Dance, the Sun God took him back to the sky with the girl he loved. When Poïa returned to the home of the Sun, the Sun God made him bright and beautiful, just like his father, Morning Star. In those days Morning Star and his son could be seen together in the east. Because Poïa appears first in the sky, the Blackfeet often mistake him for his father, and he is therefore sometimes called *Poks-o-piks-o-aks*, Mistake Morning Star.

(Reprinted from McClintock, *The Old North Trail*, 496–499.)

ORION Nebula: Blood-Clot or Smoking-Star

The Smoking-Star is thought to be the Great Nebula in ORION, the middle "star" in ORION's sword. To the unaided eye the nebula looks like a star with a hazy mass around it. The star is the mythical person Blood-Clot or, more precisely, his stone knife. In one Blackfoot ceremony, a dancer with a stone knife on his head portrays Blood-Clot; as the dancer leaps into the air, light flashes from the knife, just as light flashes from the star in the sky. The story of Blood-Clot is widespread among Plains Indians, appearing, among others, in Blackfoot, Arapaho, and Gros Ventre tales. It is the story of a hero with a miraculous birth who vanquishes evil wherever he goes.

Alice Kehoe, an archaeologist and ethnographer, suggests that the clot of blood the father picks up (and which soon becomes Blood-Clot, a man) is actually the aborted fetus of a bison. In her interpretation, Blood-Clot is not only a culture hero who rids the world of evil, but he also symbolizes the animal that provides sustenance, tools, and shelter for the Blackfoot.

A young man married the three daughters of an old man and woman. Although the old man helped his son-in-law hunt buffalo every day, the son-in-law refused to give the old couple any meat. The youngest wife, however, felt sorry for her parents and stole some meat for them.

One day when the father was on a path, he found a blood-clot and put it in his quiver, so that he and his wife could make soup from it. When his wife made the soup, she heard a baby crying and looked all around the lodge. She finally realized that there was a baby in the pot. She called the baby Blood-Clot.

The son-in-law heard the baby cry and wanted to kill it, but the old woman lied, saying the baby was a girl. The son-in-law thought that he would eventually have another wife and let the baby live.

That night, the baby told the old man to lift it up and touch its head to each of the lodge poles. As he did so, the baby grew larger and larger and became a fine young man. "I am Blood-Clot," he said, "and I will help you."

The next morning the son-in-law was once more cruel to the old man and threatened to kill him. Blood-Clot intervened, however, killing the son-in-law. Blood-Clot also killed the two older daughters, the ones who had not given meat to their parents. He gave all the dried buffalo meat in the son-in-law's lodge to the old man and women, and asked the youngest daughter to take care of her parents.

Blood-Clot set out to visit other Indians. He saved one camp from grizzly bears, killing them all except one female who was ready to give birth and who pleaded for her life. If he had not spared her there would be no grizzly bears today.

He went to another camp that was ruled by snakes. He killed them all except one female who was also ready to give birth and who pleaded for her life. Again, he spared her, and all snakes have descended from her.

As he was walking along a road, a great wind picked him up and carried him into the mouth of a huge fish. He then found other people in the fish's stomach. He tied a knife to his head and told the people that they should dance. As he jumped up and down, his knife pierced the fish's heart and the fish died. Everyone escaped through a hole he cut in the ribs.

Although he was warned not to talk with a young woman who killed people when she wrestled with them, he talked with her and agreed to wrestle with her. He saw that she had buried knives in the ground and refused to stand where she indicated. They wrestled in another place and he threw her onto the knives and killed her.

He was also warned not to ride on a swing by the river because the woman there killed everyone who did so. He tricked the woman by asking her to show him how to swing; he cut the vine and she fell into the water.

After ridding the world of these monsters, Blood-Clot returned to his parents, the old couple. One day he told them that he had to go back to the place from which he had come. "Do not be sorry for me," he said "because I will return to the sky, where I will be the Smoking-Star." Some Crow Indians later killed him, but his body disappeared. At the same time, the Smoking-Star appeared in the sky.

(Adapted from Wissler and Duvall, Mythology of the Blackfoot Indians, 53–58.)

?: The Hand in the Sky
Milky Way: Wolf Trail

Cecile Black Boy gathered narratives from the Blackfoot Reservation in Montana from 1940 to 1943 under the Montana Writers Project of the Public Works of Art Project. In her story about the creation of the Skunk Tipi design, which follows, she describes the origin of Hand in the Sky, an unidentified constellation, and the Wolf Trail, the Milky Way. Hand in the Sky could be identical to the Blackfoot Spider Man's Hand (stars in the constellation HERCULES). Or, it could be the Arapaho Hand constellation (five unidentified stars)—or it could be some other constellation entirely. Wolf's Piskun, where the story takes place, is near Calgary, Canada.

Long ago, people followed the buffalo to an area called Wolf's Piskun. After the people had gotten enough food there, both fresh and dried, they returned to their main camp, but one man and his family decided to stay where they were for a while. The man's son was a good hunter and brought the family enough to eat until winter set in and the son was no longer able to secure game. One day he went out but did not return. The father and the other son could find neither the missing man nor game. The three family members began to perish from lack of food.

A young wolf happened by and looked into the tipi. When he returned to his own camp, he told the old wolf what he had seen. The old wolf directed a group of young wolves to take the people food, being careful to turn into humans before they reached the tipi so that the family would not be frightened. The people accepted the food and eventually moved to the wolf camp.

While they were there, the father learned how to paint the design of the Skunk Tipi, and his son received a buffalo calf robe from a wolf boy. A hand had been painted in the center of the robe. "Don't ever let a woman use this robe," cautioned the wolf boy.

After a time the family returned to their own people and many years later the son got married. Although he was careful not to let his wife use the buffalo robe, she was curious and one day, when he was away hunting, she spread the robe over herself. When he returned, he was angry. "Look what you have done!" he cried. The hand had disappeared from the robe and gone into the sky. Even to this day there is a group of stars in the sky that looks just like a man's hand. There is also the Wolf Trail, along which there are groups of stars that mark where the family camped as they went to the wolf camp and then returned to their own people.

(Adapted from Oklahoma Indian Arts and Crafts Cooperative, *Painted Tipis by Contemporary Plains Indian Artists*, 77–79.

Fig. 11.11. Blackfoot tipi showing star decoration represented by circles. The six stars at the top left are the PLEIADES, the seven stars on the top right are the BIG DIPPER, and the stars around the base of the tipi are likely puffball stars. This design is called the "skunk-painted tipi." (Illustration after Oklahoma Indian Arts and Crafts Cooperative, *Painted Tipis by Contemporary Plains Indian Artists*, plate 27.)

Cheyenne

The Cheyenne moved from west of the Great Lakes onto the Great Plains by the late 1700s. As in other plains tribes, the bands lived apart during the winter and came together for the bison hunt and summertime ceremonies, including the Sun Dance. When gathered, the bands arranged their tipis in a large circular camp fashioned after the Camp Circle (NORTHERN CROWN). They decorated their tipis with designs that included the Morning Star and the PLEIADES.

To the Cheyenne, the Hanging Road (Milky Way) is the path along which spirits travel to the afterlife. The Morning Star and Evening Star merit particular esteem because of their brilliance in the heavens and their proximity to the sun and moon.

BIG DIPPER: Quillwork Girl and Her Seven Brothers

Historically, Indian women dyed porcupine quills and used them to make beautiful decorations on clothing, blankets, bags, quivers, and other skin implements. The Cheyenne story about the BIG DIPPER features Quillwork Girl, whose skill in this art was known far and wide among her people. Although Blackfoot, Assiniboin, and Crow stories about the BIG DIPPER portray the pursuer as a bear, in this account Old Bull and his herd of bison are the villains.

John Stands In Timber, a Cheyenne tribal historian, presents this tale in his book, *Cheyenne Memories*. Mothers told this tale to their children when it was time to go to sleep. The seven stars of the BIG DIPPER are the seven brothers, and as for what became of the girl, he did not know. In another version, the girl is the brightest star (Alioth, the handle star nearest the bowl), and the youngest brother is the star at the end of the handle. The small star Alcor is, apparently, the seventh brother.

Many years ago there was a girl who was well known for her quillwork. She was the only child of her parents, and though she was beautiful, she had no suitors. Thus her parents were surprised when she began making buckskin clothing for a man. It took her a whole month to make the shirt, leggings, moccasins, and other things that a man would wear. She carefully decorated each piece with brilliant designs in quillwork.

When she finished the first set, she started another, and she continued her work until she had seven sets of clothing. She told her parents, "I have no brothers or sisters, so I am going to go far away and live with seven brothers. This clothing is for them. Some day you and everyone else will know them and respect them."

Quillwork Girl and her mother placed the outfits in packs that the dogs carried, and Quillwork Girl set out. Although she had never been that way before, she seemed to know where to go. At last she came to a shallow river with a tipi on the far shore. She crossed the water and met a boy who was about ten years old. "I am the youngest," he said. "I have six older brothers who are out hunting buffalo. I knew that you were coming, but they will be surprised." The boy immediately put on his new clothing and was very pleased.

When the brothers returned, they were indeed surprised, but they admired the outfits that Quillwork Girl had made for them and ate the meal of meat, berries, and fat that she had prepared. They were overjoyed with their adopted sister and all lived content until a bison calf appeared one day.

The calf said to the youngest son, "The bison have heard of your sister and want her for themselves. She must come with me." The boy refused, saying that his brothers were away and he could not let the girl go himself. Then, a two-year-old heifer ran up to the tipi and made the same demand, but the boy again refused. Finally a bison cow appeared, but she was turned away as well. "The Old Bull himself will be next," the cow threatened as she left.

The brothers returned to the tipi and learned about what had happened and they were all afraid. Soon they heard a thunderous pounding of hooves. Old Bull, leading a herd of bison, raced toward their tipi. "Little brother," they said, "you have special powers. Save us!"

The youngest brother did indeed have unusual medicine. He had known that Quillwork Girl would be coming to live with them, and he knew what to do when the bison threatened to kill them. He picked up his quiver and shot an arrow into a nearby tree, which began to grow. The seven brothers and Quillwork Girl escaped into the tree just as the bison surrounded it, pawing and snorting. Old Bull began butting against the trunk, ripping out great chunks of wood, until the trunk began to sway. The little brother shot another arrow and the tree grew higher toward the sky. Old Bull butted again, and just as the tree began to fall, Quillwork Girl and her brothers leaped from the branches into the sky and became stars.

You can still see the seven brothers and Quillwork Girl. They are the stars called the Big Dipper. Quillwork Girl is the brightest of them all. She is busy filling the sky with her beautiful, shimmering designs.

(Adapted from Stands In Timber and Liberty, *Cheyenne Memories*, 16–19; and Erdoes and Ortiz, *American Indian Myths and Legends*, 205–209.)

Fig. 11.12. The Cheyenne modeled their camp circle after the constellation Camp Circle (NORTHERN CROWN). The circle opens to the east, as does the NORTHERN CROWN during the evenings of late summer and fall. The Council tipi was erected in the middle of the circle. (Illustration after Dorsey, The Pawnee Mythology, plate 19; Lowie, *Indians of the Plains*, fig. 63.)

PLEIADES: Possible Sack and Her Brothers
PLEIADES: Seven Puppies

The Cheyenne tell a similar story about Possible Sack, who lives with her brothers. In this story, the bull actually seizes her and carries her away. The youngest brother rescues her and transforms her, the other brothers, and himself into the PLEIADES.

In a second story about the PLEIADES, a woman gives birth to puppies that become stars. The theme of a woman giving birth to puppies is common throughout the Inuit and tribes of western Canada.

A chief had a fine-looking daughter, who had a great many admirers. At night she was visited by a young man but did not know who he was. She worried about this, and determined to discover him. She put red paint near her bed. At night he crawled on her bed, wearing a white robe. She put her hand into the paint and then on his back.

The next day she told her father to call all the young men to a dance in front of his tent. They all came, and the whole village turned out to see them. She watched all that came, looking for the mark she had made. As she turned, she saw one of her father's dogs with the mark on his back. This disheartened her so that she went straight into the tent. This broke up the dance.

The next day she went into the woods near the camp, with the dog on a string, and hit him. He eventually broke loose. She was very unhappy. Several months later she bore seven pups. She told her mother to kill them, but her mother was kind toward them and made a little shelter for them. They began to grow, and at night the old dog sometimes came to them.

After a time, the woman began to take interest in them and sometimes played with them. When they were big enough to run, the old dog came and took them away. When the woman went to see them in the morning, they were gone. She saw the large dog's tracks, and several little ones, and followed them a distance. She was sad and cried. She came back to her mother and said, "Mother, make me seven pairs of moccasins. I am going to follow the little ones, searching for them."

Her mother made seven pairs of moccasins, and she started out, tracking them all the way. Finally, in the distance, she saw a tent. The youngest one came to her and said, "Mother, father wants you to go back. We are going home. You cannot come."

"No. Wherever you go, I go," she said. She took the little one and carried him to the tent. She entered and saw a young man who, however, took no notice of her. He gave her a little meat and drink, which did not grow less however much she ate. She tied the little pup to her belt with a string. Next morning, she was left alone and the tent was gone. She followed and again came to them. Four times this happened in the same way, but the fourth time the tracks stopped. She looked up and there she saw seven pups; they were stars (the Pleiades).

(Reprinted from Kroeber, Cheyenne Tales, 181–182.)

Arapaho

The Arapaho call themselves *Hinanaeina* (Our People). They once lived near the headwaters of the Mississippi but moved to the western edge of the plains, where they served as traders to both the northern and southern plains tribes, earning them the Pawnee name *Arapaho* (He Who Trades).

The Arapaho used star symbols on many items, including leggings, moccasins, dresses, pouches, and whistles. The Morning Star appears in the shape of a Maltese cross, a diamond, and a traditional cross. The Milky Way is represented by two parallel lines with diagonal stripes. A horseshoe of six diamonds is thought to portray the NORTHERN CROWN.

Stars and Constellations

In an Arapaho story of a girl captured by bison bulls, a mole rescues the girl and the girl saves herself and her brothers by bouncing a ball into the air. They all rise with it and become the PLEIADES. The Arapaho word for the PLEIADES means "bison bulls."

The following story lists the PLEIADES and Seven Bison Bulls as separate constellations, so perhaps there were two star groups associated with bison. Or, as is sometimes the case with the BIG DIPPER and the PLEIADES—the two prominent constellations with seven stars—stories that originally described one constellation were confused with or merged with the other.

Many creation tales describe how the Earth was originally flooded and how an animal dove far beneath the waters to secure soil to make land. The following myth embraces this Earth-diver theme and describes the creation of the Sacred Wheel. The wheel is part of the Offering Lodge, a ceremony into which the Arapaho later incorporated special reverence for the sun. The rays of the wheel are considered to be rays of the sun, from which life comes.

First Man inscribed constellations—including the Old Camp (NORTHERN CROWN), Lone Star (Evening Star), Morning Star, PLEIADES, Seven Bison Bulls (the BIG DIPPER?), the Milky Way, the Hand, and the Lance—on the first Sacred Wheel. The Hand is a constellation of five stars, and nearby it is the Lance (or bow), several stars in a long row. The Hand and the Lance are unidentified.

At one time the earth was completely covered with flood waters. A man walked around on the water holding his friend, the pipe, and wondered how he might find land where he

Fig. 11.13. An Arapaho girl's legging includes celestial designs. The two triangles on the side, near the ankle, are mountains, and the Morning Star lies between them (*inset, top*). Along the back above the heel lies the Morning Star as it is rising (the diamond), and when it is high in the sky (the two crosses) (*inset, second and third from top*). The strip of leather that runs up the front of the legging incorporates the Milky Way along its outer edges (*inset, fourth from top*). The six connected diamonds ringing the foot represent a constellation, probably Camp Circle (NORTHERN CROWN) (*inset, bottom*). (Illustrations from Kroeber, *The Arapaho,* fig. 6.)

could keep his friend in safety. After fasting and thinking a great deal, he decided that there should be an earth with many kinds of animals living upon it. There, he would be able to protect his pipe.

He called all of the animals together and laid out his plan. First, they had to find soil from which to make the earth, but none had any idea where to look. Finally the turtle said, "I hear it, far below us, far below the water. There is soil there."

The animals each dove in turn, trying to locate the soil. The turnstone, the waterfowl, the kingfisher, the otter, the beaver—on and on they went, but no one was successful. At last the turtle asked to be allowed to go. He and the man, who had turned into a red-headed duck, dove together. They each brought back a bit of clay, and the man (for he had changed back) created the soil of the earth from these particles of clay. The land he made was great and strong and broad. He placed his friend the pipe on the land and was pleased. Then, animals offered parts of themselves to be kept with the pipe, the eagle his feathers and the buffalo his robe, and so on, until the sacred items were complete. At this time another man appeared and helped make mountains and hills and creeks of all sizes.

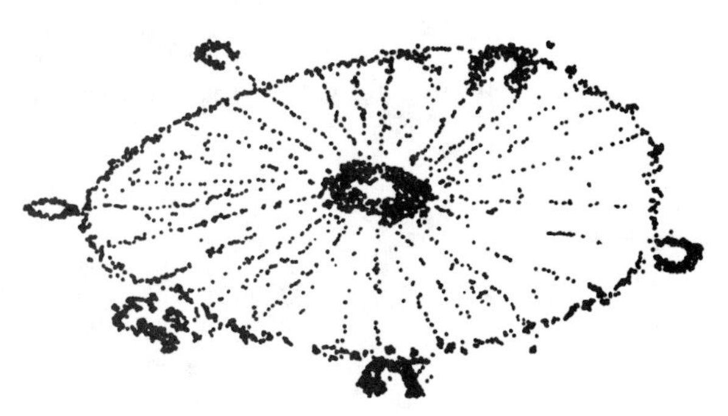

Fig. 11.14. The Big Horn Medicine Wheel in Wyoming may have been constructed to track the rising of the sun and the positions of various important stars. (Illustration after U.S. Forest Service photograph.)

Then the first man made the Wheel from a bush that has dark red bark and is very flexible. He decorated the wheel with four bunches of eagle feathers at the four quarters and inscribed the Wheel with a cross (representing the Morning Star), the Evening Star, a bunch of stars (the Pleiades), the Seven Bison Bulls, the Hand, the Lance, the Old Camp, the Sun, the Moon, and the Milky Way.

(Adapted from Dorsey, *The Arapaho Sun Dance: The Ceremony of the Offerings Lodge*, 101–205.)

The Medicine Wheel

The Big Horn Medicine Wheel in Wyoming is a circular arrangement of rocks with spokes radiating from a center and a rim that connects the spokes. At various intervals along the rim there are additional piles of rocks. Although there are several dozen such circular patterns across the northern Plains, astronomers have looked most closely at the Big Horn Medicine Wheel to gain insights into its use. One astronomer suggests that Native people built the wheel to mark the summer solstice and used it as a calendar and in ritual, and that they last used it for these purposes two hundred to seven hundred years ago. He has offered evidence that the wheel may have been constructed to sight the rising of the sun at the summer solstice, and possibly to mark the positions of Aldebaran, Rigel, and Sirius. Another astronomer has suggested an alignment for Fomalhaut, Spica, and the PLEIADES.

Although the Big Horn Medicine Wheel's exact history is unknown, the area has long been held sacred by Native Americans. It is likely that only the holy people of a tribe visited the wheel, for it is set high in the mountains and is not suitable for a large encampment. Around 1900, two Sioux said that it belonged to the Cheyenne and Arapaho, and some Crow said that it had been made by "people who had no iron," apparently people who lived before the arrival of Euro-Americans and their metal tools. Elk River, a Cheyenne, said that the Big Horn Medicine Wheel was similar in shape to a medicine lodge. The medicine wheel has long been associated with the Sun Dance Lodge.

Because the Arapaho creation myth describes how First Man made the Sacred

Wheel and decorated it with constellations, one interpretation of the Medicine Wheel may be that it is "decorated" with stars that appear along the rim in the sky at certain times of the year. The Sacred Wheel could be the prototype of the Medicine Wheel, and both then evolved to places of importance in the Sun Dance ceremony.

CONSTELLATIONS OF THE GREAT PLAINS

Culture Group	Star or Constellation	Classical Equivalent
Caddoan-Speaking Tribes		
Pawnee	Red Star	Antares?
Pawnee	Big Stretcher	BIG DIPPER
Pawnee	South Star; Spirit Star	Canopus?
Pawnee	Yellow Star	Capella?
Pawnee	Rabbit	CASSIOPEIA
Pawnee	Panther	CHARIOTEER?
Pawnee	(Second) Black Star	Deneb?
Pawnee	Bow	DOLPHIN
Pawnee	Feather Headdress Star	Great Nebula in ORION?
Pawnee	Little Stretcher	LITTLE DIPPER
Pawnee	Big Black Meteoric Star's Daughters	LYRE?
Pawnee	Great Star; Morning Star	Mars
Pawnee	Great Star's Brother	Mercury or Jupiter?
Pawnee	Bright Light in a Long Stretch Across the Heavens	Milky Way
Pawnee	Dust from Bison and Horse Race	Milky Way
Pawnee	Pathway of Departed Spirits	Milky Way
Pawnee	Star That Does Not Move	North Star
Pawnee	Council of Chiefs	NORTHERN CROWN
Pawnee	Bird's Foot	NORTHERN CROSS?
Pawnee	Deer	ORION
Pawnee	The Seven Stars	PLEIADES
Pawnee	Pahukatawa's Kneeprint	Star in PERSEUS or CEPHEUS?
Pawnee	Wind Ready to Give; Breathe	Two stars in PERSEUS?
Pawnee	Wildcat Star	Procyon?
Pawnee	Bear	SAGITTARIUS?
Pawnee	Star of Sickness	Saturn?
Pawnee	Swimming Ducks	SCORPIUS's tail
Pawnee	Real Snake	SCORPIUS

CONSTELLATIONS OF THE GREAT PLAINS *(cont.)*

Culture Group	Star or Constellation	Classical Equivalent
***Caddoan-Speaking Tribes** (continued)*		
Pawnee	Snake Not Real	SERPENT?´
Pawnee	White Star	Sirius?
Pawnee	Wolf Star	Sirius?
Pawnee	Big Black Meteoric Star	Vega?
Pawnee	Bright Star; Evening Star	Venus
Wichita	The Seven	BIG DIPPER
Wichita	South Star	Canopus?
Wichita	First Man	Morning Star
Wichita	Flint Stone Lying Down Above	Unidentified
Arikara	Invalid Being Carried	BIG DIPPER
Arikara	Girl and Brothers	PLEIADES
Arikara	Winding Trail	Unidentified
Arikara	Two Holy Fathers	Unidentified
Arikara	Star Boy	Unidentified
Siouan-Speaking Tribes		
Omaha	The Litter	BIG DIPPER
Omaha	Big Star	Morning or Evening Star
Omaha	Spirits' Path	Milky Way
Omaha	Star That Does Not Walk	North Star
Omaha	Brothers	NORTHERN CROWN?
Omaha	Large Foot of Goose	ORION's belt
Omaha	Deer's Head; Little Duck's Foot	PLEIADES
Osage	Stretcher or Litter	BIG DIPPER
Osage	Wolf That Hangs at the Side of the Heavens	BIG DOG or Sirius
Osage	Great Star or Red Star	Morning Star
Osage	Red Star	North Star
Osage	Three Deer	ORION's belt
Osage	Stars Strung Together	ORION's sword
Osage	Deer Head	PLEIADES
Dakota	Carrying the Dead Man	BIG DIPPER
Dakota	Trail of Bubbles	Milky Way
Dakota	Largest Star	Morning Star, Evening Star, others
Dakota	*Ta´maopa´*	DOLPHIN?
Dakota	Sweat Tipi	Unidentified
Lakota	Seven Stars	BIG DIPPER
Lakota	Spirit Way	Milky Way

CONSTELLATIONS OF THE GREAT PLAINS *(cont.)*

Culture Group	Star or Constellation	Classical Equivalent
***Siouan-Speaking Tribes** (continued)*		
Lakota	Star That Does Not Move	North Star
Lakota	Hand	ORION
Lakota	Bear's Lodge	Stars around GEMINI
Assiniboin	Seven Stars	BIG DIPPER
Assiniboin	Backbone of Sky	Milky Way
Assiniboin	Star Shining Daylight Chases	Morning Star
Assiniboin	North Star	North Star
Assiniboin	Wise-One and His Brothers	PLEIADES
Mandan	Woman Who Wears a Plume	Morning Star
Mandan	Woman of the West	Evening Star
Hidatsa	Ermine	BIG DIPPER
Hidatsa	Ashy Way	Milky Way
Hidatsa	Star That Does Not Move	North Star
Hidatsa	Hand Star (the hand of Long Arm)	ORION
Crow	Girl and Brothers; Seven Stars	BIG DIPPER
Crow	Old Woman's Grandchild	North Star; Morning Star
Crow	Hand Stars	ORION
Crow	Tobacco Plant	Unidentified
Algonquian-Speaking Tribes		
Blackfoot	Seven Brothers and Their Sister	BIG DIPPER
Blackfoot	Ashes Chief and Stuck-Behind	Castor and Pollux
Blackfoot	Smoking-Star	Great Nebula in ORION
Blackfoot	Spider Man's Fingers	HERCULES
Blackfoot	Poïa; Young Morning Star; Mistake Morning Star	Jupiter
Blackfoot	Big Fire Star; Big Fire Eater	Mars
Blackfoot	Wolf Trail	Milky Way
Blackfoot	Star That Stands Still	North Star
Blackfoot	Lodge of the Spider Man	NORTHERN CROWN
Blackfoot	Bunched Stars; Lost Children	PLEIADES
Blackfoot	Early Riser; Day Star; Morning Star (Poïa's father)	Venus
Blackfoot	Hand in the Sky	Unidentified
Gros Ventre	Cut-Off-Head	BIG DIPPER
Cheyenne	Quillwork Girl and Her Brothers	BIG DIPPER
Cheyenne	Hanging Road	Milky Way

CONSTELLATIONS OF THE GREAT PLAINS *(cont.)*

Culture Group	Star or Constellation	Classical Equivalent
***Algonquian-Speaking Tribes** (continued)*		
Cheyenne	Camp Circle	NORTHERN CROWN
Cheyenne	Possible Sack and Her Brothers	PLEIADES
Cheyenne	Puppies	PLEIADES
Arapaho	Seven Bison Bulls	BIG DIPPER?
Arapaho	Lone Star	Evening Star
Arapaho	Old Camp	NORTHERN CROWN
Arapaho	Girl and Brothers; Bison Bulls	PLEIADES
Arapaho	Hand	Unidentified
Arapaho	Lance	Unidentified
Uto-Aztecan-Speaking Tribes		
Kiowa	Girl and Brothers	BIG DIPPER

Chapter twelve

The Celestial Skiff

Stars of the Southeast

The Native Americans considered in this chapter—the Cherokee, Creek, Seminole, Alabama, Koasati, Caddo, and Natchez—lived in well-developed towns located in the fertile valleys of the Southeast. These Native people were predominantly farmers, cultivating corn, squash, beans, and pumpkins. Agriculture is a more efficient means of securing food per acre of land than hunting, and the Southeast was more densely settled than areas where hunting provided most of the food. The people of the Southeast did supplement crops by hunting deer, turkey, bear, and other game. They constructed their homes of mud plaster, poles, brush, and sometimes hides; many families had both summer and winter homes. They conducted religious ceremonies in council houses or the town square.

The languages spoken in the Southeast include dialects of the Iroquoian language family (Cherokee), dialects of the Muskogean language family (Creek, Seminole, Alabama, Koasati, Natchez), and a dialect of the Caddoan language family (Caddo).

Many of the tribes of the Southeast divided the cosmos into three parts, Upper World, This World, and Under World. Upper World is situated above a vault of solid rock that forms the sky. Large animals, organized into towns with chiefs and councils, live in the Upper World. These animals are able to perform remarkable deeds, such as transforming themselves into other beings. This World is a large, flat island suspended from the vault by four cords. The sun, moon, and stars are fastened on the underside of the rock vault. At some time in the past, the great animals from Upper World visited and lived in This World, but they did not remain. The animals of This World are only imitations. Under World, which lies beneath the Earth, is a place of snakes, ghosts, and monsters.

Only a limited number of star stories has been recorded from the Southeast culture area. By the time that ethnologists began work in the late 1800s and early 1900s, the tribes had sustained many years of Euro-American contact, including for most the forced removal from their homeland. What we have left are a handful of star names and stories that offer a glimpse into what must have been a much richer body of information.

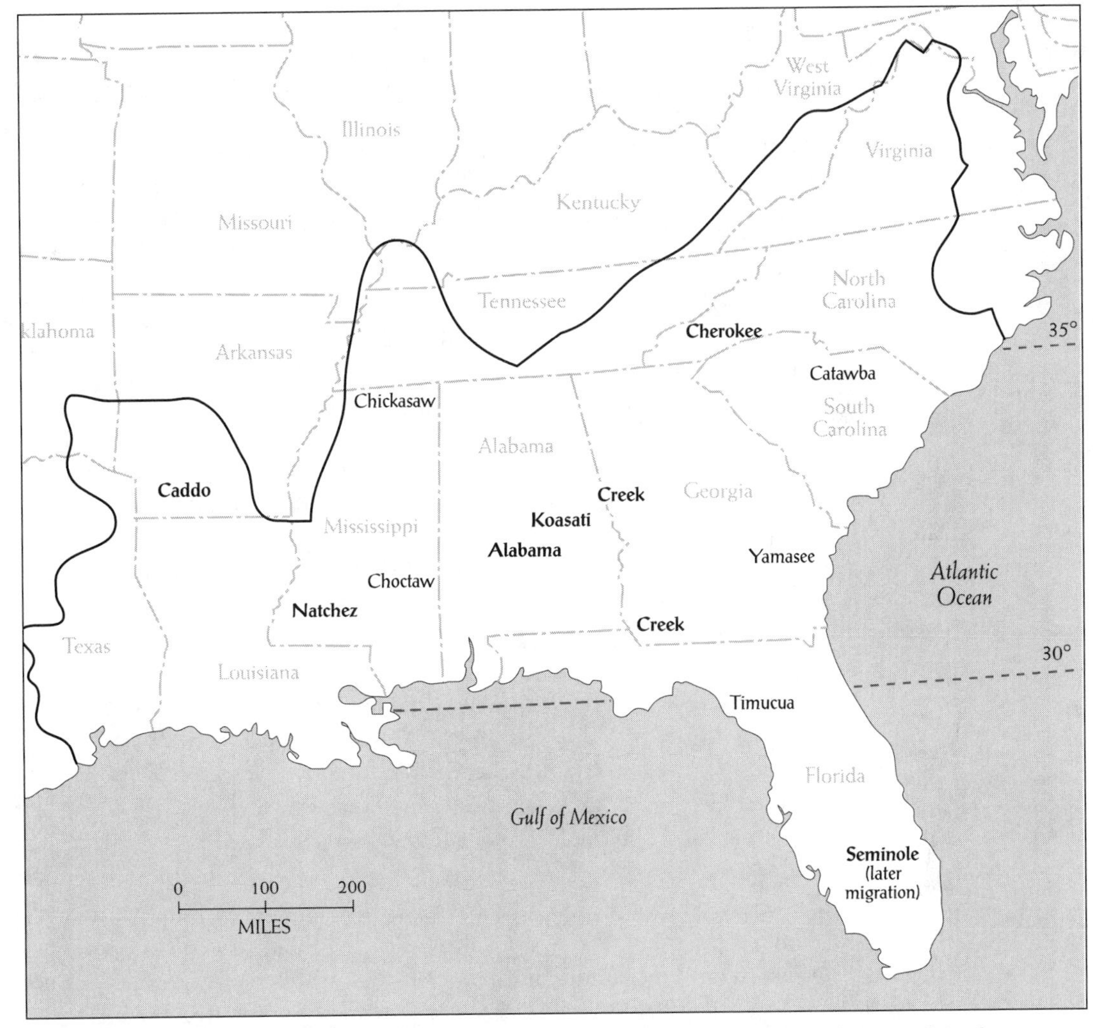

Fig. 12.1. Tribes of the Southeast culture area and their approximate territories at the time of Anglo-European contact. The tribes indicated in **boldface type** are included in this chapter. Not all tribes that inhabited this area are shown. (After Hudson, *The Southeastern Indians*, 6–7.)

Cherokee

The Cherokee, who call themselves *Ani-yun-wiya* (Real People), were loosely organized in a confederacy of about a hundred villages in the mountains of what is now western Virginia, North Carolina, South Carolina, eastern Tennessee, and northern Georgia. Today, the Eastern Cherokee still live in the Smoky Mountains of North Carolina, and the descendants of the Cherokee who were forcibly moved to Indian Territory in Oklahoma in the "Trail of Tears" live in Oklahoma.

The Cherokee raised many crops, including different varieties of corn for roasting, boiling, and making into flour. They also gathered wild foods, hunted, and fished. Each village had a White Chief (Most Beloved Man), who advised on issues involving village life, and a Red Chief, who advised on matters of war.

Sequoyah, a Cherokee, developed an alphabet in the early 1800s, enabling some cultural information to be recorded by the people of his tribe. Still, the oral tradition played a major part in passing along information. James Mooney, a late-nineteenth-century ethnologist from the Smithsonian Institution, wrote that Cherokee storytellers used to begin their tales by saying, "This is what the old men told me they had heard when they were boys."

Creation Myth

The following segment of the Cherokee creation myth involves the Great Rabbit (or the Great White Rabbit), the trickster who is prominent in many Native American tales of the eastern United States. Tribes of the Northeast variously called the rabbit *Winabojo, Nanabush, Nanabozho, Menapus, Manabozho,* or *Manabush*. The Great Rabbit is a culture hero, a worker of wonders, a being who can—like Coyote of the West—be both wise and stupid. The rabbit appears in later folktales of the Southeast, including the Uncle Remus tales of Brer Rabbit.

The Seven Stars are PLEIADES.

Before there were any stars in the sky, the Great Rabbit, who was horned, told the people to watch the eastern sky by night, and they would see the Great Star rise. Then he told them that seven days later, when the Great Star was overhead, the seven stars of the Pleiades would appear. Until then the Great Star was the only star in the sky. So the people watched this star to know when the Seven Stars would rise; and while they were waiting for them, the Great Rabbit directed them to make a great feast, with dances and all kinds of food, in honor of their coming. As they first appeared in spring, when all the people were happy, they were hailed with joy, and were greatly loved and reverenced. The Great Rabbit then told the people that the Great Star had been sent as a messenger to announce the coming of the sun; the Seven Stars, to warn them of the approach of the Deluge.

The Great Rabbit was the largest and brightest of all the rabbits. He was drowned in the Flood; but the present race of rabbits sprang up after the Flood subsided, from pieces of his body that were washed ashore.

(Reprinted from Hagar, Cherokee Star-Lore, 361.)

The Nature of the Stars

Swimmer, a Cherokee wise man, used Sequoyah's alphabet to record sacred tribal information. Swimmer also told Mooney about traditions of his people and, in the following story, described the nature of stars.

There are different opinions about the stars. Some say they are balls of light, others say they are human, but most people say they are living creatures covered with luminous fur or feathers.

One night a hunting party camping in the mountains noticed two lights like large stars moving along the top of a distant ridge. They wondered and watched until the light disappeared on the other side. The next night, and the next, they saw the lights again moving along the ridge, and after talking over the matter decided to go on the morrow and try to learn the cause. In the morning they started out and went until they came to the ridge, where, after searching some time, they found two strange creatures about *so* large (making a circle with outstretched arms), with round bodies covered with fine fur or downy feathers, from which small heads stuck out like the heads of terrapins. As the breeze played upon these feathers showers of sparks flew out.

The hunters carried the strange creatures back to the camp, intending to take them home to the settlements on their return. They kept them several days and noticed that every night they would grow bright and shine like great stars, although by day they were only balls of gray fur, except when the wind stirred and made the sparks fly out. They kept very quiet, and no one thought of their trying to escape, when, on the seventh night, they suddenly rose from the ground like balls of fire and were soon above the tops of the trees. Higher and higher they went, while the wondering hunters watched, until at last they were only two bright points of light in the dark sky, and then the hunters knew that they were stars.

(Reprinted from Mooney, *Myths of the Cherokee*, 257–258.)

Morning Star: Wicked Conjurer

The Cherokee believe that beings called Thunderers live above the sky as well as on cliffs and mountains. The sky Thunderers help people who pray to them, but the other Thunderers play pranks.

The Thunderers of the sky venerated the Morning Star, but rather as an object of fear. They say that very long ago, a wicked conjurer committed murder by witchcraft. The people combined to slay him, but divining their purpose, he gathered the shining implements of his craft around him and sprang upwards to a great height, where his possessions made him seem a star. He then became fixed in his position, and his aid is sought by all who endeavor to kill others by magic.

(Reprinted from Squier, Ne-she-kay-be-nais, or the Lone Bird: An Ojibway Legend, 256.)

PLEIADES: *Ani´tsutsă,* the Boys
PLEIADES: *Unadatsugi,* the Group

The following two stories about the PLEIADES illustrate variations on a theme. The first tale describes boys, perhaps gamblers, who disobey their mothers and forget their duties. The story ends by explaining how the pine tree comes to the Earth. The second version, which also involves obstinate children, specifically links them with the ceremonial Feather Dance used to ensure good crops.

The first story involves a game called *gatayû´stĭ,* which was popular with many Southeast tribes. The object of the game is to hit a stone disk with a special spear that is up to eight feet in length. Points are awarded for coming close to or hitting the stone. Participants and audience often gamble on who will win.

The pine tree mentioned here is likely yellow pine (shortleaf pine), which grows throughout the South.

Long ago, when the world was new, there were seven boys who used to spend all their time down by the town council-house playing the *gatayû´stĭ,* throwing a stone wheel along the ground and sliding a curved stick after it to strike it. Their mothers scolded, but it did no good, so one day they collected some *gatayû´stĭ* stones and boiled them in the pot with the corn for dinner. When the boys came home hungry their mothers dipped out the stones and said, "Since you like the *gatayû´stĭ* better than the cornfield, take the stones now for your dinner."

The boys were very angry and went down to the town council-house, saying, "As our mothers treat us this way, let us go where we shall never trouble them any more." They began a dance—some say it was the Feather Dance—and went round and round the council-house, praying to the spirits to help them. At last their mothers were afraid something was wrong and went out to look for them. They saw the boys still dancing around the townhouse, and as they watched they noticed that their feet were off the earth, and that with every round they rose higher and higher in the air. They ran to get their children, but it was too late, for they were already above the roof of the town-house—all but one, whose mother managed to pull him down with the *gatayû´stĭ* pole, but he struck the ground with such force that he sank into it and the earth closed over him.

The other six circled higher and higher until they went up to the sky, were we see them now as the Pleiades, which the Cherokee still call *Ani´tsutsă* (the Boys). The people grieved long after them, but the mother whose boy had gone into the ground came every morning and every evening to cry over the spot until the earth was damp with her tears. At last a little green shoot sprouted up and grew day by day until it became the tall tree that we call now the pine, and the pine is of the same nature as the stars and holds in itself the same bright light.

(Reprinted from Mooney, *Myths of the Cherokee*, 258–259.)

Three Cherokees—Swimmer, Ta´gwadihĭ, and Suyeta—tell nearly identical forms of this second tale of seven boys who rise into the sky. The Cherokee believe that the stars in PLEIADES command greater power than other stars and can influence the weather and crops. Tribal members can appease the star group by holding the Feather Dance, also called the Eagle Dance, in which each dancer holds seven eagle feathers.

The Great Rabbit had told the Cherokee to note the position of PLEIADES to determine the seasons and time for planting. Long ago, the Cherokee held a dance for PLEIADES and the Morning Star each month.

Stansbury Hagar, who recorded the Micmac Never-Ending Bear Hunt story, recorded many Cherokee tales, including this one about the PLEIADES. He refers to a nebula near one of the stars of the PLEIADES as the drum to which the boys danced. A nebula is a concentrated cloud of gas and dust that shows up as a hazy area in the sky.

Seven boys came together to practice shooting with bows and arrows. They used a bundle of corn-cobs as a target, sometimes penetrating as many as five cobs at a time. They would continue to shoot until all their arrows were discharged. Then, they would gather them up, move the target farther on, and renew the match. They kept this up till their mothers grew weary of it, and told them, that, if they must keep shooting at things not fit to eat, they must go elsewhere to do it.

So the boys went away around the side of a hill and disappeared. They remained away so long that their parents at length became anxious and searched for them. Soon they perceived the boys dancing the Feather Dance in a circle, accompanied by the sound of an ancient drum, *ahúlĭ*. But the parents noticed that as they danced, the boys were gradually rising higher and higher from the ground.

In alarm, they picked up poles and tried to knock the boys down, but the children were already too high to be reached. The boys continued to dance, gradually ascending higher and higher all the while, until at last they were but specks in the sky. As they were rising, their parents called to them in vain, urging them to return. They only replied, "We have found a place now where we can shoot as much as we please."

Thus the boys became the Seven Stars of the Pleiades and the drum, the nebula near one of those stars. There were seven stars for seven days; then one fell to the earth, leaving a fiery trail. All the people gathered together to know what this might mean. They found that the star had become a bearded man, who sat down and warned them of the coming of the flood. He remained on earth seven years, then disappeared, leaving his footprint on a rock. The stars are still called the Seven Stars, although there have since been but six.

(Reprinted from Hagar, Cherokee Star-Lore, 358–359.)

BIG DIPPER: The Bear and Hunters

As with many other tribes, hunting provides the basis for Cherokee BIG DIPPER stories. In the following tale, three hunters and a bear form the BIG DIPPER. Honeydew is a sweet deposit that aphids and other insects secrete on the leaves of plants. "Trying out" fat involves melting it and rendering it into a pure form.

The four stars of the bowl are a bear (*yânû*), which is pursued by three—or, as some say, seven—hunters (*aníkanátĭ*), represented by the three stars of the handle. The middle hunter carries a pot (*tu'sdĭ*), in which they are going to cook the bear when killed. The bear comes from his den (*ustágălûñĭ*) in spring. The hunters pursue him across the sky in summer and kill him in August. The honeydew that falls then comes from his fat, which they are trying out over a fire.

(Reprinted from Hagar, Cherokee Star-Lore, 357.)

In this next tale, the seven stars are seven hunters.

A party of hunters, having killed a large bear, camped, made a fire, and cooked some of the meat. After eating they started home, carrying what remained of the carcass. As they journeyed, they noticed that the landscape was gradually rising in front of them, but they kept on. In this way they reached the land of the sky.

The next evening people on earth first saw the seven stars. If you watch them closely, you will see that they are hunters, carrying a load of bear meat on their backs. They must carry it forever now, because they have lost their way, and will eternally circle around and around, but will never find their way home again.

(Reprinted from Hagar, Cherokee Star-Lore, 357–358.)

HYADES: The Arm

The HYADES is a V-shaped cluster in TAURUS. To the Cherokee, this V is a human arm bent at the elbow. This story depicts the sad fate of a hunter who could no longer hunt.

A hunter, having broken his arm, grieved much to feel that he was now useless in hunting and war. He left his home and wandered eastward to meet the rising sun. Some time after this, people on earth saw the Arm appear for the first time, and they recognized in these stars the arm of this man who had wandered into the sky-land because he was injured and felt useless on earth.

(Reprinted from Hagar, Cherokee Star-Lore, 365–366.)

Milky Way: Where the Dog Ran

The following two stories describe how spilled cornmeal becomes the Milky Way. The stories underscore the prominent place that corn had in the diet of Southeast tribes. In a present-day telling of the story, Mary Ulmer Chiltoskey depicts the dog as a very large spirit dog that can fly through the sky.

Some people in the south had a corn mill, in which they pounded the corn into meal, and several mornings when they came to fill it they noticed that some of the meal had been stolen during the night. They examined the ground and found the tracks of a dog, so the next night they watched, and when the dog came from the north and began to eat the meal out of the bowl they sprang out and whipped him. He ran off howling to his home in the north, with the meal dropping from his mouth as he ran and leaving behind a white trail where now we see the Milky Way, which the Cherokee call to this day *Gi'lĭ´-utsûñ´stănûñ´yĭ* (Where the Dog Ran).

(Reprinted from Mooney, *Myths of the Cherokee*, 259.)

The next account, in which the stolen wife symbolizes summer, explains why the south but not the north has a relatively warm climate year-round. The sequence of events portrayed here—the theft, the melting of the winter snows, and the eventual return of the woman to the south and cold to the north—occurs year after year throughout time.

There were once two hunters in the sky—one who lived in the north and hunted big game, another who lived in the south and hunted small game. The former became jealous of the latter, and one day, perceiving the southern hunter's wife grinding corn into meal, he seized her and carried her away from the corn-beating place far across the sky to his home in the north. Her dog ate the meal that was left, then followed her across the sky; and the food fell from his mouth as he ran, forming a trail of meal, the Milky Way. But when the northern hunter arrived home with his southern captive, such was the spell of her presence that the weather became warmer and warmer, until all the ice in that region began to melt. At length the northern hunter could no longer endure the heat, so he was compelled to release his prisoner. She returned home with her dog, and the weather in the north resumed its normal aspect.

(Reprinted from Hagar, Cherokee Star-Lore, 364–365.)

Sirius and Antares: Two Dog Stars

The Cherokee identify two dog stars that guard opposite ends of the path of souls. These two stars, Sirius and Antares, are not in the sky at the same time; the former appears in the winter sky, and the latter is seen in the summer sky. Both stars lie in or near the Milky Way. The path mentioned is presumably the Milky Way, and the pass is where the Milky Way touches the horizon.

Souls, after the death of the body, cross a raging torrent on a narrow pole, from which those of the evil-doers and cowardly fall off and are swept to oblivion in the waters below. Those who succeed in crossing go eastward, and then westward to the Land of Twilight. They follow a trail until they reach a pass beyond which the trail forks. There they encounter a dog (*gilĭ´*), who must be fed, otherwise he will not permit the soul to pass. Having left him behind, the soul continues to follow the trail until it encounters another dog, who must also be propitiated with food. The unfortunate soul who is insufficiently provided with food for both dogs, having passed one, will be stopped by the other. The first will not permit him to return, and he will be held a prisoner forever between the two animals.

(Reprinted from Hagar, Cherokee Star-Lore, 362–363.)

Other Stars

The sun's daughter died after being bitten by a venomous serpent. Her body was placed in a wooden box, and this box became a constellation, *Gûnesûñĭ* (Wooden Box), which has not been identified.

Creek, Alabama, Koasati, and Seminole

Historically, the Creek, Alabama, and Koasati lived in the mountains, piedmont, and coastal plain of what is now Georgia, Florida, and Alabama. Creek bands were loosely allied into the Creek Confederacy, a group that also included the Alabama, Koasati, and other tribes. Several Creek groups migrated south to Florida in the 1700s and early 1800s, becoming the Seminole. The federal government forced these tribes, like others in the Southeast, to move to Indian Territory in Oklahoma, where they are today. Some Native people resisted the move, however, and there are populations of Creek in Alabama and of Seminole in Florida.

Creek villages were separated into two sections, one for warriors and one for everyone else. Each village had both a leader and a council of advisors. The Creek honor Hisagia-imisi (Preserver of Breath, or One Sitting Above) as the creator.

The Seminole honor the same hero, calling him Hisagita misa (Breath Maker). Breath Maker made the Milky Way and gave the Seminole the pumpkin plant; another being, Little Give, presented corn to the people and showed them how to grow, store, and cook it. The Seminole lived in open-air structures with platforms called *chickee*s and used dugout canoes.

For the Creek, Seminole, and Cherokee, the Green Corn Dance—also called the *Busk*—marked the New Year. They held the ceremony, which involved four to six days of fasting, rituals, games, dancing, and feasting on corn and other foods, in July or August. They extinguished all fires and started a new fire, known as "breath master," at the center of the square and carried it to each home. The ceremony was and is a time of personal cleansing and forgiveness.

Stars and Constellations

The Creek call Morning Star *Hayàtitca* (Bringer of Daylight), North Star *Kolasniegu* (Stationary Star), and PLEIADES *Tukàbofkà* (untranslated). The Milky Way is *Poya fik-tcàlk innini* (Spirits' Road). The BIG DIPPER is *Pīlo hagi* (Image of a Canoe).

The Seminole call Morning Star *Apak k'aci* (Tomorrow Star) and Evening Star *Owaci'kidisci* (Red Star). They, like the Creek and Alabama, describe the BIG DIPPER as a boat; the stars are *Piklici k'abi* (Boat Image). The Milky Way, created by Breath Maker, shines more brightly when someone dies so that the path is well lighted.

To the Koasati, the PLEIADES are Cluster-Stars, lazy people who prefer dancing and traveling to working. They planted only pole beans, not corn, and soon after they ate their beans, they disappeared and became stars.

Another Koasati constellation, Stars in a Row, is three people who set out to cut down a bee tree. One carries an ax, another a deerskin, and the third a spoon. They get lost, however, and never return. These stars have not been identified, but the two most obvious sets of three stars, ORION's belt and Altair with its flanking stars, are possibilities; they are not otherwise identified in Native American stories of the Southeast.

BIG DIPPER: The Celestial Skiff

The following story was recorded around 1912 from members of the Alabama tribe living in Texas. Although the story mentions a celestial skiff, it does not identify any stars. A recent interpretation by astronomer Ray Williamson, however, suggests that the skiff may be the BIG DIPPER.

Williamson writes that of the few constellations that could qualify as the celestial canoe, only the BIG DIPPER fits all of the requirements. The Alabama lived at about thirty degrees north latitude, where the BIG DIPPER lies relatively low in the sky. As the BIG DIPPER moves around the North Star in the summer, it touches the horizon and then disappears for several hours (the time during which a ball game could be played during the *Busk* celebration), and then reappears as dawn breaks. The Alabama call the bowl of the BIG DIPPER *Hotci'li pi'la* (Boat Stars). Williamson suggests that the sky people descend to the Earth in these boat stars and play ball while the BIG DIPPER sinks below the horizon. After several hours the BIG DIPPER reappears, and the people climb back into the boat and return to the sky. One sky woman who has been kidnapped remains on Earth. She and her family occasionally visit the sky, using the boat and a tiny skiff (one or more stars in the handle of the BIG DIPPER).

The ball game mentioned here is similar to lacrosse.

Some people descended from above in a canoe singing and laughing. When they reached the earth they got out and played ball on a little prairie. As soon as they were through they got into the canoe again, singing and laughing continually, ascended toward the sky, and disappeared. After an interval they descended to the same place, singing and laughing, got out, and played ball again. When they were through they went back, got into the canoe, ascended toward the sky, and disappeared.

After this had gone on for some time a man came near a little while before they descended, stood on a tree concealed behind some bushes, and saw them come down, singing and laughing, and get out. While they were playing the ball was thrown so as to fall close to the man, and one woman came running toward it. When she got near he seized her, and the other people got into the canoe, ascended toward the sky, singing continually, and then disappeared.

The woman, however, he married. One time, after they had several children, the children said, "Father we want some fresh meat. Go and hunt deer for us." He started off, but he had not gotten far when he stopped and returned home. The mother said to her children, "Say, 'Father, go farther off and kill and bring back deer. We need venison very much.'" And the children said, "Father, go farther off and kill and bring back some deer. We need venison very much." When he did so, the children and their mother got into the canoe and started up, singing, but he came running back, pulled the canoe down, and laid it on the ground again.

After that the woman made a small canoe and laid it on the ground. When their father went hunting she got into one canoe and put the children into the small canoe and they started upward, singing. As they were going up the man came running back, but pulled only his children down, while their mother, singing continually, disappeared above.

But the children that the father had kept back wanted to follow their mother. They and their father got into the canoe, started off, singing continually, and vanished. Presently they came to where an old woman lived. The man said to her, "We have come because the children want to see their

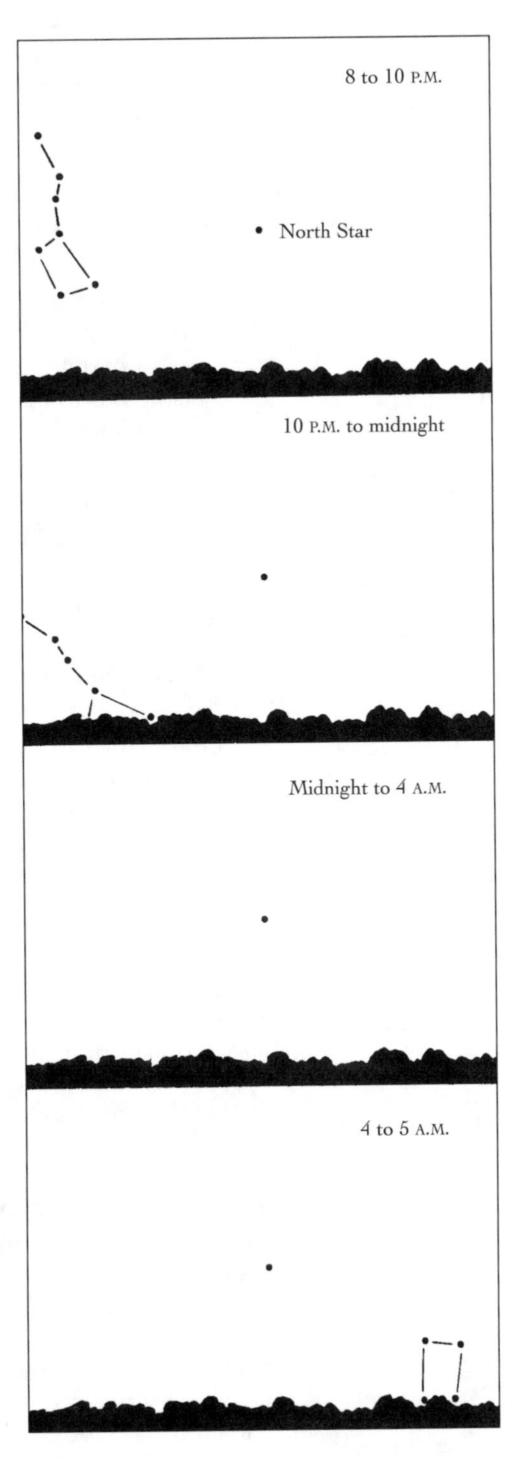

Fig. 12.2. In the southern United States the BIG DIPPER sweeps under the horizon and reappears as it circles the North Star. In the story of the Celestial Skiff, the bowl and handle of the BIG DIPPER are boats that allow travel to and from the sky. Sky People use the BIG DIPPER to come to the Earth during the summer *Busk* celebration, where they join in a game of ball. They return to the sky when the BIG DIPPER reappears above the horizon.

mother," and the old woman answered, "Their mother is dancing over there all the time, having small round squashes for breasts."

Then the old woman gave them food. She cooked some small squashes and gave pieces to each. When she set these before them, they thought, "It is too little for us." But when they took one away another appeared in the same place. When they took that one away it was as before. They ate for a long time but the food was still left. Then the old woman broke a corncob in pieces and gave a piece to each of them.

They went on and came to another person's house. This person said to them, "She stays here dancing." While they were there she went dancing around. They threw a piece of corncob at her but did not hit her. She passed through them running. The next time they threw at her when she came, she said, "I smell something," and passed through on the run. But the last one they threw hit her and she said, "My children have come," and she came running up to them. Then all got into the canoe and came back to this world.

One time after this when their father was away all got into the canoe, started up toward the sky, and disappeared. The children's father came back, and after he had remained there for a while he got into the other canoe, sang, and started upward toward the sky. He went on for a while, singing, but looked down to the ground. Then he fell back and was killed.

(Reprinted from Swanton, *Myths of the Southeastern Indians*, 138-139.)

Caddo

The Caddo are a group of tribes that lived in what is now Louisiana, southwest Arkansas, and eastern Texas. These Native peoples spoke a language similar to that of the Arikara, Pawnee and Wichita, who lived to the north on the plains. Although the Caddo were mainly villagers and farmers, those who lived on the edge of the plains also hunted buffalo. The descendants of the Caddo now live with other tribes in Oklahoma.

Stars and Constellations

The Caddo tell stories about several unidentified stars. In one tale, a young man has a dog that can talk. The dog warns the man of upcoming danger. When the two go hunting, the man shoots a deer that then jumps into a lake. Both the dog and hunter follow the deer but find themselves surrounded by monsters. The man prays to spirits, and he and his dog are lifted by a wave and set on the shore. The hunter offers deer meat as a sacrifice to the spirits that saved them. He and the dog go to the sky, which is a safer place to live, and become two bright stars in the south.

In another Caddo tale, a poor orphan is left on an island in a lake. The only way he can get off the island is to ride on the back of a horned monster. The Evening Star eventually kills the monster and takes the orphan to the sky with him. There, the Orphan-Star stands near the Evening Star.

Natchez

The Natchez are an extinct tribe that lived along the lower Mississippi River in Mississippi and Louisiana. This once-powerful tribe developed an elaborate caste society ruled by a king, called Great Sun. The king and priests resided on top of large mounds, and others lived below them. Farmers and mound builders were considered to be the lowest tier in the society.

PLEIADES: Seven Fasters

The PLEIADES are Seven Fasters, and the story about them shows how the oral tradition can incorporate historical events, like the arrival of Euro-American settlers. (The Delaware tell a similar story about Seven Prophets who become the PLEIADES, but in that story it is the Delaware people, rather than white settlers, who force the prophets to flee to the sky.)

Seven persons went apart, fasted, and took medicine for four days in order to prophesy. Then they came in and reported to the people what they had found out. Then the people said, "We will select seven [other?] persons and find out more." So they sent out seven persons who fasted and took medicine for seven days. At the end of this time the seven wondered if they should continue their fast for seven months. They fasted and took medicine until the seven months were completed. They asked one another if they could not observe their regulations for a whole year. They accomplished it, but when the time was completed they had become wild and feared to go near the rest of the people, so they went into the woods and stayed there. They asked one another what they should do, and finally said, "Let us turn ourselves into pine trees."

At that time there were no iron axes but tools made of flint with which little wood could be cut. But when the white men came and they saw them cutting down pine trees with their axes they said to one another, "That has cut us down." When the whites went on destroying pine trees they said, "Let us turn ourselves into rock. A rock lies undisturbed on top of the ground." But after they had turned themselves into rocks they saw the white people turn to the rocks and begin to use them in various ways. Then they made up their minds to go above, saying, "We cannot escape in any other manner." So they rose and went up into the air, where they became a constellation (probably the Pleiades).

(Reprinted from Swanton, *Myths and Tales of the Southeastern Indians*, 242.)

Other Stars and Constellations

The Natchez call the Milky Way *Wȧcgup ū´ic* (Dog Trail). The constellation Elbow Stars is so similar in name to the Cherokee constellation the Arm that it seems likely they are the same; the Arm is HYADES.

CONSTELLATIONS OF THE SOUTHEAST

Culture Group	Star or Constellation	Classical Equivalent
Cherokee	Hunters and Bear; Lost Hunters	BIG DIPPER
Cherokee	The Arm	HYADES
Cherokee	Where the Dog Ran (cornmeal trail)	Milky Way
Cherokee	Wicked Conjurer	Morning Star
Cherokee	The Drum	A nebula
Cherokee	Seven Stars	PLEIADES
Cherokee	The Group; the Boys	PLEIADES
Cherokee	Dog Stars	Sirius, Antares
Cherokee	Wooden Box	Unidentified
Creek	Image of a Canoe	BIG DIPPER
Creek	Spirits' Road	Milky Way
Creek	Bringer of Daylight	Morning Star
Creek	Stationary Star	North Star
Creek	*Tukȧbofkȧ*	PLEIADES
Seminole	Boat Image	BIG DIPPER
Seminole	Red Star	Evening Star
Seminole	Tomorrow Star	Morning Star
Koasati	Cluster-Stars	PLEIADES
Koasati	Stars in a Row	Unidentified
Alabama	Celestial Skiff	BIG DIPPER (bowl)
Alabama	Small Skiff	BIG DIPPER (handle)

CONSTELLATIONS OF THE SOUTHEAST *(cont.)*

Culture Group	Star or Constellation	Classical Equivalent
Caddo	Evening Star	Evening Star
Caddo	Orphan Star	Unidentified
Caddo	Man and Dog	Unidentified
Natchez	Elbow Stars	HYADES?
Natchez	Dog Trail	Milky Way
Natchez	Seven Fasters	PLEIADES

Chapter thirteen

Coming to Know the Night Sky

A Conclusion

I saw a quail in the sky one night when I was camped in the chill air of the high desert of southern California. I awoke well after midnight and stood near the tent, shivering, while I scanned the great dome above me. Then I saw it there above the southern horizon. It had a round body, a small triangular head, and a feather perched forward like a little headdress. It looked like the Gambel's quail that I'd seen scurrying along the ground earlier in the day, and it was a quail if ever there was one in the sky.

I had just done what generations of people the world over have done—related something in this world to the stars above. In my case, the quail had no special meaning other than the fact that I had watched one that day, but in other instances the act of naming has involved important and long-held beliefs.

Writing this book has been a great responsibility, because I am in fact presenting someone else's culture, or, rather, the cultures of dozens of quite distinct peoples. When dealing with long narratives like creation stories, for example, I have had to use only part of the narrative and emphasize the celestial material, sometimes (I fear) to the detriment of the original. It's similar to portraying biblical passages in this way: "In their book of creation, the Jewish people say that their God made stars on the fourth day of his seven-day creation of the universe. Genesis, the first book in a collection of books called the Bible, portrays the fourth day of creation in this way: 'God made the two great lights, the greater light to rule the day, and the lesser light to rule the night; he made the stars also' (Genesis 1:16). This book also mentions but does not identify eleven stars that the people of this time apparently held important: 'Behold, I have dreamed another dream; and behold, the sun, the moon, and eleven stars were bowing down to me' (Genesis 37:9). In later books, God is said to give names to all the stars (Psalms 146:4), and he places them in the sky (Jeremiah 31:35); unfortunately, these names are not noted in the text. Although the Jews did not generally worship the stars, at one point in their history some people did worship Käi´wän, a star-god (Amos 5:25). In the second section of the Bible, the 'New Testament,' a narrator named Matthew recorded that wise men 'from the east' saw a star in the eastern sky and interpreted it to be the star of the king of the Jews."

Of course, this passage distorts the eloquent messages found in the Bible because it pulls them out of context, emphasizes some passages that are otherwise unimportant, and gives no background information to set the ideas within the belief system of the people of that time. Although I have diligently worked to avoid this kind of presentation, there may be instances in which I did not fully succeed. In addition, an overview book like this one cannot possibly include enough detail to distinguish, properly and fully, each tribe.

I have encountered other challenges as well. Many of the original sources contain no index—a researcher's nightmare. Some material was tantalizing but sketchy. And though I followed every lead, I knew that as soon as this book was "finished," I would discover more material. In the larger sense, I consider this book a work in progress, and hope that it may serve as a nucleus for still other narratives and celestial information.

But writing this book has also been exciting. I have relished the role of detective, tracking down myths and star names, comparing constellations from different sources, and unearthing stories that have long been tucked away in ancient tomes. There is a satisfaction that comes from pulling together disparate bits of information and forming a whole.

I have gained a broader and deeper understanding of the natural world around me. Just as I enjoy knowing the names of trees, flowers, birds, and mammals and understanding how they live together, I have enjoyed coming to know the night sky in new and different ways. When I look at a map of North America, I see not only states, provinces, and countries, but a land of many cultures. Wherever I go in the Northern Hemisphere, I can see shapes in the sky that link me to the culture of the Native Americans who once lived here, and who live here still.

Likewise, when I watch the constellations, I find them richer in meaning than I did in the past. The Northern Crown is not just a semicircle of stars, it is the den where the spirit of the eternal bear lives through the long months of winter to emerge once more in the spring. When I see the Big Dipper and Cassiopeia, I think of the *hózhǫ́*, or balance and harmony, that the constellations represent. And when I see the Pleiades, I think of its Luiseño name, Hearts of the First People, an image so primal that the concerns of this life slip away, and for a fleeting moment I glimpse that time when animals could talk.

Appendix A

Classical Star Groups and their Native American Counterparts

The following chart presents classical constellations, figures within constellations (asterisms), planets, and other star features with their corresponding Native American names. The chart is divided into three sections: constellations, planets, and other sky features.

In the constellation section, classical constellations, asterisms, and individual stars are listed in alphabetical order. The stars Altair (in EAGLE), Canopus (in SHIP'S KEEL), Castor and Pollux (in GEMINI), Fomalhaut (in SOUTHERN FISH), Procyon (in LITTLE DOG), Sirius (in BIG DOG), and Spica (in VIRGO) are listed under their own names rather than by constellation, because it is the star, not the constellation, that figures in Native American myths. Likewise, the BIG DIPPER is listed under BIG DIPPER rather than GREAT BEAR. The classical constellation TAURUS is subdivided into its component parts (PLEIADES, HYADES, Aldebaran) because Native Americans did not consider it as a whole.

In the planet section, Native American stars that have been tentatively or definitely identified as Venus are listed under that name. Native American stars that are identified only as the Morning Star or Evening Star have been included under those headings. Although it is likely that many of these Native American stars are in fact Venus, it is possible that some of them are other planets or stars. The Pawnee call Mars, for example, Great Star or Morning Star.

In some cases a descriptive phrase rather than the constellation name is given; these phrases are not capitalized. In other cases, similar concepts have been combined. Ghost Road, Soul's Road and similar names, for example, have been listed together as Spirits' Path. The chart is useful for quick comparisons but does not reveal subtle differences among the tribes. Nor does the chart reflect the cultural meaning invested in each star group and the richness of the story that relates how that star group came into being. (Chapters 3 to 12 contain detailed information about individual constellations.)

Concerning tribal designations, the Seneca and Onondaga are part of the Iroquois Confederacy. In this chart, "Tewa" refers to a Tewa-speaking Pueblo when the exact village is not known; San Ildefonso is one Tewa-speaking Pueblo. Tewa-speaking Pueblos, Isleta, Taos, Picuris, Jemez, and Cochiti are among the Eastern Pueblos. Zuni, Hopi, and Laguna are Western Pueblos.

An asterisk (*) after the Native American star/constellation means that the star or constellation's identification with the classical star group is suggested or possible, but not certain. When several tribes are listed together, an asterisk is used to show that only for the tribe so indicated is the identification uncertain. When a Native American star group may be one of two classical constellations, the star group is listed under both constellations with an asterisk to denote uncertainty.

CLASSICAL STAR GROUPS AND THEIR NATIVE AMERICAN COUNTERPARTS

Constellations, Asterisms, and Stars

Altair and two stars in EAGLE (AQUILA)

	Plateau
Three Persons Running a Race	Coeur d'Alene
	California
Altair: Buzzard	Yokut, Monache, Luiseño, Tipai
Chiyi (a supernatural being)	Tipai (Kamia)
Three Steady Persons	Chumash
	Southwest
Altair: Very Cold Star	Havasupai
Altair: Feces Interrupted	Walapai
Altair: Cold's Cottonwood	Maricopa
Altair: Turkey Vulture*	Seri

Antares: See SCORPIUS

AQUARIUS

	California
Part of Cottontail Man	Northern Sierra

Arcturus: See HERDSMAN

ARROW (SAGITTA)

	California
Monsters*	Chumash

BIG DIPPER in GREAT BEAR (URSA MAJOR)

	Northeast
Bear and/or hunters	Micmac, Passamaquoddy, Penobscot, Iroquois, Seneca, Delaware, Fox
Fisher Stars	Ojibway, Menominee
	Subarctic
Fisher Star	East Cree
Grandfather Stars	Tahltan

CLASSICAL STAR GROUPS AND THEIR NATIVE AMERICAN COUNTERPARTS

BIG DIPPER in GREAT BEAR (URSA MAJOR) *(continued)*

	***Subarctic** (continued)*
Big-Headed Man	Tutchone
It Rotates Its Person	Koyukon
	Arctic
Giant Caribou	Polar Inuit
	Pacific Northwest
Yaxtê´	Tlingit
Dipper*	Tsimshian
Elk; Three Hunters and Elk	Fraser River, Snohomish0
Bear Skin	Quileute
Skate	Quinault
	Plateau
Grizzly Bear and Three Hunters	Thompson, Coeur d'Alene
Wolf Brothers and Grizzlies	Wasco
The Divers (the loons)	Klamath
	Great Basin
Rabbit Net; Men Chasing Rabbits into a Net	Northern Paiute; Western Shoshone
Mountain Sheep	Southern Paiute
	California
Stick or cane	Atsugewi, Patwin, Pomo
Looking round	Maidu
Seven Boys	Monache
Wildcat's Rump	Cahuilla, Ipai
Seven Boys Who Became Geese*	Chumash
	Southwest
Seven stars, corners, tails, or ones	Cochiti, Tewa, Zuni
Cradle	Isleta
Shield Stars	Cochiti
Lungs of one monster; head of another	Zuni
Male One Who Revolves	Navajo
Revolving Around a Central Point	Mescalero Apache
Cactus-gathering hook	Walapai, Tohono O'odham, Seri
Coyote's Fishing Net; Coyote	Maricopa, Maricopa/Pima
	Great Plains
People carrying stretcher	Pawnee, Arikara, Omaha, Osage, Dakota

	***Great Plains** (continued)*
"Seven" or seven stars	Wichita, Lakota, Assiniboin, Crow
Ermine	Hidatsa
Brothers, sometimes with girl	Crow, Blackfoot, Cheyenne, Kiowa
Cut-Off-Head	Gros Ventre
Seven Bison Bulls	Arapaho
	Southeast
Canoe or skiff	Creek, Seminole, Alabama
Lost Hunters; Hunters and Bear	Cherokee

BIG DOG (CANIS MAJOR): See Sirius

CANCER

	Southwest
Long Sash's Place of Doubt	Tewa

Canopus in SHIP'S KEEL (CARINA)

	California
Coyote's stepbrother*	Cahuilla
	Southwest
Star in South*; Gray-Haired Old Woman*	Tewa
Coyote's Star*; No-North Star	Navajo
	Great Plains
South Star	Pawnee*, Wichita*

CASSIOPEIA

	Arctic
Stones Supporting a Lamp	Polar Inuit
	Pacific Northwest
Elk Skin	Quileute
	California
Land-of-the-Widows*	Chumash
	Southwest
Star Zigzag	Zuni

CASSIOPEIA *(continued)*

	Southwest *(continued)*
Female One Who Revolves	Navajo
Spider	Maricopa/Pima
	Great Plains
Rabbit	Pawnee

Castor and Pollux in GEMINI

	Plateau
Twins (boy and girl)	Klamath
	California
Sun's Female Cousins*	Chumash
	Southwest
Place of Decision	Tewa
Two-Fighting Stars	Jicarilla Apache
	Great Plains
Ashes Chief and Stuck-Behind	Blackfoot
Stars near GEMINI: Bear's lodge	Lakota

CEPHEUS

	California
Part of Ravine after Land-of-the-Widows*	Chumash
	Great Plains
Kneeprint*	Pawnee

CHARIOTEER (AURIGA) including Capella

	Plateau
Women roasting camas roots	Thompson*, Coeur d'Alene*
Part of Bow of Twins	Klamath
	Southwest
Three Who Went Together	Mescalero Apache
	Great Plains
Panther*	Pawnee
Capella: Yellow Star*	Pawnee

CROW (CORVUS)

	Subarctic
Raven Carrying the Sun	Tahltan
	Southwest
Man with Legs Ajar	Navajo

DOLPHIN (DELPHINUS)

	California
To Roast	Maidu
Paw Print	Northern Sierras
Monsters*	Chumash
	Southwest
Sling or slingshot	Cochiti, Tewa, Zuni
	Great Plains
Bow	Pawnee
Ta´maopa´	Dakota

EAGLE (AQUILA): See Altair

Fomalhaut in SOUTHERN FISH (PISCIS AUSTRINUS)

	Plateau
Rabbit*	Klamath
	Great Basin
Big Woman*	Northern Paiute
	California
A sky chief	Luiseño

GEMINI (see also Castor and Pollux)

	Plateau
Part of Bow of Twins	Klamath

HARE (LEPUS)

	California
Part of constellation *Mech*	Chumash

CLASSICAL STAR GROUPS AND THEIR NATIVE AMERICAN COUNTERPARTS

HERCULES

	Great Plains
Spider Man's Fingers	Blackfoot

HERDSMAN (BOÖTES) including Arcturus

	Northeast
Arcturus and three stars: Hunters	Micmac
	California
Arcturus: *Nükülish* His-Hand	Luiseño
Hand*	Tipai
	Southwest
Arcturus: Left shoulder of Chief	Zuni
Hoop*	Havasupai, Yavapai
Cipa's Hand*	Maricopa

LEO

	California
Hand*	Tipai
	Southwest
Three Stars of Helpfulness	Tewa

LITTLE DIPPER (URSA MINOR)

	Northeast
Three Hunters in a Canoe	Micmac
Beaver that Spreads its Skin	Seneca
	Pacific Northwest
Skate	Quileute
	Great Basin
Mountain Sheep	Southern Paiute
	California
Wise Woman; Fox	Chumash
Wildcat's Rump*	Cahuilla, Ipai

	Southwest
Yoke	Tewa
Those to the North	Zuni
	Great Plains
Little Stretcher	Pawnee

LITTLE DIPPER: North Star

	Northeast
Star That Does Not Move	Iroquois
	Subarctic
Great Star; Guide of the People	East Cree
	Plateau
Grizzly Bear	Kootenai
	Great Basin
Star That Does Not Move	Northern Paiute, Southern Paiute
Mountain Sheep	Southern Paiute
	California
Eye of the Creator	Pomo
Sky Coyote*	Yokut, Chumash
Pretty Woman*	Cahuilla
Star That Does Not Move; Star That Never Moves	Salinan, Chumash
A sky chief; Heart of the Night Wolves	Luiseño
	Southwest
North Star on Top	Navajo
Star-That-Does-Not-Move; Not-Walking Star	Mescalero Apache, Pima
Captain	Maricopa
Pole	Maricopa/Pima
	Great Plains
Star That Does Not Move (or Walk); Star That Stands Still	Pawnee, Omaha, Lakota, Hidatsa, Blackfoot
Red Star	Osage
North Star	Assiniboin
Old Woman's Grandchild	Crow
	Southeast
Star That Does Not Move	Creek

CLASSICAL STAR GROUPS AND THEIR NATIVE AMERICAN COUNTERPARTS

LITTLE DOG: **see Procyon**

FOX (VULPECULA)

	Great Basin
Part of Grizzly Bear	Shoshone

LIZARD (LACERTA)

	California
Part of Ravine after Land-of-the-Widows*	Chumash

LYRE (LYRA) Including Vega

	California
Part of Scorpion Woman*	Chumash
Vega: Scorpion Woman's Stinger*	Chumash
Vega: Buzzard's Right Hand	Luiseño
	Southwest
Part of Heart of the Night*	Zuni
One of two men with Hats*	Maricopa
	Great Plains
Vega: Big Black Meteoric Star*	Pawnee
Big Black Meteoric Star's Daughters*	Pawnee

NORTHERN CROWN (CORONA BOREALIS)

	Northeast
Bear's Den; Bear's Net; Bear's Head	Micmac, Iroquois, Delaware
Celestial Sisters	Shawnee
	California
Hand*	Tipai
	Southwest
Meal-Drying Bowl	Tewa
Red Star	Zuni
Hoop*	Havasupai, Yavapai
Cipa's Hand*	Maricopa
Council	Maricopa/Pima

	Great Plains
Council of Chiefs	Pawnee
Brothers*	Omaha
Lodge of Spider Man	Blackfoot
Camp Circle; Old Camp	Cheyenne, Arapaho

ORION including various elements

	Northeast
She is Sitting	Seneca
	Subarctic
Belt: Three Chiefs; Three Hunters	Montagnais/Cree/Naskapi, Chilcotin*
	Arctic
Belt: The Great Stretchers	Alaskan Inuit
Belt: Three Lost Hunters	Alaskan Inuit
Hunters, Sledge, and Bear	Canadian Inuit
Belt: Three Steps Cut in a Snowbank	Polar Inuit
	Pacific Northwest
Belt: Three Men in a Line	Tlingit
Harpooner-of-Heaven; Paddlers	Kwakiutl
Two Canoes and a Fish*	Snohomish
	Plateau
Bark Canoe; Canoe; Two Canoes	Thompson, Coeur d'Alene*, Wasco
Five Canoe Builders	Flathead*
	Great Basin
Belt: Mountain Sheep	Northern Paiute, Southern Paiute
Mountain Sheep and Arrow	Western Shoshone
	California
Coyote Carries on Head	Patwin
Belt: Coyote's Arrow	Yana
Belt: Strong Sucker*	Pomo
Six Young Men*	Yokut, Monache
Belt: Crane and Her Sons	Yokut
Belt: Bear Chasing Antelope	Yokut
Men (or Coyote) chasing PLEIADES	Yokut
Three Babies	Yokut
Belt: Three in a Row	Salinan
Belt: Bear*	Chumash
Hunter, Arrow, Mountain Sheep (sometimes also Deer, Antelope)	Cahuilla, Ipai, Tipai
Hulaish	Luiseño

CLASSICAL STAR GROUPS AND THEIR NATIVE AMERICAN COUNTERPARTS

ORION including various elements *(continued)*

	Southwest
Belt: In a Row	Tewa, Zuni
Belt: Pot-Rest Stars*	Cochiti
Rows	Jemez
Belt: Fawns	Isleta, Laguna
Long Sash	San Ildefonso
Belt: Right arm of Chief	Zuni
Belt: Strung Together	Hopi
First Slim One	Navajo
Belt: Three-Vertebrae Stars	Jicarilla Apache
Hunter, Arrow, Mountain Sheep (sometimes also Deer, Antelope)	Yavapai, Cocopa, Maricopa, Maricopa/Pima, Seri
	Great Plains
Deer	Pawnee, Osage
Nebula: Feather Headdress Star*	Pawnee
Belt: Large Foot of Goose	Omaha
Sword: Stars Strung Together	Osage
The Hand; Hand Star	Lakota, Hidatsa, Crow
Nebula: Smoking Star	Blackfoot

PEGASUS

	Southwest
GREAT SQUARE OF PEGASUS: Four Big Ones; Square	Zuni
Part of Cottontail Man	Northern Sierra

PERSEUS

	Great Plains
Kneeprint*, Wind Ready to Give*, Breathe*	Pawnee

Procyon in LITTLE DOG (CANIS MINOR)

	Great Plains
Wildcat Star*	Pawnee

SAGITTARIUS

	Southwest
Left hand of Chief; Eight Ones	Zuni
Fox	Maricopa/Pima

	Great Plains
Bear*	Pawnee

SCORPIUS including Antares

	California
Antares: A sky chief	Luiseño
Boy with Bow and Arrow	Ipai
Shooting*	Tipai
	Southwest
Antares: Left forearm of Chief	Zuni
Big First One and Cane	Navajo
Tail: Rabbit Tracks	Navajo
Antares: Red Star*, Coyote's Star*	Navajo
Coyote Carrying a Pole; Pole-Carrier	Havasupai, Walapai, Seri
Scorpion	Maricopa
Earth Doctor	Maricopa/Pima
Antares: One of Two Men with Hats*	Maricopa
	Great Plains
Tail: Swimming Ducks	Pawnee
Real Snake	Pawnee
Antares: Red Star*	Pawnee
	Southeast
Antares: Dog Star	Cherokee

SERPENT (SERPENS)

	Great Plains
Snake Not Real*	Pawnee

SHIP'S KEEL (See Canopus)

Sirius in BIG DOG (CANIS MAJOR)

	Arctic
Moon Dog	Alaskan Inuit
	Great Basin
The Chaser	Northern Paiute

CLASSICAL STAR GROUPS AND THEIR NATIVE AMERICAN COUNTERPARTS

Sirius in BIG DOG (CANIS MAJOR) *(continued)*

	California
Part of constellation *Mech*	Chumash
Sirius: Two Hunters*	Tipai
	Southwest
Mountain Sheep Pursuer; hunter	Maricopa, Cocopa*
Dog	Seri
	Great Plains
Sirius: White Star*	Pawnee
BIG DOG or Sirius: Wolf Star Wolf that Hangs at the Side of the Heavens	Pawnee*, Osage
	Southeast
Dog Star	Cherokee

SOUTHERN FISH (see Fomalhaut)

Spica in VIRGO

	California
A sky chief	Luiseño
	Southwest
Left elbow of Chief	Zuni

SWAN (CYGNUS) including NORTHERN CROSS and Deneb

	Great Basin
Part of Grizzly Bear	Shoshone
	California
Part of Scorpion Woman*	Chumash
Monsters*	Chumash
NORTHERN CROSS: Cross Star	Tipai
	Southwest
Turkey Foot*	Tewa
Part of Heart of the Night*	Zuni

	Great Plains
Deneb: Second Black Star*	Pawnee
Bird's Foot*	Pawnee

TAURUS

	Northeast
Horned Head of Wild Animal	Iroquois

TAURUS: Aldebaran

	Arctic
Part of Sharing-Out of Food	Alaskan Inuit
	Plateau
Dog*	Thompson
Badger*	Coeur d'Alene
Coyote's Eyeball	Spokane
	California
Coyote	Yokut, Cahuilla, Chumash, Luiseño, Ipai
	Southwest
Bull's Eye*	Tewa
Broad Star	Hopi
Red Star*	Navajo
Queeto	Seri

TAURUS: HYADES

	California
Six Young Men*	Yokut, Monache
Girl Who Is Planting*	Tipai
	Southwest
Doubtful or Pinching Stars	Navajo
Wrestlers; Hard Flint Women	Navajo
	Southeast
The Arm; Elbow Stars	Cherokee, Natchez*

TAURUS: PLEIADES

	Northeast
Dancers (brothers, children, women)	Iroquois, Seneca, Onondaga, Delaware, Huron/Wyandot, Chippewa

CLASSICAL STAR GROUPS AND THEIR NATIVE AMERICAN COUNTERPARTS

TAURUS: PLEIADES *(continued)*

	Northeast *(continued)*
Bunched Up Stars (holy men or prophets)	Delaware
Hole in the Sky	Ojibway
	Subarctic
Herd of Caribou	Carrier
	Arctic
Little Foxes	Alaskan Inuit
Part of Sharing-Out of Food	Alaskan Inuit
Dogs after a Bear	Polar Inuit
	Pacific Northwest
Sculpin	Tlingit
Seven Brothers in a Boat	Haida
Otter	Kwakiutl
Crying Children	Fraser River
	Plateau
Bunch; Cluster; Bunch of Stars	Thompson, Coeur d'Alene, Wasco
Sisters	Nez Perce
Children who Announce Dawn	Klamath
Aqkitl kanka	Kootenai
	Great Basin
Women Fighting*	Northern Paiute
Coyote's Daughters; Coyote's Family; Women with Children	Northern and Southern Paiute, Western Shoshone, Ute*
	California
Many; Little Stars; Seven Sisters at Puberty Dance	Yurok, Karok, Atsugewi
Six People in a Boat	Wiyot
Raccoon's Children	Shasta
Grizzly Sisters	Northern Sierras
Buckeyes Bunched Up	Pomo
Young women	Yokut, Monache, Salinan, Cahuilla, Luiseño
Hearts of the First People	Luiseño
Children Shaking	Costanoan
Eight Wise Men	Chumash
Xatca; Hecha	Tipai

	Southwest
In a Bunch; Clustered	Tewa, Hopi
Jumbled	Isleta
Deer	Taos
Seven Stars	Jemez
Seeds; Seed Stars	Zuni
Dilyéhé	Navajo
Six Hunters and Girl	Jicarilla Apache
Hecha or Hestah	Yavapai, Cocopa
Gopher	Maricopa/Pima
Running Women	Pima, Tohono O'odham
Women and Children	Seri
	Great Plains
Seven Stars	Pawnee
Deer Head	Omaha, Osage
Duck's Foot	Omaha
Wise One (or Girl) and Brothers	Assiniboin, Arikara, Cheyenne, Arapaho
Bison Bulls	Arapaho
Puppies	Cheyenne
Lost Children; Bunched Stars	Blackfoot
	Southeast
Seven Stars; Group; Boys	Cherokee
Tukâbofkâ	Creek
Cluster–Stars	Koasati
Seven Fasters	Natchez

Vega: see LYRE

VIRGO: see Spica

Planets

Evening Star

	Plateau
Big Star	Kootenai
	California
Seining in the Slough; West Star Now Comes	Patwin
Night Woman	Pomo

Evening Star *(continued)*

	Southwest
Yellow-Going Old Woman	Tewa
Asking Star; Prayer Star	Isleta
One Following the Sun	Zuni
Liver of monster	Zuni
Big Star	Navajo
Star that Travels Through the Sun	Maricopa
	Great Plains
Big Star; Largest Star	Omaha, Dakota
Woman of the west	Mandan
Lone Star	Arapaho
	Southeast
Red Star	Seminole
Evening Star	Caddo

Jupiter

	Pacific Northwest
Marten Month*	Tlingit
	Great Plains
Great Star's Brother*	Pawnee
Mistake Morning Star; Poïa	Blackfoot

Mars

	California
Xolxol; Condor*	Chumash
	Great Plains
Great Star, Morning Star	Pawnee
Big Fire Star	Blackfoot

Mercury

	Great Plains
Great Star's brother*	Pawnee

Morning Star

	Northeast
Chief M´Sūrtū	Passamaquoddy
Star Woman	Iroquois
It Brings the Day	Seneca
Great Star, Red Star	Delaware
A *manitou*	Fox
	Subarctic
Old Woman with Torch	Chilcotin
	Arctic
Bunch of Glowing Grass	Alaskan Inuit
	Northwest
Sun's Daughter	Tsimshian
	Plateau
Big Star	Kootenai
Bringing the Day	Thompson, Coeur d'Alene
	Great Basin
Qaŋa	Southern Paiute
	California
Day Woman; Morning Eye Fire; Big Star	Pomo
Young Woman who Announces the Sun	Yuki
Teammate of Sky Coyote	Chumash
	Southwest
Big Star; Bright Star; Great Star	Tewa, Isleta, Zuni, Hopi, Maricopa
Dark Star Man	Tewa
Heart of monster; Gambler's Eyeball	Zuni
Broken Water Jug	Zuni
Woman	Tohono O'odham
	Great Plains
Big Star; Greatest Star; Largest Star	Omaha, Osage, Dakota
First Man	Wichita
Red Star	Osage
Old Woman's Grandchild	Crow
Star Shining Daylight Chases	Assiniboin
Woman Wearing a Plume	Mandan

Morning Star *(continued)*

	Southeast
Wicked Conjuror	Cherokee
Bringer of Daylight	Creek
Tomorrow Star	Seminole

Saturn

	Great Plains
Star of Sickness*	Pawnee

Venus

	Arctic
He Who Stands and Listens	Polar Inuit
	Pacific Northwest
Morning Round Thing	Tlingit
Eye*	Quileute
	California
Thunderers*	Chumash
Eagle*; Chief of Land-of-the-Dead*	Chumash
That Which is Left Over	Luiseño
	Southwest
Big Star; Morning Star	Navajo
Bluebird; Visible Star	Pima
	Great Plains
Bright Star; Evening Star	Pawnee
Early Riser; Day Star; Poïa's father	Blackfoot

Other Sky Features

Milky Way

	Northeast
Spirits' Path; Great Sky Road; Soul's Road; Road of the Dead	Passamaquoddy, Iroquois, Delaware, Ojibway, Menominee
Path of Migrating Birds	Ojibway

	Northeast (continued)
Spray from Turtle's Tail	Ojibway
White River	Fox
	Subarctic
Ghost Road	Montagnais/Cree/Naskapi
Snowshoers in the Sky	Tahltan
Flight of the Loon	Tutchone
	Arctic
Raven's Snowshoe Tracks	Alaskan Inuit
	Pacific Northwest
Tracks of Tlingit hero	Tlingit
Cannibal Pole	Kwakiutl
River	Snohomish
	Plateau
Tracks of the Dead; a path, trail, or road	Kootenai, Thompson, Coeur d'Alene, Wasco, Klamath
River	Klamath
	Great Basin
Ghost Road; Sky Path; Dusty or People's Trail	Northern Paiute, Southern Paiute, Shoshone, Ute
Smoke; Sky Backbone; Ice Crystal Trail	Shoshone; Western Shoshone
	California
Road to Land-of-the-Dead; Path of Soul; Devil's Trail	Karok, Shasta, Atsugewi, Chumash
Ocean (foam?)	Wiyot
Tracks of animal	Patwin, Pomo, Yokut, Cahuilla
Ashes	Patwin
Morning Star's Trail	Maidu
Journey of Piñon Gatherers	Chumash
Night's Backbone; Sky-Its-Backbone	Chumash, Tipai
Piwish	Luiseño
	Southwest
Arch Across the Sky	Cochiti
Backbone of the Universe; Sky Backbone	Tewa, Taos, Jemez
Endless Trail	San Ildefonso
Whitishness	Tewa
Ashes	Zuni
Entrails of monster	Zuni
Awaits the Dawn	Navajo

Milky Way *(continued)*

Scattered Stars	Mescalero Apache
Deer and Antelope Tracks	Maricopa
Spider's Web	Maricopa/Pima
Flour; Ashes	Pima
White Beans	Tohono O'odham
	Great Plains
Bright Light in a Long Stretch Across the Heavens	Pawnee
Dust from Bison and Horse Race	Pawnee
Spirits' Path; Spirits' Way	Pawnee, Omaha, Lakota
Trail of Bubbles	Dakota
Backbone of Sky	Assiniboin
Ashy Way; Wolf Trail; Hanging Road	Hidatsa, Blackfoot, Cheyenne
	Southeast
Where the Dog Ran; Dog Trail	Cherokee, Natchez
Spirits' Road	Creek

Appendix B

The Twenty Brightest Stars

1. Sirius in BIG DOG (CANIS MAJOR)
2. Canopus in SHIP'S KEEL (CARINA)
3. Rigel Kent in CENTAUR (CENTAURUS)
4. Arcturus in HERDSMAN (BOÖTIES)
5. Vega in LYRE (LYRA)
6. Capella in CHARIOTEER (AURIGA)
7. Rigel in ORION
8. Procyon in LITTLE DOG (CANIS MINOR)
9. Achernar in RIVER ERIDANUS (ERIDANUS)
10. Hadar (also called Beta Centauri) in CENTAUR (CENTAURUS)
11. Altair in EAGLE (AQUILA)
12. Betelgeuse in ORION
13. Aldebaran in BULL (TAURUS)
14. Acrux in CROSS (CRUX)
15. Spica in VIRGIN (VIRGO)
16. Antares in SCORPION (SCORPIUS)
17. Pollux in TWINS (GEMINI)
18. Fomalhaut in SOUTHERN FISH (PISCIS AUSTRINUS)
19. Deneb in the asterism NORTHERN CROSS, which is in SWAN (CYGNUS)
20. Becrux in CROSS (CRUX)

Notes

Preface

page xii. "The Parts of this weed": Wissler and Duvall 1909, 5.

1. Stars of the First People

page 1. Definition of myth: Gill and Sullivan 1992, xi.

page 4. "The old men used to study": Hooper 1920, 362.

2. Greek Constellations

page 8. Callisto and Arcas: Hamilton 1942, 290–291.

page 10. Story of CASSIOPEIA, CEPHEUS, PERSEUS, and ANDROMEDA: Hamilton 1942, 146–148; Graves 1957, 1:240–241.

page 12. Castor and Pollux: Hamilton 1942, 41–42; Graves 1947, 1:247–248.

page 14. ORION: Graves 1957, 1:151.

page 14. Orthrus as Sirius: Graves 1957, 1:30.

page 15. Disappearance of Electra and Merope: Graves 1957, 1:154, 218.

page 15. Arion and his lyre: Graves 1957, 1:290–291.

page 16. Cycnus becoming a swan: Hamilton 1942, 326.

page 18. Crab becomes part of the zodiac: Graves 1957, 2:108–109.

page 18. Icarius, Erigone, and Maera: Graves 1957, 1:262.

page 18. The NORTHERN CROWN: Graves 1957, 1:106; Hamilton 1942, 56.

page 20. "You will have to pass": Hamilton 1942, 132–133.

page 20. Chiron: Graves 1957, 2:114.

page 22. Asclepius: Graves 1957, 1:175.

page 22. Nemean lion as constellation: Although many star books state that LEO is the Nemean lion, a search of two standard Greek mythology texts (Graves 1957 and Hamilton 1942) does not reveal the link. In a book about constellations from around the world, Allen (1899) does say that the Nemean lion becomes LEO, but gives no citation; Allen may be the source of subsequent star-book stories about the Nemean lion.

Only if there are myths in Graves or Hamilton that specifically describe people or animals being placed in the sky are those events included here.

page 22. Crow: Allen (1963, 178–180) cites Ovid as his source.

page 24. Ganymede: Graves 1957, 1:116.

3. The Celestial Bear Hunt (Northeast)

page 36. "The sky is just the same" and rebirth of the celestial bear: Hagar 1900, 94–95.

page 40. "The three guardians" and Micmac LITTLE DIPPER: Ibid., 99.

page 40. Identification of stars in the LITTLE DIPPER as bear hunters in a canoe: Because Father Le Clerq uses the phrase "three guardians of the North Star," it is likely that the "three guardians" (like the North Star) are his own words and not those of the Micmac. Allen (1899, 459) reports that the English and the Spanish called Kochab and

Pherkad the Guards or Guardians as early as the 1500s, and the French may also have used these terms. In *Othello* (2.1.14–15), Shakespeare writes that the swell of the Mediterranean Sea "seems to cast water on the Burning Bear and quench the guards of the ever-fixed Pole." Even today, stargazers call Kochab and Pherkad the Guardians. In addition, Kochab is the brightest star near the North Star (Pole Star); Kochab, Pherkad, and a third star form a line that might be seen as three men in a canoe.

pages 40–41. Mrs. Brown's version: Brown 1890, 213–214.

page 43. Penobscot BIG DIPPER: Spotted Elk 1938, Story 14.

page 43. League as longhouse: Tooker, 418 in Trigger 1978.

page 44. PLEIADES at dusk: Fenton, 300 in Trigger 1978.

page 44. Hunters addressing the stars: McElwain 1992, 261, 268.

page 44. "The women are the most experienced star-gazers" and "a horned head": Wassenaer in O'Callaghan 1850, 3:20, 22.

page 44. Sky Woman: Parker 1923, 69.

page 45. Iroquois creation: Beauchamp cites Hewitt (1903, 227–228) as his source for this material.

page 45. The Dancing Stars: Converse 1908, 53–54.

page 49. Seneca BIG DIPPER story: Curtin and Hewitt 1918, 276–277.

page 49. Stone monster: Smith 1883, 81.

page 54. Delaware stars: Speck 1931, 23, 48–49, 105–107, 171–173.

page 55. Earth-sky connection of Bear Sacrifice: Speck and Moses 1945, 32, 55.

page 55. "At a certain season": Speck 1931, 48.

page 55. Only shamans can know: Ibid., 48.

pages 55–56. "A shaman took": Nekatcit (Nicodemus Peters), a Delaware-Mahican, narrated two versions of this myth in the late 1930s (Speck and Moses 1945, 79–81).

page 56. "We know this is true" and accompanying story: Judson 1917, 74–76.

page 58. PLEIADES and shamans' star material: Conway 1992, 253.

page 60. Ojibway Milky Way stories: Jenness 1935, 28.

page 60. The land of the dead: Cappel (1928, 63) notes that Panguk, the Death Spirit, escorts spirits along the Milky Way until they reach Chebiabos, keeper of the Shadowland. There the spirits receive new clothing made of white deerskin trimmed with fringe and beads.

page 61. For another version of the Chippewa Fisher Stars, see Carson 1917, 492–493.

page 63. Chippewa PLEIADES: Cappel 1928, 51–53.

page 64. Jones's respect for religious material: Margaret Welpley Fisher in Jones 1939, vii–ix.

page 64. "You saw the stars": Rideout 1912, 201.

page 65. "Some of these *manitous*": Jones 1939, 13.

page 65. The bear returns every year: Ibid., 22.

page 66. Menominee Milky Way and payment for tales: Skinner and Satterlee 1915, 229, 237.

page 67. "This is a sacred story": Skinner and Satterlee 1915, 471.

4. The Fisher Stars (Subarctic)

page 75. "The long period of darkness": Nelson 1983, 39. McClellan (1975, 78) observes that the relatively short nights in summer as well as the bright northern lights (aurora borealis) that overpower the stars and the frigid conditions in winter contribute to the lack of star stories.

page 76. "Good health, good hunting": Rogers and Leacock, 184 in Helm 1981.

page 80. Milky Way and ORION: Speck 1935, 65. Speck uses the vague term "Montagnais-Naskapi" when referring to Montagnais, Cree, or Naskapi constellations.

page 80. Carrier PLEIADES: Hamilton 1905, 48.

page 81. Identity of the Great Bear: Several Plateau tribes to the immediate south call the BIG DIPPER a grizzly bear and hunters, and so it could be for the Chilcotin. The Chilcotin story relates, however, that the Great Bear "appeared" in the sky. Because the BIG DIPPER circles the North Star, which is quite high at the latitude of the Chilcotin, the BIG DIPPER is in the sky all night and therefore would not "appear." The Great Bear likens its path to that of the sun; conceivably, the bear could be the moon.

pages 82–83. Big-Raven as the constellation CROW: The Tahltan translate Big-Raven as "crow" when speaking English. Crows and ravens belong to the same genus, *Corvus,* which is also the Latin name for the classical constellation CROW. The curious reader may wonder whether identification of Big-Raven with the western constellation of the same name is purely coincidental, a cultural parallel, or perhaps a result of Euro-American influence.

page 83. Raven carrying the sun: Lawrence Hall of Science 1978.

page 83. Tahltan BIG DIPPER: McClellan (1975, 78) reports that the Tagish, a tribe just south of the Tutchone, relate a similar tale. The constellation tells his grandchildren on Earth that when he stops going around, the "last day" will arrive.

page 84. Tutchone Milky Way and BIG DIPPER: McClellan 1975, 78.

5. Raven's Snowshoe Tracks (Arctic)

page 88. "While he was standing close" and Bering Strait constellations: Nelson 1899, 449, 462.

page 88. Point Barrow constellations: Boas 1895, 121.

page 90. Polar Inuit creation of stars: Rasmussen 1921, 16–17.

page 90. Identification of PLEIADES: Rasmussen does not identify the tale as involving the PLEIADES, but Kroeber (1899, 173) writes that the Smith Sound/Polar Inuit identify as PLEIADES a group of dogs that pursue *nanuq* (bear) along the ice and into the sky.

page 92. Polar Inuit BIG DIPPER, CASSIOPEIA, and ORION: Burland 1965, 24. George Gibbs (in Swan 1870, 90) writes in a footnote: "The Eskimo call the Great Bear the Cariboo." Gibson did not cite the source of his knowledge or identify the Inuit group about which he spoke.

6. The Harpooner-of-Heaven (Pacific Northwest)

page 96. Elder sky watchers in the Northwest: Miller 1992, 193, 197. Miller cites Dauenhauer and Dauenhauer (1987, 95), who present text from a Tlingit clan history that includes stars.

page 96. "Most, if not all": Swan 1870, 90.

page 97. Tlingit constellations: Swanton 1905, 427; Swanton 1909, 107. Swanton says that Three Men in a Line are "probably" ORION's belt, and he tentatively identifies the Evening Star as Jupiter.

page 97. BIG DIPPER as clock: de Laguna 1972, 802.

page 97. "So nowadays": Swanton 1909, 107.

page 98. "When a child was lazy": Ibid.

page 99. Tsimshian constellations: Boas 1916, 113–116, 454. Boas edited the stories that Henry Tate, a Tsimshian, recorded over a period of twelve years.

page 99. Harpooner-of Heaven: Boas 1910, 165; Boas 1966, 306; Boas and Hunt 1902, 383, 387.

page 100. Cannibal pole: Alexander 1916, 248–249.

page 100. Fraser River constellations: Boas 1916, 604; Lowie 1908, 125.

page 102. Makah constellations: Swan 1870, 90.

page 102. Quileute and Quinault constellations: Clark 1953, 151–153, 158–160.

page 105. "The handle is the elk's tail": Reagan and Walters do not explain the curious use of "handle," which would seem to apply to the BIG DIPPER rather than CASSIOPEIA.

7. Bringing the Day (Plateau)

page 110. About James Teit: Maud 1982, 63–65.

page 111. Lillooet and Carrier constellations: Ray 1945, 189.

page 111. Kootenai constellations: Chamberlain, A. F. 1895, 69. Boas (1918, 247–249) presents a Star Husband story not included here.

page 113. Thompson constellations: Boas 1917, 16, 26; Teit 1898, 22, 91; Teit 1094, 341–342; Teit 1930, 179–180. Unless otherwise noted, identification or possible identification of Thompson and Coeur d'Alene stars is by Teit.

page 113. "Dog following Bunch's trail": Since no other qualifying information is given, the dog could be one of many stars. Aldebaran, however, is the brightest star that appears in the vicinity of PLEIADES and also follows it. The Arabic word for Aldebaran, *Al Dabarān,* means "the follower," that is, the follower of PLEIADES. Perhaps the Thompson, like the Arabs, viewed the reddish star Aldebaran as a follower of the bunched stars, PLEIADES.

page 114. Coeur d'Alene constellations: Boas 1917, 125–126; Teit 1930, 179–180.

page 114. Badger as Aldebaran: Teit (1930, 179) writes that a small star, Coyote's child, lies to the side of the PLEIADES, and that "behind" the small star is the red star, Badger. If "behind" means "following," then Aldebaran fits the description of being a red star that follows the PLEIADES.

page 114. "Women Cooking Camas Roots": Boas (1917, 125), who edited a collection of Teit's Coeur d'Alene stories, writes, "I think [the stars] are the large stars of Auriga and Perseus." Teit (1930, 179) says that the constellation is "a group of stars forming a circle, with one to the side (probably Auriga)." Because Teit actually spoke with the Coeur d'Alene, it is reasonable to suppose that he had better knowledge of the identity of their constellations.

page 117. Coyote's eyeball: Coffin (1961, 146–147) presents a Cheyenne tale about a man who juggles and loses his eyeballs, but the tale does not include the element of the eyeballs becoming stars.

page 118. Knife stars: McWhorter writes that the Knife is "a line of six stars in close proximity to the Pleiades" that "has not been identified." The most likely line of stars is one in PERSEUS, directly above the PLEIADES, that could be interpreted as a knife.

page 121. Star as evil sign: Voegelin 1942, 147.

pages 121–122. Klamath constellations: Spier 1930, 221.

page 122. Klamath Milky Way: Voegelin 1942, 146–147.

8. Chasing Rabbits into a Heavenly Net (Great Basin)

pages 124–125. Storytelling and mythical characters: Liljeblad, 650–656 in D'Azevedo 1986.

page 125. Movement of four planets: Ibid., 642.

page 127. Milky Way: Curtis 1926, 15, 82, 134; Fowler and Liljeblad, 453 in D'Azevedo 1986; Kelly 1932, 154; Stewart 1941, 417.

page 127. BIG DIPPER: Kelly 1932, 154; Liljeblad, 642 in D'Azevedo 1986.

page 128. Great Basin hunting petroglyphs related to ritual: Heizer and Baumhoff 1962, 239.

page 128. Surprise Valley Paiute's identification of ORION's belt: Kelly 1932, 154. Northern Paiute identification of belt and Sirius: Liljeblad, 657 in D'Azevedo 1986. Lowie (1924, 232–234) does not identify the three stars in a row being chased by a fourth star, but the stars he describes are visible in the winter, as are the stars in ORION. Because the story Lowie presents is similar to the one that Curtis presents, it is likely that they describe the same constellation, ORION.

page 129. PLEIADES as Coyote's Daughters: Liljeblad, 642 in D'Azevedo 1986.

page 129. PLEIADES as Women Fighting: Kelly 1932, 154.

page 129. North Star: Liljeblad, 642 in D'Azevedo 1986.

page 129. "Large single star" and unidentified constellations: Kelly 1932, 154.

page 130. Stars hiding: Palmer 1946, 11.

page 130. Milky Way as road: Stewart 1942, 324; Steward 1941, 325.

page 130. Morning Star and ORION's belt: Sapir 1930, 581, 631.

page 130. Coyote as seducer: Kroeber 1901, 268–270; Lowie 1909, 248–251; Lowie 1924, 28–30. Palmer (1946, 11–14) does not include the sequence in which Coyote covets his daughters but does show the family fleeing to the sky.

page 130. Identification of the BIG DIPPER: Sapir (1930, 463) records a tale in which Coyote's family becomes the BIG DIPPER, but the original narrator was unsure of the exact identity of the "seven stars." Other Great Basin stories, including Palmer's Southern Paiute tale, identify Coyote's family as PLEIADES.

page 132. Shoshone constellations: Lawrence Hall of Science 1978; Steward 1941, 325; Steward 1943, 353; Stewart 1942, 324.

page 132. Panamint constellations: Driver 1937, 86–87.

page 132. Ute constellations: Stewart 1942, 324.

9. Knocking off Acorns (California)

page 136. Use of datura: Datura is highly poisonous. It can cause hallucinations, coma, and death. Historically, carefully trained shamans strictly supervised the taking of datura so that it would not be fatal, and they guided people through the frightening visions it caused.

page 137. Yurok, Karok, and Wiyot constellations: Kroeber 1976, 35–36; Kroeber and Gifford 1980, 16, 98, 205; Driver 1937, 402.

page 138. Karok Evening Star: Bright, 187 in Heizer 1978.

page 140. Shasta Milky Way: Silver, 220 in Heizer 1978.

page 140–141. Atsugewi cultural information: Garth, 236–243 in Heizer 1978.

page 141. Atsugewi constellations: Garth 1960, 195; Voegelin 1942, 147.

page 141. Constellations from northern Sierra: Lawrence Hall of Science 1978.

page 141. Patwin constellations: Kroeber 1932, 285.

page 142. Yana constellation: Sapir and Spier 1943, 283.

page 142. Maidu constellations: Dixon 1905, 264.

page 143. Cahto creation of stars: Gifford and Block 1930, 153.

page 143. Pomo constellations and "three stars under the Pleiades": Loeb 1926, 228–229.

page 144. Pomo Morning Star stories: Clark and Williams 1954, 11–14, 47–48.

page 144. Yuki Morning Star: Foster 1944, 207.

page 145. PLEIADES and the salmon run: Gayton 1948, 162, 165.

page 145. American Joe's comments: Gayton and Newman 1940, 50.

page 149. Yokut ORION: Driver 1937, 86–87. ORION's belt as bear and antelope: Latta 1949, 240.

page 149. Falcon and Duck's race: Latta 1936, 174. Other descriptions of the Milky Way: Driver 1937, 86–87.

page 150. Altair and Aldebaran: Hudson and Underhay 1978, 102, 121.

page 150. Yokut Sky Coyote: Latta 1949, 240.

page 150. Monache BIG DIPPER: Driver 1937, 86–87.

page 150. Unidentified Yokut and Monache constellations: Driver 1937, 86–87; Hudson 1988b, 19; Kroeber 1907, stories 15 and 39.

page 150. Costanoan and Salinan constellations: Harrington 1942, 29–30.

page 151. Cahuilla creation of stars: Curtis 1926, 109.

page 151. Cahuilla constellations: Hooper 1920, 363–371. Cahuilla Coyote, Wildcat, identification of ORION: Hudson 1988b, 15–17.

page 151. Patencio's story: Patencio 1943, 50–52.

pages 151–152. The identity of the Cahuilla Pretty Woman is puzzling. Chief Patencio says that the woman is the North Star and her "necklace of jewels" guides the world at night. The chief's identification does not square with Garcia's description of the woman who is wearing a pin that "shines very brightly." The North Star, which is some distance away from the PLEIADES, is not among the twenty brightest stars in the sky. Other stars—such as nearby Capella, the sixth brightest star in the sky—are closer. Perhaps, as is sometimes the case, different bands attributed stories to different stars, and Garcia's Pretty Woman is indeed another star or set of stars.

pages 151–152. Garcia says that the two men, Isilihnup and Holinach, are brothers who are represented by two stars. One star, the burning stick, appears at the beginning of the "first month." The second star, the bodies of both men, appears one month later. Chief Patencio relates that the older brother, Isilihnup, is a red star that appears in January, and the younger brother is a white star that appears in February. Since Isilihnup is a cognate of Isil (Coyote), and the Cahuilla associate Coyote with Aldebaran, it seems likely that the Cahuilla suitors are Aldebaran (as either the older brother or the burning stick) and another star, perhaps one of the stars in ORION, because ORION follows Aldebaran into the sky.

page 153. "Comes up at night": Hooper 1920, 371. Canopus is seen only in very southern latitudes, and in the area in which the Cahuilla live it rises only a short distance above the southern horizon in January and February. Canopus appears when the PLEIADES and Aldebaran, stars that are also involved in these Cahuilla stories, are in the sky.

page 153. "This star twinkles": Hooper 1920, 363.

page 154. Chumash beliefs and game of *peon:* Blackburn 1975, 30–37.

page 154. *ʔAntap* cult: Hudson and Underhay 1978, 36–38; Bean and Vane, 669 in Heizer 1978.

page 155. Chumash constellations: Hudson and Underhay 1978, 99–125; Hudson 1988a, 97–108. Fernando Librado's Chumash constellations: Hudson et al., 1977, 35–37.

page 155. "Long ago, when animals were people": Blackburn, 1975, 245.

page 155. To Separate in the Middle: Hudson and Underhay 1978, 101.

page 156. María Solares's Chumash stories: Blackburn 1975, 91–93, 234–236.

page 161. "In the old times" and religion of fear: DuBois 1908, 162.

page 161. Luiseño constellations: Ibid., 97, 162–164; Kroeber 1925, 682.

page 161. The Luiseño specify that Buzzard is Altair and that Buzzard's right hand is Vega. His headdress would likely be a star or star group "above" Buzzard. DOLPHIN, a small but distinctive constellation that lies near Altair and would be considered above it (given the position of Buzzard's right hand), is one candidate, while another is Deneb, a bright star nearby. All of the stars representing Sky Chiefs except the North Star are among the twenty brightest in the sky. If brightness is the key concern, then Deneb, which is the nineteenth brightest and is also located near Altair and Vega, might be the better candidate for Buzzard's headdress.

page 161. Heart of the Night Wolves: Harrington 1933, 200–210; Kroeber 1925, 682.

page 161. "This is the reason" and "Albañas's grandfather": DuBois 1908, 163–164.

page 162. "Starting from the North Star": DuBois 1906, 54.

page 163. Hearts of the First People: Harrington 1933, 154, 186.

page 164. Luiseño Milky Way in ground painting: Sparkman 1908, plate 20.

page 164. "The Milky Way glows brilliantly": DuBois 1908, 86.

page 164. Tipai ground paintings: Waterman 1910, 301–302, 350–354; Kroeber 1925, 662–664.

page 164. Ipai constellations: DuBois 1908, 165; Hudson 1988b, 15–17.

page 165. Identification of Hand constellation: Spier 1928, 167.

page 165. Tipai constellations: Spier 1923, 319–320; Kroeber 1925, 662–664.

page 165. Anyihai and *Hecha:* Gifford (1931, 66–67) says that *Anyihai,* a "cluster of stars," is "evidently the Pleiades or the Hyades." Narpai, Gifford's informant, says that *Hecha* (PLEIADES) rises before *Anyihai,* which would make *Anyihai* the HYADES.

page 165. Chiyi: Gifford (1931, 68–71) says that Chiyi is three large stars in a row, with the center star largest; these stars are "evidently" the same as the Tipai constellation Buzzard (Altair).

page 165. "This apt name," illus.: DuBois 1908, 165.

page 165. Mountain sheep symbol, illus.: Heizer and Whipple 1951, 197.

page 167. Identity of two hunters: The Cahuilla, to the north of the Tipai, say that ORION's belt is a group of three mountain sheep, the sword is an arrow, and the bright star Rigel is the hunter who has shot the arrow. The Maricopa, situated relatively nearby in the Southwest, say that ORION is Mountain Sheep and Sirius is Mountain Sheep Pursuer, a hunter.

10. Chief of the Night (Southwest)

page 175. 1054 supernova, illus.: Brandt and Williamson 1979.

page 176. "The sign that the Indians" and Cochiti stories: Benedict 1931, 4–5.

page 176. Pot-Rest Stars: The second Cochiti story says that the Three Stars are also called the Pot-Rest Stars. Although the three stones would be set down in a triangle to hold the pot, it is possible that the constellation simply contains three stars and does not reflect the shape formed by the stones. If so, then the Three Stars of the Cochiti could be the same as the "three stars" of the Zia, which are identified as ORION's belt. Inhabitants of both Cochiti and Zia Pueblos spoke a language of the eastern division of the Keresan language family, so some similarities might be expected.

page 176. Identification of Long Sash figures: Leonidas Romero de Vigil, from Nambe, and Maria Martinez and Antonio Da, from San Ildefonso (as was Verlarde), told the Long Sash story to Alice Marriott, who published it in 1968. Marriott's storytellers either could not or would not identify the stars involved; Hopi storytellers with whom she spoke knew the myth but said that they could not tell it to non-Hopis and that religious taboos prevented them from identifying the stars.

page 178. Constellations of Tewa-speakers: Harrington 1916, 49–51; Parsons 1926, 4, 16, 170–171. Parsons names constellations but does not translate all of the names.

page 178. Isleta constellations: Parsons 1932, 342.

page 178. Taos constellations: Curtis 1926, 271.

page 178. Various Pueblo constellations: Parsons 1939, table 2.

page 179. Bull's Eye: The Tewa word for the star in question translates as Bull's Eye. Also, according to Harrington (1916, 50), Spanish-speaking people in the Pueblos called Aldebaran *Ojo del Toro* (Bull's Eye). The early English name for Aldebaran is Bull's Eye, a term that was used at least until 1899 (Allen 1899, 384). It seems likely that the Pueblo name is borrowed from the Spanish, and that the star is Aldebaran.

page 179. "An easily learned constellation": Harrington 1916, 51.

page 179. Identification with Twin War Gods: Parsons 1926, 4, 23.

page 179. Olivella Flower and Yellow Corn Girl: For a similar story, see Erdoes and Ortiz (1984, 173–175), in which Deer Hunter and Corn Maiden become two new stars in the western sky.

page 181. Importance of Morning Star: Ellis 1975, 84–85.

page 181. Stars as higher powers: Stevenson 1904, 23.

page 181. Seven stars as gourd: Cushing 1896, 392.

page 181. Rain ceremony: Stevenson 1904, 194.

page 181. Role of War God Twins in Zuni ritual: Young 1992, 76–84.

page 181. "Henceforth two stars": Cushing 1901, 378–379.

page 182. Cloud Swallower: Benedict 1935, 51.

page 182. Bush Man: Parsons 1930, 10–11.

page 182. Tale of the gambler: Parsons 1930, 45–46. For a similar story, see Boas 1922, 69–70.

page 183. Stick Game Ones: Young and Williamson 1981, 189. The authors note that the constellation is near LIBRA (which is between SCORPIUS and VIRGO), but the group has not been identified to their satisfaction.

page 183. Zuni constellations: Young and Williamson 1981, 183–191; Bunzel 1932, 487. Young and Williamson include unpublished material by John P. Harrington and discuss the possible identities of some Zuni constellations.

page 183. Great White Bear: Personal statement by Frank Cushing to Stansbury Hagar (Hagar 1900, 99).

page 183. Milky Way as snowdrift: Cushing 1901, 381.

page 183. Zuni Milky Way: Benedict 1935, 38, 51.

page 183. Acoma Milky Way: Stirling 1942, 19.

page 183. "This is the most majestic constellation": Harrington n.d., in Young and Williamson 1981, 187. Identification of Chief of the Night, 187–189.

page 184. "Some part of the body": Ibid., 189.

pages 184–185. Hopi constellations: Parsons 1936, 25, 87, 860, 1209, 1320; Parsons 1939, table 2.

page 185. Coyote creating stars: Cushing 1923, 166.

page 186. Navajo tribal government: Griffin-Pierce 1992a, 20–21.

page 186. Difference between Navajo and Pueblo ceremonials: Wyman, 537 in Ortiz 1983.

page 186. Celestial motifs originating in Pueblos: Chamberlain 1994, 13.

page 186. Laws "are written in the stars": Newcomb 1967, 83, 88. Stars as symbols, Britt 1975, 95; Griffin-Pierce 1992a, 142–173; Griffin-Pierce 1992b, 111.

page 186. Pictographs: Britt 1975, 95.

page 186. Sandpaintings: Wyman 1952; Kluckhohn and Wyman 1940; Newcomb, Fishler, and Wheelwright 1956; Griffin-Pierce 1992a.

page 187. Accuracy in sandpaintings: Griffin-Pierce 1992a, 104–105.

page 187. Son-of-the-late-Cane: Haile 1947, star chart.

page 187. Primary and secondary constellations: Griffin-Pierce 1992a, table 4.3, 80–82.

page 187. Male One Who Revolves and Man with Legs Ajar: Griffin-Pierce (personal communication) recommends "Male One Who Revolves" and "Man with Legs Ajar" over other translations of the Navajo names for the BIG DIPPER and CROW.

page 187. Navajo painting: Brewer 1950, 134–136.

page 187. Navajo star books: Anthropologist Trudy Griffin-Pierce's book, *Earth Is My Mother, Sky Is My Father,* presents information on Navajo constellations as they appear in sandpaintings and discusses the interpretations of current Navajo chanters. Griffin-Pierce worked with six chanters and seven other consultants who are closely associated with chanters and ceremonials. Father Berard Haile, a Catholic priest who lived among the Navajo for many years, collected material from the chanter Son-of-the-late-Cane in 1908 and published this information in *Starlore Among the Navajo* in 1947.

page 188. Navajo creation: Gill, 502–504 in Ortiz 1983.

page 188. Creation of stars: Goddard 1934, 137–138; Haile 1947, 1–4; Matthews 1883, 213–214. Wyman (1952, 110–111) records that in the Beauty Chant story, which is part of the Beadway ceremonial, First Man and First Woman rather than Black

God make the stars. The constellations are described as shadows of people or objects. Big First One (part of SCORPIUS) is the shadow of a chief who died of old age, a desirable way to die. This chief carries a walking stick and a basket of seeds, from which he eats. He also eats "rabbit tracks" (the tail of SCORPIUS). A bluebottle fly (CROW) flies through the sky spreading news. Ants (PLEIADES) come from underneath the Earth and go to the sky. Other constellations shown in the sandpainting made during this part of the ceremonial include Wife of the God of Night (CASSIOPEIA), their fire or ignitor (North Star), and the God of Night (BIG DIPPER).

page 188. Spelling of Navajo words: Most spellings are from Young and Morgan, 1980.

page 188. Female One Who Revolves: This constellation is generally identified as CASSIOPEIA, though some chanters say it is the LITTLE DIPPER. For more information, see O'Bryan 1956, 21; Brewer 1950, 135.

page 189. Warrior with His Bows and Arrows: O'Bryan 1956, 21.

page 189. Littlesalt: Brewer 1950, 133–136.

page 189. Broken leg: Griffin-Pierce 1992a, 156.

page 189. Stars as hogan: Griffin-Pierce 1992a, 153; 1992b, 111. As a married couple: Newcomb 1966, 156. As leaders and as First Man and First Woman: Griffin-Pierce 1992a, 153–156.

page 189. Hózhǫ́: Griffin-Pierce 1992b, 113–114.

page 189. Bows: Griffin-Pierce 1992a, 155–156.

page 189. Translation of *Dilyéhé:* Brewer 1950, 135; Griffin-Pierce 1992a, 157.

page 189. Dice game: Griffin-Pierce 1992a, 157.

pages 189–190. "Before the Navahos had guns" and boys dodging: Brewer 1950, 135. Hard Flint Boys: Haile 1947, 33.

page 190. Dilyéhé representing all stars: Griffin-Pierce 1992a, 158.

page 190. String figures: Griffin-Pierce 1992b, 124. The string figures are to be practiced only in winter, when spiders are not present. For directions for making string figures, see Jayne 1906.

page 190. Two women as Doubtful Stars: Haile 1947, 16–19. *Baalchini* as central pairs of stars and *Dilyéhé* marrying: O'Bryan 1956, 20–21.

page 190. Hard Flint Women: Haile 1947, 14; Wyman 1952, 47. Wrestling boys and "fire of the twin stars": Brewer, 1950, 136. Griffin-Pierce (1992a, table 4.3) cites Hosteen Klah, who identifies the red star as Antares, but Littlesalt says that the red star is near the two wrestlers in HYADES, which would make it Aldebaran.

page 190. Month feathers: O'Bryan 1956, 16–17.

page 191. Cane: Brewer 1950, 136; Griffin-Pierce 1992a, 166, 180.

page 191. Milky Way as symbol: Griffin-Pierce 1992b, 124. As spilled meal: Kluckhohn and Wyman 1940, 183. As ashes: Griffin-Pierce 1992a, 169.

page 191. Rabbit Tracks: Although some sources identify Rabbit Tracks as a cluster of stars located beneath or within the constellation BIG DOG, Griffin-Pierce's chanters (1992a, 80) say that the constellation Rabbit Tracks consists of stars in the tail of SCORPIUS.

page 191. Honoring the rabbit and hunting rabbits: Ibid., 168–169.

page 191. Actual rabbit tracks and petroglyphs: Chamberlain 1994, 12–19.

page 191. First Slim One: Brewer 1950, 136. The stick is not identified, but one possibility is that it is ORION's shield, a curved line of stars that lies above when ORION is rising and to the right when ORION is high in the sky. Griffin-Pierce's list of stars and constellations does not include the stick and seed basket of First Slim One. She does include a stick for First Big One, part of SCORPIUS.

page 193. Stars and planets: Griffin-Pierce 1992a, table 4.3.

page 193. Coyote Star as Antares: In most Navajo texts, Coyote Star is identified as Canopus, though Navajo medicine man Hosteen Klah clearly identified it as Antares (Chamberlain 1983, 52). Williamson (1984, 164, 332), an astronomer, points out that Canopus appears at only two to three degrees above the horizon at the mean latitude of the Navajo and cannot be seen in their region. He believes Coyote Star is more likely Antares. Chamberlain, also an astronomer, writes, "Canopus should be barely visible from Navajo country" (1983, 54).

page 193. Unidentified constellations and stars: Griffin-Pierce 1992a, table 4.3; Newcomb, Fishler, and Wheelwright 1956, 26; Newcomb and Reichard 1937, 55–59. Chamberlain (1983, 48–58) presents an overview of all Navajo constellations in literature, art, artifacts, and rock art.

page 193. Mescalero Apache constellations and stars as guides: Farrer 1987, 223–236; 1989, 486; 1991, 23, 48, 55–56; 1992, 71–73.

page 194. Translation of *Nahakus:* Farrer 1991, 48.

page 194. Jicarilla Apache creation myth: Tiller, 446 in Ortiz 1983.

page 197. Havasupai constellations: Spier 1928, 166–172. Yavapai constellations: Gifford 1932, 248. For all three tribes: Spier 1955, 20–21.

page 197. Identity of Hoop: Although the circlet of PISCES matches the shape of the Hoop constellation recorded by Spier, PISCES rises in the evening, rather than before dawn, in the late fall. Spier (1928, 170) records that the Hoop rises before dawn in the first Havasupai month, whereas Gifford (1932, 248) says that the constellation rises before dawn in the third Yavapai month. This discrepancy complicates identification.

page 199. "The stars change" and "the first crops": Spier 1933, 146.

page 199. Maricopa constellations: Spier 1933, 146–148; Spier 1955, 27. Spier discusses the possible identifications of various constellations.

page 199. "When an old person is dying": Spier 1933, 148.

page 199. Quail: Gambel's quail is a small chickenlike bird that lives in the desert of the Southwest. The bird has a short black plume that arches forward from the top of the head. Both the Pima and the Tohono O'odham tell stories about quail, and the Seri name but do not identify a Quail constellation. Spier (1933, 147) describes the constellation as "a ring of faint stars forming the body, three larger ones the head, surmounted by a single tiny star, the plume."

page 200. Maricopa stars in ORION: Spier (1933, 146–147) identifies Sirius and suggests that ORION's sword is the shaft.

page 200. Cocopa ORION and quote, "Star people set an example": Gifford 1933, 286.

page 200. Sobar as Sirius: Gifford (1933, 286) says that Sobar is "standing to S. of Amuh." If you are looking east just after ORION and BIG DOG have risen, Sirius will appear below and somewhat to the south (as opposed to the north) of ORION. Although another star, such as Rigel, which lies to the south of ORION's belt when ORION is high in the sky, could be suggested as Sobar, it is more likely that the Cocopa Sobar is the same as the Maricopa Mountain Sheep Pursuer.

page 201. "When the yellow blossoms" and "That is why": Spier 1933, 147–148.

page 202. Identity of Scorpion: Culin (1907, 202) writes that the Scorpion is composed of five stars and notes: "As [SCORPIUS] rises in the east about August, the three stars of the body are nearly horizontal. The two claws point toward the south, upward and downward." The description is a bit ambiguous, but if the forepart of SCORPIUS is Earth Doctor, then the hind part must be the Scorpion. The two stinger stars, Shaula and Lesath, along with a small star to the east, do form a line that is nearly horizontal. If these three stars are the Scorpion's body, then two stars to the south would be the claws.

page 203. Identification of Earth Doctor: Culin (1907, 203) notes that Earth Doctor is composed of stars in "Scorpio and the others." If Antares is the head and other stars in SCORPIUS form the braid, then perhaps stars above SCORPIUS form the two feathers.

page 205. Identity of prostitutes and Bluebird: Hoskinson 1992, 134–137. According to Russell (1908, 198) prostitution was said to have been rare in the "old days."

page 205. Pima constellations: Russell 1908, 252.

page 206. Tohono O'odham Cactus Hook: Spier 1955, 21; Densmore 1929, 151.

page 208. Tohono O'odham Morning Star: See Wright (1929, 143–151) for another version of this myth.

page 209. Seri constellations: Felger and Moser 1985, 57; Kroeber 1931, 12; Spier 1955, 21.

page 209. Seri Vulture: Kroeber (1931, 12) says that Altair "may be" Buzzard and cites the Luiseño Buzzard.

11. A Bright Light Across the Heavens (Great Plains)

page 215. Two distinct cultures: Josephy 1968, 110–123.

page 218. Skidi Pawnee background information: Murie 1981, 4–11.

page 218. Supervision of a star: Weltfish 1965, 19.

page 218. Sacred bundles: Murie 1981, 13.

page 218. Patron stars: Dorsey 1904c, xx.

page 219. Taboo about snakes: Ibid., xxii–xxiii.

page 221. "To you I give power": Chamberlain (1982, 54), quoting George A. Dorsey's undated original manuscripts and notes on file in the Anthropology Archives, Field Museum of Natural History, Chicago.

page 221. "A great warrior": Fletcher 1903, 11.

page 222. "Oh Morning Star": Fletcher 1904, 323.

page 222. Mars as Great Star in conjunction with Venus: Chamberlain (1982, 84–90) compared the known dates of the human sacrifice, which was conducted to commemorate the union of Bright Star and Great Star, with events in the sky by running a planetarium instrument for a two-hundred-year period. The human sacrifice was not performed every year, and it is thought that the practice ended sometime before 1850. The Bright Star ceremony was sometimes performed without a sacrifice.

page 222. "You four shall be known": Dorsey 1904c, 4.

page 222. Identification of four-quarter stars: Chamberlain 1982, 101.

page 224. Milky Way as bright light: Fletcher 1903, 13.

page 225. BIG DIPPER and LITTLE DIPPER: Dorsey 1906, 135.

page 225. "The people took their way": Fletcher 1903, 15.

page 225. Regarding Swimming Ducks, "It is believed": Murie 1981, 41.

pages 225–226. Identification of Real Snake, Snake Not Real, Rabbit, Deer, Bird's Foot: Fletcher 1903, 15; Chamberlain 1982, 82, 114, 132–136.

page 226. "Each warrior places": Murie 1981, 110.

page 226. Identification of Bear, Panther, and Wildcat: Chamberlain 1982, 128–130.

page 226. Wolf Star: Ibid.; Murie 1981, 39.

page 227. "Look as they rise": Fletcher 1904, 330.

page 227. Basket dice game: Culin 1907, 101.

page 228. Rattling skull: Dorsey 1906, 119–122.

page 228. Identification of Great Star's brother and Star of Sickness: Chamberlain 1982, 90–92.

page 228. Pahukatawa: Chamberlain (1982, 111–112) suggests Beta Persei (Algol) or Gamma Cephei as the Kneeprint Star, but adds that there are other reasonable choices as well.

page 228. "By their breaths": Murie 1981, 30. Identity of Wind Ready to Give and Breathe: Chamberlain 1982, 111–112.

page 229. Lone-Chief's story: Murie 1981, 40; Dorsey 1904c, 49.

page 229. Identification of Bow and Feather Headdress Star: Chamberlain 1982, 114, 124–127, 142.

page 229. Wichita creation, First Man and First Woman: Dorsey 1904b, 20.

page 229. Wichita stories about the BIG DIPPER and unidentified stars: Dorsey 1904a, 153–160; Dorsey 1904b, 69–74.

page 229. South Star as Canopus: The Wichita South Star is likely the same as the Pawnee South Star, which Chamberlain suggests is Canopus.

page 229. Identity of Flint Stone Lying Down Above: There is not enough information to identify this star. One wonders, however, if there is a connection between the Flint Stone star and the flint knife that a Blackfoot hero uses to kill monsters; the knife becomes Smoking-Star, the Great Nebula in ORION. There are not enough recorded Wichita star stories to establish a pattern of similarities between the two tribes, but the Wichita tale about the BIG DIPPER (seven brothers and a sister chased by a bear) does parallel that of the Blackfoot. ORION is in the sky when Canopus—thought to be the South Star—appears in the southern sky.

page 230. Arikara constellations: Parks 1991, 148–161, 213–235, 247–259, 575–592.

page 231. Star Husband: Grinnell (1894, 197–200) relates a similar myth from the Pawnee about Star Boy.

page 231. Kiowa BIG DIPPER: Momaday 1969, 8.

page 233. "Ho! Sun, Moon, Stars": Fletcher and La Flesche 1911, 115.

page 233–234. Omaha constellations: Ibid., 110, 177, 178, 514, 588; Dorsey 1888, 378.

page 234. Osage constellations: Dorsey 1888, 378; Dorsey 1894, 379; La Flesche 1928, 73–75; La Flesche 1930, 658; La Flesche 1932, 91, 139, 183, 205, 206.

page 234. "I am not" and Osage creation: Dorsey 1888, 377–379.

page 235. Dakota constellations: Wallis 1923, 44–45.

pages 235–236. Lakota star lore including Hand constellation: Goodman 1992, 215–220; Walker 1980, 114–115.

page 236. Winter Count, illus.: see Chamberlain (1984, 1–53) for more information about Winter Counts.

page 237. Assiniboin constellations: Denig 1930, 417.

page 238. A Comanche story about seven brothers who decide to become stars (St. Clair 1909, 282) also parallels the Assiniboin PLEIADES story.

page 240. Mandan constellations: Dorsey 1894, 508; Squier 1848, 255.

page 240. Hidatsa constellations: Dorsey 1894, 517; Densmore 1923, 115–116.

page 242. Crow storytelling: Lowie 1918, 13.

page 242. Crow constellations: Ibid., 210–211; Lowie 1935, 152–153, 274; Simms 1903b, 309–312.

page 243. Puffball stars: Kehoe 1992, 207; Wissler and Duvall 1909, 40.

page 243. Mars: Grinnell 1892, 99.

pages 243–244. The Seven Brothers: McClintock 1910, 488–490; Wissler and Duvall 1909, 68–70; Wissler 1936, 9–10; Wilson 1893, 200–203.

pages 243–244. Woman changing into a bear: Kroeber (1907b, 105–108) gives a similar Gros Ventre tale about the BIG DIPPER; the constellation is called *Täbiitciçaa*[n] (Cut-Off-Head).

page 246. "Ever since those days": McClintock 1910, 500.

page 246. Medicine woman organizing the Sun Dance: Leitch 1979, 66.

page 246. In Wolf-head's telling of the story of a woman-who-marries-a-star (Wissler and Duvall 1909, 60), the woman's child dies and becomes the Fixed Star, the North Star.

page 246. "Whenever you see": McClintock 1910, 500.

page 246. "long ago, when": Ibid., 491–492.

page 250. "I see them in the eastern sky": Ibid., 499.

page 252. Smoking-Star as stone knife and Blackfoot ceremony: Wissler 1936, 16.

page 252. Blod clot as bison fetus: Kehoe 1992, 207–214.

page 255. Star designs on Cheyenne tipis: Grinnell 1923, 233–234.

page 255. Cheyenne constellations: Harrod 1987, 144; Hoebei 1978, 88.

page 257. Possible Sack stories: Grinnell 1926, 220–231; Kroeber 1900, 182–183.

page 259. Arapaho constellations: Dorsey 1903, 205; Dorsey and Kroeber 1903, 338.

page 259. Arapaho Hand constellation: If this constellation is composed of stars in ORION, as it is in Lakota and Hidatsa myth, then the Lance might be stars nearby, such as the stars of ORION's shield.

page 261. Stellar alignments: Eddy 1974, 1035–1043; Eddy 1977, 149–152; Robinson 1980, 15–19; Robinson 1989, 497–498.

page 261. Use and creators of medicine wheel: Simms 1903a, 107; Williamson 1984, 209.

page 261. Elk River: Grinnell 1922, 307.

12. The Celestial Skiff (Southeast)

page 266. The cosmos: Swanton 1928, 477–478; Mooney 1900, 231.

page 268. "This is what the old men told me": Mooney 1900, 232.

page 270. Cherokee PLEIADES: For a contemporary narration of this story by Mary Ulmer Chiltoskey, see Galloway 1990, 45–47.

page 271. Nebula as the drum: It is unclear whether Hagar is referring to the faint nebulosity in which the PLEIADES are bathed (a nebulosity that is invisible to the naked eye), or to some other nebula. Another Cherokee story, preserved by John Howard Payne in manuscript and quoted in Squier (1848, 256) also says that the boys seized a drum and beat it as they ascended to the sky.

page 273. Spirit Dog: Galloway 1990, 62–63.

page 275. Wooden Box: Hagar 1906, 365.

page 275. Creek constellations: Swanton 1928, 478–480.

page 275. Seminole constellations: Greenlee 1945, 138–139.

page 275–276. Koasati constellations: Swanton 1929, 166.

page 276. Celestial skiff: Monroe and Williamson 1987, 111–115; Williamson 1992, 52–66.

page 278. Caddo constellations: Dorsey 1905, 25–27.

References

Works frequently cited are identified by the following abbreviations:

BAE	Bureau of American Ethnology, Smithsonian Institution
JAFL	*Journal of American Folk-Lore*
AR/BAE	Annual Report of the Bureau of American Ethnology, Smithsonian Institution
Earth and Sky/Williamson and Farrer	Williamson, Ray A., and Claire R. Farrer, eds. 1992. *Earth and Sky: Visions of the Cosmos in Native American Folklore* Albuquerque: University of New Mexico Press.
The North American Indian	Curtis, Edward S. [1926] 1970. *The North American Indian: Being a Series of Volumes Picturing and Describing the Indians of the United States, and Alaska.* Edited by Frederick W. Hodge. Norwood, Mass.: Plimpton Press. Reprint, New York: Johnson Reprint.
Native American Constellations	Lawrence Hall of Science. 1978. Native American Constellations. Berkeley: University of California. (This resource is a listing of Native American constellations printed in a round "star wheel" format.

General References

Gill, Sam D., and Irene F. Sullivan. 1992. *Dictionary of Native American Mythology*. New York and Oxford: Oxford University Press.

Leitch, Barbara A. 1979. *A Concise Dictionary of Indian Tribes of North America*. Algonac, Mich.: Reference Publications, Inc.

Waldman, Carl. 1988. *Encyclopedia of Native American Tribes*. New York: Facts on File.

Preface

Wissler, Clark, and D. C. Duvall. 1909. Mythology of the Blackfoot Indians. Part 1. *Anthropological Papers of the American Museum of Natural History* 2.

Chapter 1: Stars of the First People

Gill, Sam D., and Irene F. Sullivan. 1992. *Dictionary of Native American Mythology*. New York and Oxford: Oxford University Press.

Hooper, Lucile. 1920. The Cahuilla Indians. *Publications in American Archaeology and Ethnology* (University of California) 16:315–380.

Chapter 2: Greek Constellations

Allen, Richard H. [1899] 1963. *Star-Names and Their Meanings*. Reprint, New York: Dover Publications.

Burnham, Robert Jr. [1966–1973] 1978. *Burnham's Celestial Handbook: An Observer's Guide to the Universe and Beyond the Solar System*. Flagstaff, Ariz.: Celestial Handbook Publications. Reprint, New York: Dover Publications.

Graves, Robert. [1955] 1957. *The Greek Myths*. 2 vols. Reprint, New York: George Braziller, Inc.

Hamilton, Edith. [1940] 1942. *Mythology*. Reprint, New York and Toronto: New American Library.

Chapter 3: The Celestial Bear Hunt

Alger, Abby L. 1897. *In Indian Tents*. Boston: Roberts Brothers.

Allen, Richard H. [1899] 1963. *Star-Names and Their Meanings*. Reprint, New York: Dover Publications.

Beauchamp, William M. 1895. Onondaga Tale of the Pleiades. *JAFL* 13:281–282.

———. 1922. *Iroquois Folk Lore*. Syracuse: The Dehler Press.

Brown, Mrs. W. Wallace. 1890. Wa-Ba-Ba-Nal, or Northern Lights. *JAFL* 3:213–214.

Cappel, Jeanne L'strange. 1928. *Chippewa Tales*. Los Angeles: Wetzel Publishing Company.

Carson, William. 1917. Ojibwa Tales. *JAFL* 30:492–493.

Connelley, William E. 1899. *Wyandot Folk-Lore*. Topeka, Kans.: Crane & Company, Publishers.

Converse, Harriet M. [1908] 1974. *Myths and Legends of the New York State Iroquois*. New York State Museum Bulletin 125. Reprint, Albany: State University of New York.

Conway, Thor. 1992. The Conjurer's Lodge: Celestial Narratives from Algonkian Shamans. Chapter 15 in *Earth and Sky*/Williamson and Farrer.

Curtin, Jeremiah. 1922. *Seneca Indian Myths*. New York: E. P. Dutton & Company.

Curtin, Jeremiah, and J. N. B. Hewitt. 1918. *Seneca Fiction, Legends, and Myths*. AR/BAE (1910–1911) 32:37–813.

Hagar, Stansbury. 1900. The Celestial Bear. *JAFL* 13:92–103.

Hewitt, J. N. B. 1903. *Iroquoian Cosmology*. AR/BAE (1899–1900) 21:127–339.

Jenness, Diamond. 1935. *The Ojibwa Indians of Parry Island, Their Social and Religious Life*. Canada Department of Mines Bulletin 78, Anthropological Series 17.

Jones, William. 1907. Fox Texts. Edited by Margaret Welpley Fisher. *Publications* (American Ethnological Society) 1.

———. 1939. Ethnography of the Fox Indians. BAE Bulletin 125.

Judson, Katherine Berry. 1917. *Myths and Legends of British North America*. Chicago: A. C. McClurg & Co.

Leland, C. G. 1884. *The Algonquin Legends of New England; or, Myths and Folk Lore of the Micmac, Passamaquoddy, and Penobscot Tribes*. Boston: Houghton Mifflin.

McElwain, Thomas. 1992. Asking the Stars: Seneca Hunting Ceremonial. Chapter 16 in *Earth and Sky*/Williamson and Farrer.

O'Callaghan, E. B. 1850. *Documentary History of the State of New York*. Vol. 3. Albany: Weed, Parsons & Co.

Parker, Arthur. 1923. *Seneca Myths and Folk Tales*. Buffalo Historical Society Publications 27.

Rideout, Henry M. 1912. *William Jones: Indian,*

Cowboy, American Scholar, and Anthropologist in the Field. New York: Frederick A. Stokes Company.

Schoolcraft, Henry Rowe. 1856. *The Myth of Hiawatha and other Oral Legends, Mythologic and Allegoric, of the North American Indians.* Philadelphia: J. B. Lippincott & Co.

Skinner, Alanson, and John V. Satterlee. 1915. Folklore of the Menomini Indians. *Anthropological Papers of the American Museum of Natural History* 13.

Smith, Erminnie A. [1883] 1983. *Myths of the Iroquois.* AR/BAE (1880–1881) 2:51–112. Reprint, Ohsweken, Ontario: Iroqrafts Ltd.

Speck, Frank G. 1915. *Myths and Folklore of the Timiskaming Algonquin and Temagami Ojibwa.* Geological Survey of Canada Anthropology Series 9.

———. 1931. *A Study of the Delaware Indian Big House Ceremony.* Vol 2. Harrisburg: Pennsylvania Historical Commission.

Speck, Frank G., and Jesse Moses. 1945. *The Celestial Bear Comes Down to Earth.* Reading (Pennsylvania) Public Museum and Art Gallery Scientific Publications 7.

Spotted Elk, Molly. 1938. Penobscot Indian Legends. Orono: University of Maine Archives. Duplicated.

Trigger, Bruce G., ed. 1978. *Handbook of North American Indians.* Vol. 15, *Northeast.* Washington, D.C.: Smithsonian Institution. The following chapters were consulted for background and cultural information: Brock, Philip K., Micmac; Callender, Charles, Fox and Shawnee; Erickson, Vincent O., Maliseet-Passamaquoddy; Fenton, William N., Northern Iroquoian Culture Patterns; Goddard, Ives, Delaware; Ritzenthaler, Robert E., Southwestern Chippewa; Rogers, E. S., Southeastern Chippewa; Spindler, Louise, Menominee; Tooker, Elisabeth, The League of the Iroquois: Its History, Politics, and Ritual; Trigger, Bruce G., Introduction.

Chapter 4: The Fisher Stars

Farrand, Livingston. 1900. Traditions of the Chilcotin Indians. *Memoirs* (American Museum of Natural History) 4:1–54.

Hamilton, J. C. 1905. Stellar Legends of American Indians. *Transactions* (Royal Astronomical Society of Canada) 47–50.

Helm, June, ed. 1981. *Handbook of North American Indians.* Vol. 6, *Subarctic.* Washington, D.C.: Smithsonian Institution. The following chapters were consulted for background and cultural information: Honigmann, John J., Expressive Aspects of Subarctic Indian Culture; Lane, Robert B., Chilcotin; MacLachlan, Bruce B., Tahltan; McClellan, Catharine, Tutchone; Rogers, Edward S., and Eleanor Leacock, Montagnais-Naskapi.

Judson, Katherine Berry. 1917. *Myths and Legends of British North America.* Chicago: A. C. McClurg & Co.

Lawrence Hall of Science. 1978. Native American Constellations.

McClellan, Catharine. 1975. *My Old People Say: An Ethnographic Survey of Southern Yukon Territory.* Part 1. Ottawa: National Museums of Canada.

Nelson, Richard K. 1983. *Make Prayers to the Raven.* Chicago: The University of Chicago Press.

Speck, Frank G. 1925. Montagnais and Naskapi Tales from the Labrador Peninsula. *JAFL* 38:1–32.

Teit, James A. 1919. Tahltan Tales. *JAFL* 31:198–250.

Chapter 5: Raven's Snowshoe Tracks

Boas, Franz. [1888] 1964. *The Central Eskimo.* AR/BAE (1894–1895) 6:399–669. Reprint, Lincoln: University of Nebraska Press.

———. 1895. Literary Notes. *The American Antiquarian* 17:121.

Burland, Cottie. 1985. *North American Indian Mythology.* 2d ed. New York: Peter Bedrick Books.

Damas, David, ed. 1984. *Handbook of North American Indians.* Vol. 5, *Arctic.* Washington, D.C.: Smithsonian Institution. The following chapters were consulted for background and cultural information: Damas, David, Introduction; Kemp, William B., Baffinland Eskimo; Ray, Dorothy Jean, Bering Strait Eskimo; Spencer, Robert F., North Akaska Coast Eskimo.

Kroeber, Alfred Lewis. 1899. Tales of the Smith Sound Eskimo. *JAFL* 12:166–182.

Nelson, Edward. [1899] 1983. *The Eskimo About Bering Strait.* AR/BAE (1896–1897) 18:3–518. Reprint, Washington, D.C.: Smithsonian Institution.

Nuttall, Zelia. 1901. Fundamental Principles of Old and New World Civilizations. *Archaeological and Ethnological Papers* (Peabody Museum of American Archaeology and Ethnology) 2.

Rasmussen, Knud. 1908. *The People of the Polar North.* Edited by G. Herring. London: Kegan Paul, Trench, Trubner & Co. Ltd.

———. 1921. *Eskimo Folk-Tales.* Edited and translated by W. Worster. London: Gyldendal.

Swan, James G. 1870. The Indians of Cape Flattery. *Contributions to Knowledge* (Smithsonian Institution) 220.

Turner, Lucien M. 1894. *Ethnology of the Ungava District, Hudson Bay Territory.* AR/BAE (1889–1890) 11:159–316.

Chapter 6: The Harpooner-of-Heaven

Alexander, Hartley B. 1916. *Mythology of All Races.* Vol. 10. *North American Mythology.* Boston: Marshall Jones Co.

Boas, Franz. 1901. *Kathlamet Texts.* BAE Bulletin 26.

———. 1910. Kwakiutl Tales. *Contributions to Anthropology* (Columbia University) 2.

———. 1916. *Tsimshian Mythology.* AR/BAE (1909–1910) 31:29–1037.

Boas, Franz. 1966. *Kwakiutl Ethnography.* Edited by Helen Codere. Chicago and London: The University of Chicago Press.

Boas, Franz and George Hunt. 1902. Kwakiutl Texts. *Publications of the Jesup North Pacific Expedition* (American Museum of Natural History) 2.

Clark, Ella E. 1953. *Indian Legends of the Pacific Northwest.* Berkeley: University of California Press.

Dauenhauer, Nora Marks, and Richard Dauenhauer. 1987. *Haa Shuka, Our Ancestors: Tlingit Oral Narratives.* Seattle: University of Washington Press.

de Laguna, Frederica. 1972. Under Mount Saint Elias: The History and Culture of the Yakutat Tlingit 2. *Contributions to Anthropology* (Smithsonian Institution) 7.

Haeberlin, Hermann K. 1924. Mythology of Puget Sound. *JAFL* 37:371–438.

Hamilton, J. C. 1905. Stellar Legends of American Indians. *Transactions* (Royal Astronomical Society of Canada).

Lowie, Robert H. 1908. The Test-Theme in North American Mythology. *JAFL* 21:97–148.

Mallery, Garrick. 1893. *Picture-Writing of the American Indians.* AR/BAE (1888–1889) 10:103–822.

Miller, Jay. 1992. North Pacific Ethnoastronomy: Tsimshian and Others. Chapter 11 in *Earth and Sky*/Williamson and Farrer.

Reagan, Albert, and L. V. W. Walters. 1933. Tales from the Ho and Quileute. *JAFL* 46:297–364.

Shelton, William. 1935. *The Story of the Totem Pole.* Everett, Wash: N.p.

Suttles, Wayne, ed. 1990. *Handbook of North American Indians.* Vol. 7, *Northwest Coast.* Washington, D.C.: Smithsonian Institution. The following chapters were consulted for background and cultural information: de Laguna, Frederica, Tlingit; Halpin, Marjorie M., and Margaret Seguin, Tsimshian Peoples: Southern Tsimshian, Coast Tsimshian, Nishga, and Gitksan; Renker, Ann M., and Erna Gunther, Makah; Suttles, Wayne, Introduction.

Swan, James G. 1870. The Indians of Cape Flattery. *Contributions to Knowledge* (Smithsonian Institution) 220.

Swanton, John R. 1908. *Social Condition, Beliefs, and Linguistic Relationship of the Tlingit Indians.* AR/BAE (1904–1905) 26:391–485.

———. 1909. *Tlingit Myths and Texts.* BAE Bulletin 39.

Chapter 7: Bringing the Day

Boas, Franz. 1918. *Kutenai Tales.* BAE Bulletin 59.

Boas, Franz, ed. 1917. Folk-Tales of Salishan and Sahaptin Tribes. *Memoirs of the American Folklore Society* 11.

Chamberlain, A. F. 1895. Notes on the Kootenay Indians. *The American Antiquarian* 17:68–72.

Clark, Ella E. 1953. *Indian Legends of the Pacific Northwest.* Berkeley: University of California Press.

———. 1960. *Indian Legends of Canada.* Toronto: McClelland and Stewart Limited.

———. 1966. *Indian Legends from the Northern Rockies*. Norman: University of Oklahoma Press.
Coffin, Tristan P., ed. 1961. *Indian Tales of North America*. Philadelphia: American Folklore Society.
Lawrence Hall of Science. 1978. Native American Constellations.
Maud, Ralph. 1982. *A Guide to B.C. Indian Myths and Legends*. Vancouver: Talonbooks.
McWhorter, Lucullus V. Papers. The L. V. McWhorter Archives, Holland Library, State College of Washington. Manuscript Collection, folders 2 and 3.
Ray, Verne. 1945. Cultural Element Distributions, XXII: Plateau. *Anthropological Records* (University of California) 8.
Spier, Leslie. 1930. Klamath Ethnography. *Publications in American Archaeology and Ethnography* (University of California) 30.
Teit, James. 1898. Traditions of the Thompson River Indians of British Columbia. *Memoirs of the American Folklore Society* 6.
———. 1904. The Thompson Indians of British Columbia. *Publications of the Jesup North Pacific Expedition* (American Museum of Natural History) 4.
———. 1930. *The Salishan Tribes of the Western Plateaus*. AR/BAE (1927–1928) 45:23–439.
Voegelin, Erminie. 1942. Cultural Element Distributions, XX: Northeast California. *Anthropological Records* (University of California) 7:47–252.

Chapter 8: Chasing Rabbits into a Heavenly Net

Curtis, Edward S. [1926] 1970. *The North American Indian*. Vol. 15.
D'Azevedo, Warren L., ed. 1986. *Handbook of North American Indians*. Vol. 11, *Great Basin*. Washington, D.C.: Smithsonian Institution. The following chapters were consulted for background and cultural information as well as celestial information: Fowler, Catherine S., Subsistence; Fowler, Catherine S., and Sven Liljeblad, Northern Paiute; Kelly, Isabel T., and Catherine S. Fowler, Southern Paiute; Liljeblad, Sven, Oral Tradition: Content and Style of Verbal Arts.
Driver, Philip. 1937. Cultural Element Distributions, VI: Southern Sierra Nevada. *Anthropological Records* (University of California) 1:53–154.
Heizer, Robert F., and Martin A. Baumhoff. 1962. *Prehistoric Rock Art of Nevada and Eastern California*. Berkeley and Los Angeles: University of California Press.
Heizer, Robert F., and C. W. Clewlow, Jr. 1973. *Prehistoric Rock Art of California*. Ramona, Calif.: Ballena Press.
Kelly, Isabel T. 1932. Ethnography of the Surprise Valley Paiute. *Publications in American Anthropology and Ethnology* (University of California) 31:67–210.
Kroeber, Alfred L. 1901. Ute Tales. *JAFL* 14: 252–285.
Lawrence Hall of Science. 1978. Native American Constellations.
Lowie, Robert H. 1909. The Northern Shoshone: Part 2, Mythology. *Anthropological Papers of the American Museum of Natural History* 2.
———. 1924. Shoshonean Tales. *JAFL* 37:1–242.
Mayer, Dorothy. 1975. Star-Patterns in Great Basin Petroglyphs. Chapter 6 in *Archaeoastronomy in Pre-Columbian America*, edited by Anthony F. Aveni. Austin: University of Texas Press.
Palmer, William R. 1946. *Why the North Star Stands Still and Other Indian Legends*. Springdale, Utah: Zion Natural History Association.
Sapir, Edward. 1930. The Southern Paiute Language. *Proceedings of the American Academy of Arts and Sciences* 65.
Steward, Julian H. 1941. Cultural Element Distributions, XIII: Nevada Shoshoni. *Anthropological Records* (University of California) 4:209–360.
———. 1943. Cultural Element Distributions, XXIII: Northern and Gosiute Shoshone. *Anthropological Records* (University of California) 8:263–392.
Stewart, Omer C. 1941. Cultural Element Distributions, XIV: Northern Paiute. *Anthropological Records* (University of California) 4:361–446.
———. 1942. Cultural Element Distributions, XVIII: Ute-Southern Paiute. *Anthropological Records* (University of California) 6:231–356.

Chapter 9: Knocking off Acorns

Blackburn, Thomas C., ed. 1975. *December's Child: A Book of Chumash Oral Narratives*. Berkeley: University of California Press.
Clark, Cora, and Texa Bowen Williams. 1954.

Pomo Indian Myths. New York: Vantage Press, Inc.

Curtis, Edward S. [1926] 1970. *The North American Indian*. Vol. 15.

Dixon, Roland B. 1905. *Maidu Myths*. American Museum of Natural History Bulletin 17.

Driver, Philip. 1937. Cultural Element Distributions, X: Northwest California. *Anthropological Records* (University of California) 1:297–433.

DuBois, Constance G. 1906. Mythology of the Mission Indians. *JAFL* 19:52–60.

———. 1908. The Religion of the Luiseño and Diegueño Indians of Southern California. *Publications in American Archaeology and Ethnology* (University of California) 8 (3):69–186.

Farrand, Livingston. 1915. Shasta and Athapascan Myths from Oregon. Edited by Leo J. Frachtenberg. *JAFL* 28:207–242.

Foster, George M. 1944. A Summary of Yuki Culture. *Anthropological Records* (University of California) 5 (3):155–244.

Garth, Thomas R. 1960. Atsugewi Ethnography. *Anthropological Records* (University of California) 14 (2):129–212.

Gayton, Anna H. 1948. Yokuts and Western Mono Ethnography. *Anthropological Records* (University of California) 10 (1–2):1–302.

Gayton, Anna H. and Stanley S. Newman. 1940. Yokuts and Western Mono Myths. *Anthropological Records* (University of California) 5 (1):1–110.

Gifford, Edward W. 1931. *The Kamia of Imperial Valley*. BAE Bulletin 97.

Gifford, Edward W., and Gwendoline Harris Block. 1930. *Californian Indian Nights Entertainments*. Glendale, California: Arthur H. Clark Company.

Grant, Campbell. 1965. *The Rock Paintings of the Chumash*. Berkeley: University of California Press.

Harrington, John. 1933. Annotations to *Chinigchinich*, edited by P. T. Hanna. Santa Ana, Calif.: Fine Arts Press.

———. 1942. Cultural Element Distributions, XIX: Central California Coast. *Anthropological Records* (University of California) 7:1–46.

Heizer, Robert F., ed. 1978. *Handbook of North American Indians*. Vol. 8, *California*. Washington, D.C.: Smithsonian Institution. The following chapters were consulted for background and cultural information and star myths: Baumhoff, Martin A., Environmental Background; Bean, Lowell John, Cahuilla; Bean, Lowell John and Florence C. Shipek, Luiseño; Bean, Lowell John, and Dorothea Theodoratus, Western Pomo and Northeastern Pomo; Bean, Lowell John, and Sylvia B. Vane, Cults and their Transformation; Bright, William Karok; Elasser, Albert B., Wiyot; Gart, Thomas R., Atsugewi; Grant, Campbell, Eastern Coast Chumash; Heizer, Robert F., Natural Forces and Native World View; Johnson, Patti J., Patwin; Luomala, Katharine, Tipai and Ipai; McLendon, Sally and Michael J. Lowy, Eastern Pomo and Southeastern Pomo; Pilling, Arnold, Yurok; Riddell, Francis A., Maidu and Konkow; Silver, Shirley, Shastan Peoples; Spier, Robert F. G., Foothill Yokuts; Wallace, William J., Northern Valley Yokuts and Southern Valley Yokuts.

Heizer, Robert F., and M. A. Whipple, eds. 1951. *The California Indians: A Source Book*. Berkeley and Los Angeles: University of California Press

Hooper, Lucile. 1920. The Cahuilla Indians. *Publications in American Archaeology and Ethnology* (University of California) 16 (6):315–380.

Hudson, Travis. 1988a. The "Classical Assumption" in Light of Chumash Astronomy. In *Visions of the Sky*, edited by Robert A. Schiffman. Salinas, Calif.: Coyote Press.

———. 1988b. The Nature of California Indian Astronomy. In *Visions of the Sky*, edited by Robert A. Schiffman. Salinas, Calif.: Coyote Press.

Hudson, Travis, Thomas Blackburn, Rosario Curletti, and Janice Timbrook. 1977. *The Eye of the Flute: Chumash Traditional History and Ritual as Told by Fernando Librado Kitsepawit to John P. Harrington*. Santa Barbara, Calif.: Santa Barbara Museum of Natural History.

Hudson, Travis, and Ernest Underhay. 1978. *Crystals in the Sky: An Intellectual Odyssey Involving Chumash Astronomy, Cosmology and Rock Art*. Menlo Park, Calif.: Ballena Press and Santa Barbara Museum of Natural History.

Kroeber, Alfred L. 1907. Indian Myths of South Central California. *Publications in American Archaeology and Ethnology* (University of California) 4 (4):169–245.

———. 1925. *Handbook of the Indians of California*. BAE Bulletin 78.

———. 1932. The Patwin and Their Neighbors. *Publications in American Archaeology and Ethnology* (University of California) 29 (4):253–423.

———. 1976. *Yurok Myths*. Berkeley: University of California Press.

Kroeber, Alfred L., and E. W. Gifford. 1980. *Karok Myths*. Berkeley: University of California Press.

Latta, Frank. F. 1936. *California Indian Folklore*. Shafter, Calif.: Latta.

———. 1949. *Handbook of Yokuts Indians*. Oildale, Calif.: Bear State Books.

Lawrence Hall of Science. 1978. Native American Constellations.

Loeb, Edwin M. 1926. Pomo Folkways. *Publications in American Archaeology and Ethnology* (University of California) 19 (2):149–405.

Patencio, Francisco. 1943. *Stories and Legends of the Palm Springs Indians*. Edited by Margaret Boynton. Los Angeles: Times-Mirror.

Sapir, Edward, and Leslie Spier. 1943. Notes on the Culture of the Yana. *Anthropological Records* (University of California) 3 (3):239–298.

Sparkman, Philip S. 1908. The Culture of the Luiseño Indians. *Publications in American Archaeology and Ethnology* (University of California) 8 (4):187–234.

Spier, Leslie. 1923. Southern Diegueño Customs. *Publications in American Archaeology and Ethnology* (University of California) 20 (16):295–358.

———. 1928. Havasupai Ethnography. *Anthropological Papers of the American Museum of Natural History* 29 (3):83–392.

Voegelin, Erminie. 1942. Cultural Element Distributions, XX: Northeast California. *Anthropological Records* (University of California) 7:47–252.

Waterman, Thomas, T. 1910. The Religious Practices of the Diegueño Indians. *Publications in American Archaeology and Ethnology* (University of California) 8 (6):271–358.

Chapter 10: Chief of the Night

Allen, Richard H. [1899] 1963. *Star-Names and Their Meanings*. Reprint, New York: Dover Publications.

Benedict, Ruth. 1931. *Tales of the Cochiti Indians*. BAE Bulletin 98.

———. 1935. *Zuni Mythology*. Vol. 1. New York: Columbia University Press.

Boas, Franz. 1922. Tales of Spanish Provenience from Zuñi. *JAFL* 35:63–98.

Brandt, John C., and Ray A. Williamson. 1979. The 1054 Supernova and Native American Rock Art. *Archaeoastronomy* 1:1–38.

Brewer, Sallie P. 1950. Notes on Navaho Astronomy. In *For the Dean: Essays in Anthropology in Honor of Byron Cummings*, edited by Erik K. Reed and Dale S. King. Tucson, Ariz.: Hohokam Museums Association; and Santa Fe, N.M.: Southwestern Monuments Association.

Britt, Claude Jr. 1975. Early Navajo Astronomical Pictographs in Canyon de Chelly, Northeastern Arizona, U.S.A. Chapter 5 in *Archaeoastronomy in Pre-Columbian America*, edited by Anthony Aveni. Austin and London: University of Texas Press.

Bunzel, Ruth L. 1932. *Introduction to Zuñi Ceremonialism*. AR/BAE (1929–1930) 47:467–544.

Chamberlain, Von Del. 1983. Navajo Constellations in Literature, Art, Artifact and a New Mexico Rock Art Site. *Archaeoastronomy* 6:48–58.

———. 1994. Tracking Rabbits in the Sky. *Proceedings of the International Planetarium Society 1994 Conference*.

Culin, Stewart. [1907] 1975. *Games of the North American Indians*. Reprint, New York: Dover Publications.

Curtis, Edward S. [1926] 1970. *The North American Indian*. Vol. 16.

Cushing, Frank Hamilton. 1896. *Outlines of Zuñi Creation Myths. 1896*. AR/BAE (1891–1892) 13: 321–447.

———. [1901] 1931. *Zuñi Folk Tales*. New York: G. P. Putnam's Sons. Reprint, New York: Alfred A. Knopf.

———. 1923. Origin Myth from Oraibi. *JAFL* 36:163–170.

Densmore, Frances. 1929. *Papago Music*. BAE Bulletin 90.

Dorsey, George A., and Henry R. Voth. 1901. *The Oraibi Soyal Ceremony*. Field Columbian Museum Publication 55, Anthropological Series 3 (1):5–59.

Ellis, Florence Hawley. 1975. A Thousand Years of the Pueblo Sun-Moon-Star Calendar. Chapter 4 in *Archaeoastronomy in Pre-Columbian Amer-*

ica, edited by Anthony F. Aveni. Austin and London: University of Texas Press.

Erdoes, Richard, and Alfonso Ortiz. 1984. *American Indian Myths and Legends*. New York: Pantheon Books.

Farrer, Claire R. 1987. Star Clocks: Mescalero Apache Ceremonial Timing. *Canadian Journal of Native Studies* 7:223–236.

———. 1989. Star Walking: The Preliminary Report. Chapter 38 in *World Astronomy*, edited by Anthony F. Aveni. New York and Cambridge: Cambridge University Press.

———. 1991. *Living Life's Circle: Mescalero Apache Cosmovision*. Albuquerque: University of New Mexico Press.

———. 1992. ". . . by you they will know the directions to guide them": Stars and Mescalero Apaches. Chapter 4 in *Earth and Sky*/Williamson and Farrer.

Felger, Richard S., and Mary Beck Moser. 1985. *People of the Desert and Sea: Ethnobotany of the Seri Indians*. Tucson: University of Arizona Press.

Fewkes, J. Walter. 1904. *Two Summers' Work in Pueblo Ruins*. AR/BAE Part 1 (1900–1901) 22:3–195.

Gifford, E. W. 1932. The Southeastern Yavapai. *Publications in American Archaeology and Ethnology* (University of California) 29:177–252.

———. 1933. The Cocopa. *Publications in American Archaeology and Ethnology* (University of California) 31:257–334.

Goddard, Pliny Earle. 1934. Navajo Texts. *Anthropological Papers of the American Museum of Natural History* 34:1–179.

Griffin-Pierce, Trudy. 1992a. *Earth Is My Mother, Sky Is My Father: Time, Space and Astronomy in Navajo Sandpaintings*. Albuquerque: University of New Mexico Press.

———. 1992b. The *Hooghan* and the Stars. Chapter 7 in *Earth and Sky*/Williamson and Farrer.

Hagar, Stansbury. 1900. The Celestial Bear. *JAFL* 13:92–103.

Haile, Berard. 1947. *Starlore Among the Navajo*. Santa Fe: Museum of Navajo Ceremonial Art.

Handy, Edward L. 1918. Zuñi Tales. *JAFL* 31:451–471.

Harrington, John P. 1916. *Ethnogeography of the Tewa Indians*. AR/BAE (1907–1908) 29:29–636.

———. 1928. Picuris Children's Stories, with Texts and Songs. AR/BAE (1925–1926) 43:289–447.

Hoskinson, Tom. 1992. Saguaro Wine, Ground Figures, and Power Mountains: Investigations at Sears Point, Arizona. Chapter 8 in *Earth and Sky*/Williamson and Farrer.

Jayne, Caroline Furness. [1906] 1962. *String Figures*. Reprint, New York: Dover Publications.

Kluckhohn, Clyde, and Leland C. Wyman. 1940. An Introduction to Navaho Chant Practice. *Memoirs of the American Anthropological Association* 53.

Kroeber, Alfred. 1931. *The Seri*. Southwest Museum Paper 6.

Marriott, Alice, and Carol Rachlin. 1968. *American Indian Mythology*. New York: Thomas Y. Crowell Co.

Matthews, Washington. 1883. A Part of the Navajo's Mythology. *American Antiquarian* 5:207–224.

Newcomb, Franc Johnson. 1966. *Navaho Neighbors*. Norman: University of Oklahoma Press.

———. [1967] 1990. *Navaho Folk Tales*. Santa Fe: Museum of Navajo Ceremonial Art. Reprint, Albuquerque: University of New Mexico Press.

Newcomb, Franc Johnson, Stanley Fishler, and Mary C. Wheelwright. 1956. A Study of Navajo Symbolism. *Papers* (Peabody Museum of Archaeology and Ethnology) 32 (3).

Newcomb, Franc Johnson, and Gladys Reichard. [1937] 1975. *Sandpaintings of the Navajo Shooting Chants*. New York: J. J. Augustin. Reprint, New York: Dover Publications.

O'Bryan, Aileen. 1956. *The Dîné: Origin Myths of the Navaho Indians*. BAE Bulletin 163.

Opler, Morris Edward. 1938. Myths and Tales of the Jicarilla Apache Indians. *Memoirs of the American Folklore Society* 31.

Ortiz, Alfonso, ed. 1979. *Handbook of North American Indians*. Vol. 9, *Southwest*. Washington, D.C.: Smithsonian Institution.

———. 1983. *Handbook of North American Indians*. Vol. 10, *Southwest*. Washington, D.C.: Smithsonian Institution. The following chapters were consulted for background and cultural information: Bowen, Thomas, Seri; Gill, Sam D., Navajo Views of Their Origin; Harwell, Henry O., and Marsha C. S. Kelly, Maricopa; Jorgenson, Joseph G., Comparative Traditional Economies and Ecological Adapations; Khera, Sigrid, and Patricia S. Mariella, Yavapai; McGuire, Thomas R., Walapai; Schwartz, Douglas W., Havasupai; Stewart, Kenneth M., Yumans: Introduction; Tiller, Veronica, E., Jicarilla Apache; Wyman, Leland C., Navajo Ceremonial System.

Parsons, Elsie Clews. 1926. Tewa Tales. *Memoirs of the American Folklore Society* 19.

———. 1930. Zuñi Tales. *JAFL* 43:1–58.

———. 1932. *Isleta, New Mexico*. AR/BAE (1929–1930) 47:193–466.

———. 1939. *Pueblo Indian Religion*. 2 vols. Chicago: University of Chicago Press.

Parsons, Elsie Clews, ed. 1936. Hopi Journal of Alexander M. Stephen. 2 parts. *Contributions to Anthropology* (Columbia University Press) 23.

Russell, Frank. 1908. *The Pima Indians*. AR/BAE (1904–1905) 26:3–389.

Saxton, Dean, and Lucille Saxton. 1973. *O'othham Hoho'ok A'agitha: Legends and Lore of the Papago and Pima Indians*. Tucson: University of Arizona Press.

Schaafsma, Polly. 1992. *Rock Art in New Mexico*. Santa Fe: Museum of New Mexico Press.

Spier, Leslie. 1928. Havasupai Ethnography. *Anthropological Papers of the American Museum of Natural History* 29.

———. 1933. *Yuman Tribes of the Gila River*. Chicago: University of Chicago Press.

———. 1955. *Mohave Culture Items*. Flagstaff: Northern Arizona Society of Science and Art, Inc.

Stevenson, Matilda (Coxe). 1894. *The Sia*. AR/BAE Part 1 (1889–1890) 11:3–157.

———. 1904. *The Zuñi Indians: Their Mythology, Esoteric Fraternities, and Ceremonies*. AR/BAE (1901–1902) 23:3–634.

Stirling, Matthew W. 1942. *Origin Myths of Acoma and Other Records*. BAE Bulletin 135.

Verlarde, Pablita. 1960. *Old Father, The Story Teller*. Flagstaff, Ariz.: Northland Press.

Williamson, Ray A. 1984. *Living the Sky: The Cosmos of the American Indian*. Boston: Houghton Mifflin.

Wright, Harold B. 1929. *Long Ago Told*. New York: D. Appleton & Company.

Wyman, Leland C. 1952. *The Sandpaintings of the Kayenta Navaho*. Albuquerque: University of New Mexico Press.

Young, M. Jane. 1992. Morning Star, Evening Star: Zuni Traditional Stories. Chapter 5 in *Earth and Sky*/Williamson and Farrer.

Young, M. Jane, and Ray A. Williamson. 1981. Ethnoastronomy: The Zuni Case. Chapter 17 in *Archaeoastronomy in the Americas*, edited by Ray A. Williamson. Los Altos, Calif.: Ballena Press; and College Park, Md.: Center for Archaeoastronomy. Quoting John P. Harrington, unpublished manuscripts and papers pertaining to fieldwork at Zuni Indian Reservation held in the National Anthropological Archives, Smithsonian Institution.

Young, Robert, and William Morgan. 1980. *The Navajo Language: A Grammar and Colloquial Dictionary*. Albuquerque: University of New Mexico Press.

Chapter 11: A Bright Light Across the Heavens

Beckwith, Martha Warren. 1938. Mandan-Hidatsa Myths and Ceremonies. *Memoirs of the American Folklore Society* 32.

Chamberlain, Von Del. 1982. *When Stars Came Down to Earth: Cosmology of the Skidi Pawnee Indians of North America*. Menlo Park, Calif.: Ballena Press; and College Park, Md.: Center for Archaeoastronomy.

———. 1984. Astronomical Content of North American Plains Indian Calendars. *Archaeoastronomy* 6:1–54.

———. 1992. The Chief and His Council: Unity and Authority from the Stars. Chapter 14 in *Earth and Sky*/Williamson and Farrer.

Culin, Stewart. [1907] 1975. *Games of the North American Indians*. New York: Dover Publications.

Curtis, Natalie. 1907. *The Indians' Book*. New York: Harper and Brothers Publishers.

Denig, Edwin T. 1930. *Indian Tribes of the Upper Missouri*. AR/BAE (1928–1929) 46:375–626.

Densmore, Frances. 1923. *Mandan and Hidatsa Music*. BAE Bulletin 80.

Dorsey, George A. 1903. *The Arapaho Sun Dance: The Ceremony of the Offerings Lodge*. Field Columbian Museum Publication 75, Anthropological Series 4.

———. 1904a. Wichita Tales. *JAFL* 17:153–160.

———. 1904b. Mythology of the Wichita. *Publications* (Carnegie Institution of Washington) 21.

———. 1904c. Traditions of the Skidi Pawnee. *Memoirs of the American Folklore Society* 8.

———. 1906. The Pawnee Mythology. Part 1. *Publications* (Carnegie Institution of Washington) 59.

Dorsey, George A., and Alfred L. Kroeber. 1903. *Traditions of the Arapaho*. Field Columbian Museum Publication 81, Anthropological Series 5.

Dorsey, James Owen. 1888. *Osage Traditions*. AR/BAE (1894–1895) 6:377–397.

———. 1894. *A Study of Siouan Cults*. AR/BAE Part 3 (1889–1890) 11.

Eddy, John A. 1974. Astronomical Alignment of the Big Horn Medicine Wheel. *Science* 184 (June 7): 1035–1043.

———. 1977. Medicine Wheels and Plains Indian Astronomy. Chapter 11 in *Native American Astronomy*, edited by Anthony F. Aveni. Austin and London: University of Texas Press.

Erdoes, Richard, and Alfonso Ortiz. 1984. *American Indian Myths and Legends*. New York: Pantheon Books.

Fletcher, Alice C. 1903. Pawnee Star Lore. *JAFL* 16:10–15.

———. 1904. *The Hako: A Pawnee Ceremony*. AR/BAE Part 2 (1900–1901) 22.

Fletcher, Alice C., and Francis La Flesche. 1911. *The Omaha Tribe*. AR/BAE (1905–1906) 27.

Goodman, Ronald. 1992. On the Necessity of Sacrifice in Lakota Stellar Theology as Seen in "The Hand" Constellation, and the Story of "The Chief Who Lost His Arm." Chapter 13 in *Earth and Sky*/Williamson and Farrer.

Grinnell, George Bird. [1892] 1962. *Blackfoot Lodge Tales: The Story of a Prairie People*. New York: Forest & Stream. Reprint, Lincoln: University of Nebraska Press.

———. 1894. A Pawnee Star Myth. *JAFL* 7:197–200.

———. 1922. The Medicine Wheel. *American Anthropologist* n.s. 24.

———. [1923] 1972. *The Cheyenne Indians: Their History and Ways of Life*. 2 vols. New Haven: Yale University Press. Reprint, Lincoln: University of Nebraska Press.

———. [1926] 1971. *By Cheyenne Campfires*. New Haven: Yale University Press. Reprint, Lincoln: University of Nebraska Press.

Harrod, Howard L. 1987. *Renewing the World: Plains Indian Religion and Morality*. Tucson: University of Arizona Press.

Hoebei, E. Adamson. 1978. *The Cheyennes: Indians of the Great Plains*. New York: Holt, Rinehart and Winston.

Josephy, Alvin M. Jr. 1968. *The Indian Heritage of America*. New York: Alfred A. Knopf.

Kehoe, Alice B. 1992. Clot-of-Blood. Chapter 12 in *Earth and Sky*/Williamson and Farrer.

Kroeber, Alfred L. 1900. Cheyenne Tales. *JAFL* 13:161–190.

———. 1902. *The Arapaho*. Part 1. American Museum of Natural History Bulletin 18.

———. 1907a. *The Arapaho*. Part 4. American Museum of Natural History Bulletin 18.

———. 1907b. Gros Ventre Myths and Tales. *Anthropological Papers of the American Museum of Natural History* 1.

La Flesche, Francis. 1928. *The Osage Tribe: Two Versions of the Child-Naming Rite*. AR/BAE (1925–1926) 43:23–165.

———. 1930. *The Osage Tribe: Rite of the Wa-Xo'-Be*. AR/BAE (1927–1928) 45.

———. 1932. *A Dictionary of the Osage Language*. BAE Bulletin 109.

Lowie, Robert H. 1910. The Assiniboine. *Anthropological Papers of the American Museum of Natural History* 4 (1).

———. 1918. Myths and Traditions of the Crow Indians. *Anthropological Papers of the American Museum of Natural History* 25 (1).

———. 1935. *The Crow Indians*. New York: Holt, Rinehart and Winston.

———. 1982. *Indians of the Plains*. Lincoln and London: University of Nebraska Press.

Mallery, Garrick. 1893. *Picture-Writing of the American Indians*. AR/BAE (1888–1889) 10:103–822.

McClintock, Walter. [1910] 1968. *The Old North Trail*. London: MacMillan and Co., Limited. Reprint, Lincoln: University of Nebraska Press.

Momaday, N. Scott. 1969. *The Way to Rainy River*. Albuquerque: University of New Mexico Press.

Murie, James R. 1914. Pawnee Societies. *Anthropological Papers of the American Museum of Natural History* 11:543–644.

———. 1981. *Ceremonies of the Pawnee*. Edited by Douglas R. Parks. Washington, D.C.: Smithsonian Institution Press.

Oklahoma Indian Arts and Crafts Cooperative. 1973. *Painted Tipis by Contemporary Plains Indian Artists*. Anadarko, Okla.: Oklahoma Indian Arts and Crafts Cooperative.

Parks, Douglas R. 1991. *Traditional Narratives of the Arikara Indians*. Vols. 3 and 4. Lincoln: University of Nebraska Press.

Robinson, Jack H. 1980. Fomalhaut and Cairn D at the Big Horn and Moose Mountain Medicine Wheels. *Archaeoastronomy* 3 (4):15–19.

———. 1989. Medicine Wheels: Testing the Astronomical Theory. In *World Archaeoastronomy*, edited by Anthony F. Aveni. Cambridge and New York: Cambridge University Press.

Simms, S. C. 1903a. A Wheel-Shaped Monument in Wyoming. *American Anthropologist* n.s. 5:107–110.

———. 1903b. *Traditions of the Crows*. Field Columbian Museum Publication 85, Anthropological Series 2 (6).

Squier, Ephriam G. 1848. Ne-she-kay-be-nais, or the Lone Bird. *American Review* n.s. 2 (3):255–259.

Stands In Timber, John, and Margot Liberty. 1967. *Cheyenne Memories*. New Haven: Yale University Press.

St. Clair, H. H. 1909. Shoshone and Comanche Tales. *JAFL* 22:262–282.

Walker, James R. 1980. *Lakota Belief and Ritual*. Edited by Raymond J. DeMallie and Elaine A. Jahner. Lincoln: University of Nebraska Press.

Wallis, Wilson D. 1923. Beliefs and Tales of the Canadian Dakota. *JAFL* 36:36–101.

Weltfish, Gene. 1965. *The Lost Universe*. New York: Basic Books, Inc.

Williamson, Ray A. 1984. *Living the Sky: The Cosmos of the American Indian*. Boston: Houghton Mifflin.

Wilson, R. N. 1893. Blackfoot Star Myths. *The American Antiquarian* 15:200–203.

Wissler, Clark. 1936. *Star Legends Among the American Indians*. American Museum of Natural History Science Guide 91.

Wissler, Clark, and D. C. Duvall. 1909. Mythology of the Blackfoot Indians. Part 1. *Anthropological Papers of the American Museum of Natural History* 2.

Chapter 12: The Celestial Skiff

Dorsey, George A. 1905. Traditions of the Caddo. *Publications* (Carnegie Institution of Washington) 41.

Galloway, Mary, ed. 1990. *Aunt Mary, Tell Me a Story: As Told by Mary Ulmer Chiltoskey*. Cherokee, N.C.: Cherokee Communications.

Greenlee, R. F. 1945. Folk Tales of the Florida Seminole. *JAFL* 58:138–144.

Hagar, Stansbury. 1906. Cherokee Star-Lore. Pages 354–366 in *Boas Anniversary Volume*. New York: G. E. Stechert.

Hudson, Charles. 1976. *The Southeastern Indians*. Knoxville: University of Tennessee Press.

Monroe, Jean Guard, and Ray A. Williamson. 1987. *They Dance in the Sky: Native American Star Myths*. Boston: Houghton Mifflin Company.

Mooney, James. 1900. *Myths of the Cherokee*. AR/BAE Part 1 (1897–1898) 19.

Squier, Ephriam George. 1848. Ne-she-kay-be-nais; or the Lone Bird: An Ojibway Legend. *American Review* n.s. 2 (3):255–259.

Swanton, John R. 1928. *Religious Beliefs and Medical Practices of the Creek Indians*. AR/BAE (1924–1925) 42:23–857.

———. 1929. *Myths and Tales of the Southeastern Indians*. BAE Bulletin 88.

Williamson, Ray A. 1992. The Celestial Skiff: An Alabama Myth of the Stars. Chapter 3 in *Earth and Sky*/Williamson and Farrer.

Index

Note: Page references ending in "m" refer to a map on the cited page; those ending in "f" refer to a figure or illustration on the cited page.

Classical constellation names are in ALL UPPERCASE LETTERS.

Classical star names are identified with "(star)" after the name.

Tribal constellation and star names are identified with the name of the tribe in parentheses after the subject name.